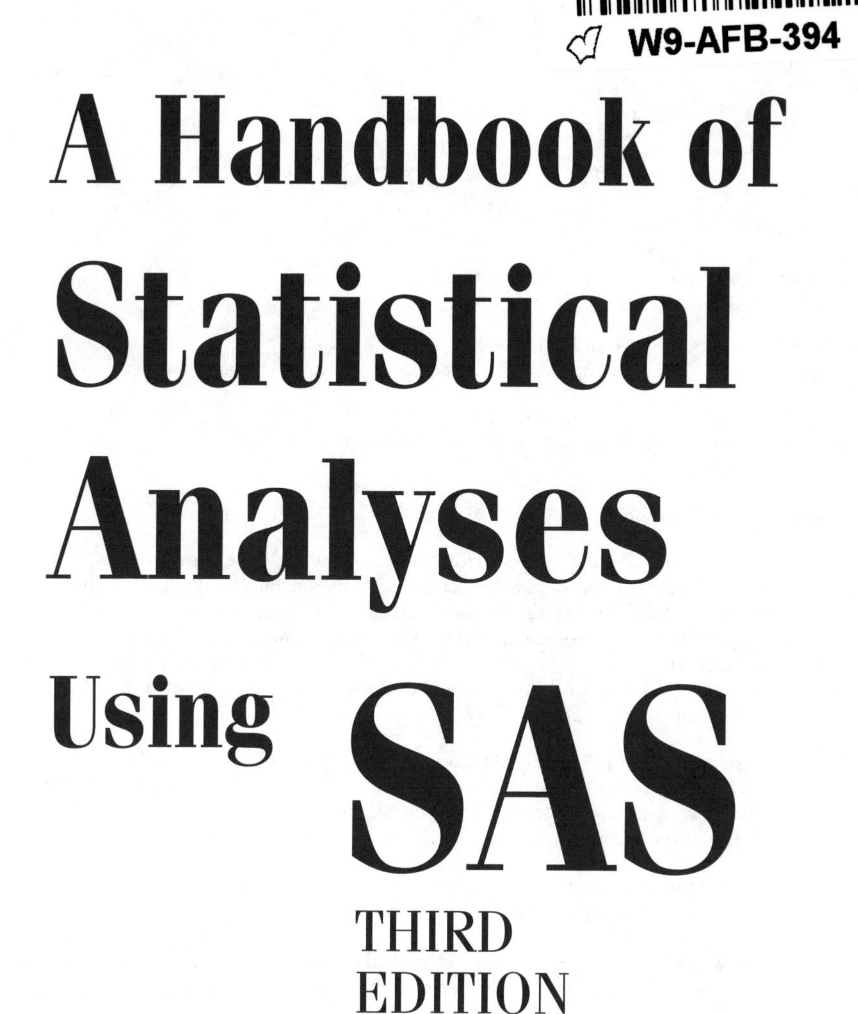

Geoff Der and Brian S. Everitt

CRC Press is an imprint of the
Taylor & Francis Group, an **informa** business
A CHAPMAN & HALL BOOK

Chapman & Hall/CRC
Taylor & Francis Group
6000 Broken Sound Parkway NW, Suite 300
Boca Raton, FL 33487-2742

Printed in the United States of America on acid-free paper
10 9 8 7 6 5 4 3 2

International Standard Book Number-13: 978-1-58488-784-3 (Softcover)

Library of Congress Cataloging-in-Publication Data

Der, Geoff.
A handbook of statistical analysis using SAS / authors, Geoff Der [and] Brian S. Everitt. -- 3rd ed.
p. cm.
Includes bibliographical references and index.
ISBN 978-1-58488-784-3 (hardback : alk. paper)
1. SAS (Computer file) 2. Mathematical statistics--Data processing. I. Everitt, Brian. II. Title.

QA276.4.E88 2008
519.5--dc22 2008021184

Visit the Taylor & Francis Web site at
http://www.taylorandfrancis.com

and the CRC Press Web site at
http://www.crcpress.com

Contents

Preface

SAS may well be the most widely used statistical package in the world. From its inception in 1976, both the software and the company behind it have grown consistently, so, too, has the documentation. When we wrote the first edition of this book in 1996, we were able to point to a set of manuals running to some 10,000 pages. At the time of the second edition, we could point our web browser at a single CD. Today we could point it to http://support.sas.com/. But, as with previous incarnations, this excellent website is not easy to use without prior acquaintance with the SAS system.

Our aim, again in this edition, has been to give a brief and straightforward description of how to conduct a range of statistical analyses using the latest version of SAS, version 9.2. We hope the book will provide students and researchers with a self-contained means of using SAS to analyse their data, and that it will also serve as a "stepping stone" to using the wider resources available to the SAS user.

The power and flexibility of SAS for manipulating data prior to analysis have long been recognized as one of the greatest strengths of the system. In version 9.2, this front end capability is equally matched at the other end of the process—that of preparing the results for publication. The Output Delivery System (ODS), the new graphics procedures, and ODS graphics can be used together to render the results of statistical analyses in a format ready for publication, whether that publication is on paper or online.

The SAS programs and the data used in this book are all available online at http://support.sas.com/HandbookofStatisticalAnalyses.

This book contains answers to some of the exercises at the end of each chapter in the form of the necessary SAS code. A full set of answers providing not only SAS code, but also relevant SAS output and comments, will be available as an accompanying solutions manual.

Geoff Der

Brian S. Everitt

Authors

Geoff Der works as a consulting statistician at the Medical Research Council's Social & Public Health Sciences Unit, in Glasgow. He advises research staff in the unit on study design and statistical analysis and also conducts his own research.

Professor Brian S. Everitt retired from being head of the Department of Biostatistics and Computing, Institute of Psychiatry, King's College, London, in 2005. In retirement, he continues to write and is currently working on his 60th book on statistics; he also acts as statistical consultant to a number of companies.

Chapter 1

Introduction to SAS

1.1 Introduction

The name SAS began as an acronym for Statistical Analysis System, reflecting the origins of the software and the company behind it. Today, the software covers such a broad range of products that the acronym is no longer considered appropriate. The majority of these products comprise a set of modules that can be added to the basic system, known as BASE SAS. Here we concentrate on the SAS/STAT and SAS/GRAPH modules in addition to the main features of the base system. Any installation of SAS intended for statistical analysis should include these two modules at the very least.

At the heart of SAS is a programming language made up of statements that specify how data are to be processed and analysed. The statements correspond to operations to be performed on the data or instructions about the analysis. A SAS program consists of a sequence of SAS statements grouped together into blocks, referred to as "steps". These fall into two types: data steps and procedure (proc) steps. A data step is used to prepare data for analysis. It creates a SAS data set and may re-organize the data and modify it in the process. A proc step is used to perform a particular type of analysis, or statistical test, on the data in a SAS data set.

A simple program might comprise a data step to read in some raw data followed by a series of proc steps analysing that data. If, in the course of the analysis, the data need to be modified, a second data step would be used to do this. Although the emphasis of this book is on the statistical analysis, one of the great strengths of SAS is the power and flexibility it gives the user to perform the data manipulation that is so often a large part of the overall task of analysing the data.

The SAS system is available for a wide range of different computers and operating systems, and the way in which SAS programs are entered and run differs somewhat according to the computing environment. We describe the Microsoft Windows interface, as this is by far the most popular, although other windowing environments, such as X-windows, are quite similar.

At the time of writing, the latest version of SAS is version 9.2, and all the examples have used version 9.2 running under Microsoft Windows XP.

1.2 User Interface

Figure 1.1 shows how SAS version 9.2 appears running under Windows XP. When SAS is started, there are five main windows open, namely the Editor, Log, Output, Results and Explorer windows. In Figure 1.1, the editor, log and results windows are visible. The explorer window is hidden behind the results window, and the output window is hidden behind the program editor and log windows.

At the top are the SAS title bar, the menu bar and the tool bar with the command bar at its left end. The buttons of the tool bar change, depending on which window is active. The command bar allows less frequently used commands to

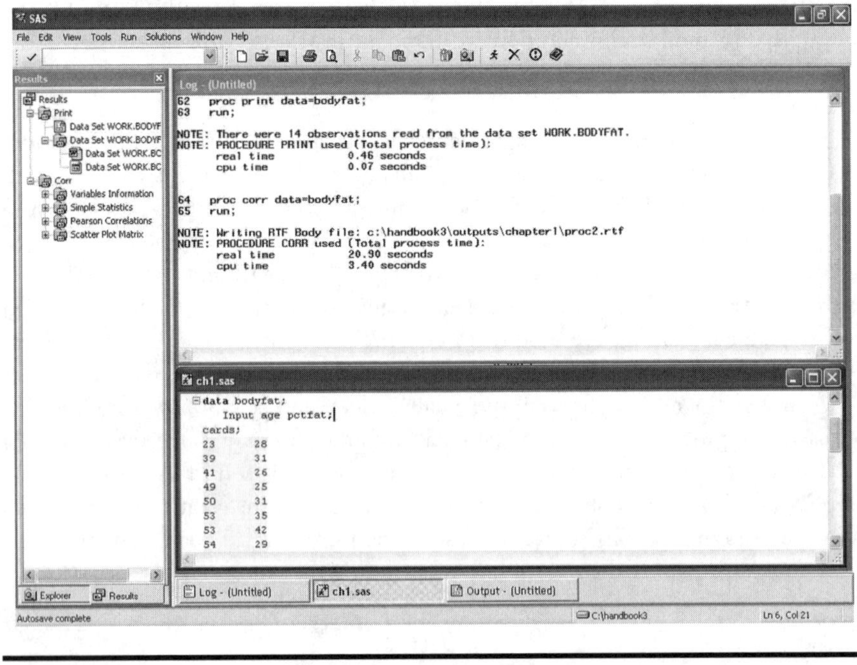

Figure 1.1 SAS user interface.

be typed in. At the bottom, the status line comprises a message area with the current directory and editor cursor position at the right. Double clicking on the current directory allows it to be changed. Above the status line is a series of tabs, which allows a window to be selected when it is hidden behind other windows.

Briefly, the purposes of the main windows are as follows:

Editor: The editor window is for typing in programs, editing them and running them. When a SAS program is run, two main types of output, the log and the procedure output, are generated and displayed in the log and output windows.

Log: The log shows the SAS statements that have been submitted together with information about the execution of the program, including warning and error messages.

Output: The output window shows the printed results of any procedures. It is here that the results of any statistical analyses are shown.

Results: The results window provides easy access to SAS output.

Explorer: The explorer window accesses files, SAS data sets and libraries.

When some graphical procedures are run, a *Graph* window is opened to display the resulting graphs. When ODS output is generated, a *Results Viewer* window may open to display the results.

Managing the windows, for example, moving between windows, resizing them and re-arranging them, can be done with the normal windows controls, including the Windows menu and the tabs at the bottom of the screen. If a window has been closed, it can be re-opened using the view menu.

1.2.1 Editor Window

SAS has two editors, a newer version, referred to as the enhanced editor, and an older version, known as the program editor. The program editor has been retained for reasons of compatibility but is not recommended. Here we describe the enhanced editor and may refer to it simply as "the editor." It may be distinguished from the program editor by the + in its icon. If SAS starts up using the program editor rather than the enhanced editor, then from the Tools menu select Options; Preferences then the Edit tab and select the use Enhanced Editor box.

The enhanced editor is essentially a built-in text editor specifically tailored to the SAS language with additional facilities for running SAS programs.

Some aspects of the editor window will be familiar as standard features of Windows applications. The File menu allows programs to be read from a file, saved to a file or printed. The File menu also contains the command to exit from SAS. The Edit menu contains the usual options for cutting, copying and pasting text, and those for finding and replacing text.

The program currently in the editor window may be run by choosing the Submit option from the Run menu. The Run menu is specific to the editor window and will

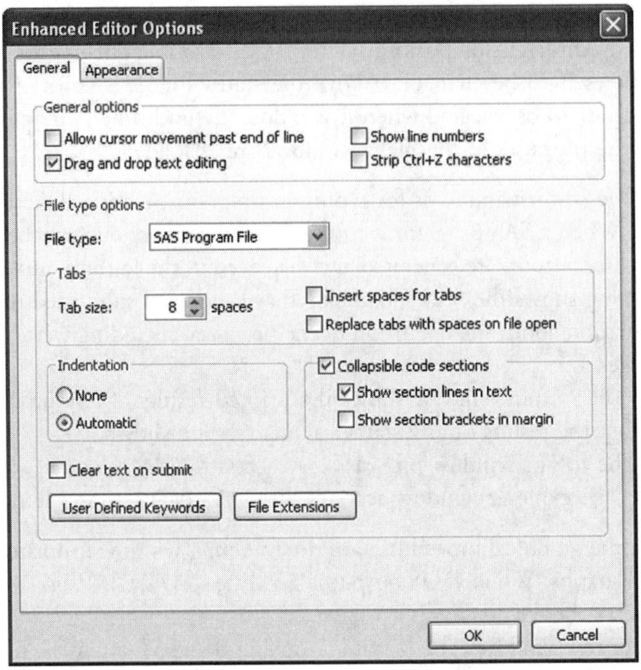

Figure 1.2 Enhanced Editor Options window.

not be available if another window is the active window. Submitting a program may remove it from the editor window. If so, it can be retrieved by choosing Recall Last Submit from the Run menu.

It is possible to run part of the program in the editor window by selecting the text and then choosing Submit from the Run menu. With this method, the submitted text is not cleared from the editor window. When running parts of programs in this way, make sure that a full step has been submitted. The easiest way to do this is to include a run statement as the last statement.

The Options sub-menu within Tools allows the editor to be configured. When the enhanced editor window is the active window (View, Enhanced Editor will ensure that it is), Tools; Options; Enhanced Editor will open a window similar to that in Figure 1.2. This shows the setup we recommend, in particular, that the options for Collapsible code sections and automatic indentation are selected, and that Clear text on submit is not.

1.2.2 Log Window

The log window is the main source of feedback to the user about the program or statements that have been submitted. It shows the statements themselves, together with notes, warnings and error messages. Although it is tempting to assume that if some output is generated the program has "worked", this is by no means the case. It is a good discipline to check the log every time. Although error messages and warnings obviously demand attention, the notes can also be important. For example, when a SAS data set is created, a note in the log gives the number of observations and variables it contains and, if these are not as anticipated, there may be an error in the program.

The Clear all option in the Edit menu, or the New button on the toolbar, will empty the window. This is useful if a program has been run several times as errors were corrected.

1.2.3 Output Window

The output window works in tandem with the results window. The entire contents of the window can be cleared in the same way as the log window, or sections can be deleted via the results window. The output window is designed for viewing plain text. When specially formatted output is generated, such as rich text format (rtf) or html, a separate window may open with an appropriate viewer.

1.2.4 Results Window

The results window provides a graphical index to the various procedure results, including the contents of the output window, the graph window and any output in other formats that may have been generated. It is useful for navigating around large amounts of output. Right clicking on a procedure, or section of output, allows that portion of the output to be viewed, printed, deleted or saved to file. Double clicking on a section of output opens the appropriate window with that section of the output visible. The level of detail that it shows can be controlled by expanding or collapsing sections or the entire output (right click and select Expand All or Collapse All).

1.2.5 Explorer Window

The explorer window performs much the same functions as the Windows Explorer, but with the added advantage of being able to view the contents of a SAS data set or a list of the variables it contains (right click, Open and right click, View Columns).

1.2.6 Some Other Menus

The View menu is useful for re-opening a window that has been closed.

The Solutions menu allows access to built-in SAS applications but these are beyond the scope of this text.

The Help menu tends to become more useful as experience of SAS is gained, although there may be access to some tutorial materials, if they have been licensed from SAS. There are also links to the main SAS web site and the customer support web site.

Context-sensitive help can be invoked with the F1 key. Within the editor, when the cursor is positioned over the name of a SAS procedure, the F1 key brings up the help for that procedure. This is very useful for quickly checking the syntax of the procedure.

For version 9, pdf files of the documentation are available online at http://support.sas.com/documentation/.

1.3 SAS Language

Learning to use the SAS language is largely a question of learning the statements that are needed to do the analysis required and of knowing how to structure them into steps. There are a few general principles that are useful to know.

Most SAS statements begin with a keyword that identifies the type of statement. (The most important exception is the assignment statement that begins with a variable name.) The enhanced editor recognises keywords as they are typed and changes their colour to blue. If a word remains red, this indicates a problem. The word may have been mistyped or is invalid for some other reason.

All SAS statements must end with a semicolon.

The most common mistake for new users is to omit the semicolon, and the effect is to combine two statements into one. Sometimes, the result will be a valid statement, albeit one that has unintended results. If the result is not a valid statement, there will be an error message in the SAS log when the program is submitted. However, it may not be obvious that a semicolon has been omitted before the program is run, as the combined statement will usually begin with a valid keyword.

Statements may extend over more than one line, and there may be more than one statement per line. However, keeping to one statement per line, as far as possible, helps avoid errors and to identify those that do occur.

SAS statements fall into four broad categories according to where in a program they can be used. These are

1. Data step statements
2. Proc step statements
3. Statements that can be used in both data and proc steps
4. Global statements that apply to all following steps

Because the functions of the data and proc steps are so different, it is perhaps not surprising that many statements are only applicable to one type of step.

1.3.1 Program Steps

Data and proc steps begin with a data or proc statement, respectively, and end at the next data or proc statement, or the next run statement. When a data step has the data included within it, the step ends after the data. Understanding where steps begin and end is important because SAS programs are executed in whole steps. If an incomplete step is submitted, it will not be executed. The statements that were submitted will be listed in the log, but SAS will appear to have stopped at that point without explanation. In fact, SAS will simply be waiting for the step to be completed before running it. For this reason, it is good practice to explicitly mark the end of each step by inserting a run statement and especially important to include one as the last statement in the program.

The enhanced editor offers several visual indicators of the beginning and end of steps. The data, proc and run keywords are colour coded in navy blue, rather than the standard blue used for other keywords. If the enhanced editor options for collapsible code sections have been selected, as shown in Figure 1.2, each data and proc step will be separated by lines in the text and indicated by brackets in the margin. This gives the appearance of enclosing each data and proc step in its own box.

Data step statements must be within the relevant data step, that is, after the data statement and before the end of the step. Likewise, proc step statements must be within the proc step.

Global statements may be placed anywhere. If they are placed within a step, they will apply to that step and all subsequent steps until reset. A simple example of a global statement is the title statement that defines a title for procedure output and graphs. The title is then used until it is changed or reset.

1.3.2 Variable Names and Data Set Names

In writing a SAS program, names must be given to variables and data sets. These may contain letters, numbers and underline characters, and may be up to 32 characters long but cannot begin with a number. Variable names may be in upper or lower case, or a mixture, but differences in case are ignored. So Height, height and HEIGHT would all refer to the same variable.

1.3.3 Variable Lists

When a list of variable names is needed in a SAS program an abbreviated form can often be used. A variable list of the form sex--weight refers to the variables sex and weight and all the variables positioned between them in the data set. A second form of variable list may be used where a set of variables has names of the form score1, score2, ... score10. That is, there are 10 variables with the root, score, in common and ending in the digits 1–10. In this case, they can be referred to by the variable list score1-score10, and do not need to be contiguous in the data set.

Before looking at the SAS language in more detail, the short example shown in Table 1.1 can be used to illustrate some of the preceding material. The data are adapted from Table 17 of *A Handbook of Small Data Sets* (Hand & Ostrowski, 1994; *SDS*) and show the age, percentage body fat and sex for 18 subjects. The program consists of three steps: a data step followed by two proc steps. Submitting this program results in the log and procedure output shown in Tables 1.2 and 1.3.

From the log we can see that the program has been split into steps and each step is run separately. Notes on how the step ran follow the statements that comprise the step and in the case of the data step show that the bodyfat data set contains the correct number of observations and variables.

The reason the log refers to the SAS data set as "WORK.BODYFAT," rather than simply "bodyfat," will be explained later.

Table 1.1 Simple SAS Program

```
data bodyfat;
input age pctfat sex $;
cards;
23   9.5  M
23  27.9  F
27   7.8  M
27  17.8  M
39  31.4  F
41  25.9  F
45  27.4  M
49  25.2  F
50  31.1  F
53  34.7  F
53  42.0  F
54  29.1  F
56  32.5  F
57  30.3  F
58  33.0  F
58  33.8  F
60  41.1  F
61  34.5  F
;
proc print data=bodyfat;
run;
proc corr data=bodyfat;
run;
```

Table 1.2 SAS Log after Submitting the Program in Table 1.1

```
27    data bodyfat;
28     input age pctfat sex $;
29    cards;

NOTE: The data set WORK.BODYFAT has 18 observations and 3 variables.
NOTE: DATA statement used (Total process time):
        real time     0.00 seconds
        cpu time      0.00 seconds


48    ;
49    proc print data=bodyfat;
50    run;

NOTE: There were 18 observations read from the data set WORK.BODYFAT.
NOTE: PROCEDURE PRINT used (Total process time):
        real time     0.00 seconds
        cpu time      0.00 seconds


51    proc corr data=bodyfat;
52    run;

NOTE: PROCEDURE CORR used (Total process time):
        real time     0.01 seconds
        cpu time      0.01 seconds
```

1.4 Reading Data – The Data Step

Before data can be analysed in SAS they need to be read into a SAS data set. Creating a SAS data set for subsequent analysis is the primary function of the data step. The data may be "raw" data or may come from a previously created SAS data set. A data step is also used to manipulate or re-organize the data. This can range from relatively simple operations, such as transforming variables, to more complex operations, such as restructuring of the data. In many practical situations, organizing and pre-processing the data take up a large proportion of the overall time and effort. The power and flexibility of SAS for such data manipulation is one of its great strengths.

We begin by describing how to create SAS data sets from raw data and store them on disc before turning to data manipulation. Each of the subsequent chapters includes the data step used to prepare the data for analysis, and several of them illustrate features not described in this chapter.

Table 1.3 Procedure Output of the Program in Table 1.1

Obs	Age	pctfat	Sex
1	23	9.5	M
2	23	27.9	F
3	27	7.8	M
4	27	17.8	M
5	39	31.4	F
6	41	25.9	F
7	45	27.4	M
8	49	25.2	F
9	50	31.1	F
10	53	34.7	F
11	53	42.0	F
12	54	29.1	F
13	56	32.5	F
14	57	30.3	F
15	58	33.0	F
16	58	33.8	F
17	60	41.1	F
18	61	34.5	F

The CORR Procedure

2 Variables: age pctfat

Simple Statistics

Variable	N	Mean	Std Dev	Sum	Minimum	Maximum
age	18	46.33333	13.21764	834.00000	23.00000	61.00000
pctfat	18	28.61111	9.14439	515.00000	7.80000	42.00000

Pearson Correlation Coefficients, N = 18
Prob > |r| under H0: Rho = 0

	age	pctfat
age	1.00000	0.79209 <.0001
pctfat	0.79209 <.0001	1.00000

*1.4.1 Creating SAS Data Sets from Raw Data**

Table 1.4 shows some hypothetical data on members of a slimming club, giving the membership number, team, starting weight and current weight. The following data step could be used to create a SAS data set:

```
data SlimmingClub;
  infile 'c:\handbook3\datasets\slimmingclub.dat';
  input idno team $ startweight weightnow;
run;
```

Table 1.4 Hypothetical Data for a Slimming Club

1023	Red	189	165
1049	Yellow	145	124
1219	Red	210	192
1246	Yellow	194	177
1078	Red	127	118
1221	Yellow	220	.
1095	Blue	135	127
1157	Green	155	141
1331	Blue	187	172
1067	Green	135	122
1251	Blue	181	166
1333	Green	141	129
1192	Yellow	152	139
1352	Green	156	137
1262	Blue	196	180
1087	Red	148	135
1124	Green	156	142
1197	Red	138	125
1133	Blue	180	167
1036	Green	135	123
1057	Yellow	146	132
1328	Red	155	142
1243	Blue	134	122
1177	Red	141	130
1259	Green	189	172
1017	Blue	138	127
1099	Yellow	148	132
1329	Yellow	188	174

* A "raw" data file may also be referred to as a text file or ASCII file. Such files include only the printable characters plus tabs, spaces and end-of-line characters. The files produced by database programs, spreadsheets and word processors are not normally "raw" data, although such programs usually have the ability to "export" their data to such a file.

1.4.2 Data Statement

The data statement often takes this simple form where it merely names the data set being created, in this case SlimmingClub.

1.4.3 Infile Statement

The infile statement specifies the file where the raw data are stored. The full path name of the file is given. If the file is in the current directory, that is, the one specified at the bottom right of the SAS window, the file name could have been specified simply as 'SlimmingClub.dat'. The name of the raw data file must be in single quotes. In many cases, the infile statement will only need to specify the filename as in this example.

In some circumstances, additional options on the infile statement will be needed. One such instance is where the values in the raw data file are not separated by spaces. Common alternatives are files where the data values are separated by tabs or commas. The expandtabs option changes tab characters into a number of spaces. The delimiter option can be used to specify a separator. For example, delimiter=',' could be used for files where the data values are separated by commas. More than one delimiter can be specified. Tab- and comma-separated data are discussed in more detail later.

Another situation where additional options may be needed is to specify what happens when the program requests more data values than a line in the raw data file contains. This can happen for a number of reasons, particularly where character data are read. Often, the solution is to use the pad option which adds spaces to the end of each data line as it is read.

There is one situation where an infile statement is not needed. This is when the data are contained within the SAS program itself. This is referred to as "instream" data. If data are instream, an infile statement is only needed when additional options are required. When data are instream, SAS automatically expands tabs according to the tab size setting for the editor (see Figure 1.2) so the expandtabs option is not needed.

1.4.4 Input Statement

The input statement in the example specifies that four variables are to be read in from the raw data file: idno, team, startweight and weightnow, and the dollar sign after team indicates that it is a character variable. SAS has only two types of variables: numeric and character.

The function of the input statement is to name the variables, specify their type as numeric or character and indicate where in the raw data the corresponding data values are. Where the data values are separated by spaces, as they are here, a simple form of the input statement is possible in which the variable names are merely listed in order and character variables are indicated by a dollar sign after their names. This is the so called "list" form of input. SAS has three main modes of input:

1. List – for data separated by spaces
2. Column – for data arranged in columns
3. Formatted – for data in non-standard formats

There is a fourth form – named input – but data suitable for this form of input occur so rarely that its description can safely be omitted.

1.4.4.1 List Input

In practice, there is often a choice of which mode of input to use and it is a question of which mode is more convenient for the data at hand. As list input is the simplest, it is usually preferred for that reason. However, the requirement that the data values be separated by spaces has some important implications. The first is that missing values cannot be represented by spaces in the raw data; a period (.) should be used instead. In the example, the value of weightnow is missing for member number 1221. The second is that character values cannot contain spaces. With list input it is also important to bear in mind that the default length for character variables is 8.

When using list input always examine the SAS log. Check that the correct number of variables and observations have been read in. The message "SAS went to a new line when INPUT statement reached past the end of a line" often indicates a problem in reading the data. If so, the pad option on the infile statement may be the answer.

With small datasets, it is advisable to print them out with proc print, or open the data set via the explorer window, and check that the raw data have been read in correctly.

1.4.4.2 Column Input

If list input is problematic and the data are arranged in columns, column input may be simpler. Table 1.5 shows the slimming club data with members' names instead of their membership numbers. This version of the data set is in the file SlimmingClub2.dat. To read in the data in the column form of input the statement would be

```
input name $ 1-18 team $ 20-25 startweight 27-29 weightnow 31-33;
```

As can be seen, the difference between the two forms of input statement is simply that the columns containing the data values for each variable are specified after the variable name, or after the dollar in the case of a character variable. The start and finish columns are separated by a hyphen, but for single column variables it is only

Table 1.5 Hypothetical Slimming Data with Members' Names

David Shaw	Red	189	165
Amelia Serrano	Yellow	145	124
Alan Nance	Red	210	192
Ravi Sinha	Yellow	194	177
Ashley McKnight	Red	127	118
Jim Brown	Yellow	220	
Susan Stewart	Blue	135	127
Rose Collins	Green	155	141
Jason Schock	Blue	187	172
Kanoko Nagasaka	Green	135	122
Richard Rose	Blue	181	166
Li-Hwa Lee	Green	141	129
Charlene Armstrong	Yellow	152	139
Bette Long	Green	156	137
Yao Chen	Blue	196	180
Kim Blackburn	Red	148	135
Adrienne Fink	Green	156	142
Lynne Overby	Red	138	125
John VanMeter	Blue	180	167
Becky Redding	Green	135	123
Margie Vanhoy	Yellow	146	132
Hisashi Ito	Red	155	142
Deanna Hicks	Blue	134	122
Holly Choate	Red	141	130
Raoul Sanchez	Green	189	172
Jennifer Brooks	Blue	138	127
Asha Garg	Yellow	148	132
Larry Goss	Yellow	188	174

necessary to give the one column number. Note also that Jim Brown's current weight is missing, but the blank columns are treated as a missing value so the period is not needed as it would be with list input.

1.4.4.3 Formatted Input

With formatted input, each variable is followed by its input format, referred to as its informat. Alternatively, a list of variables in parentheses is followed by a format list also in parentheses. Formatted input is the most flexible, partly because a wide range of informats is available. To read the above data using formatted input the following input statement could be used:

```
input name $19. team $7. startweight 4. weightnow 3.;
```

The informat for a character variable consists of a dollar sign, the number of columns occupied by the data values and a period. The simplest form of informat for numeric data is simply the number of columns occupied by the data and a period. Note that the spaces separating the data values have been taken into account in the informat.

Formatted input must be used if the data are not in a standard numeric format. Such data are rare in practice. The most common use of special SAS informats is likely to be the date informats. When a date is read using a date informat the resultant value is the number of days from January 1, 1960, to that date. The following data step illustrates the use of the ddmmyy*w*. informat. The width *w* may be from 6 to 32 columns. There is also the mmddyy*w*. informat for dates in American format. (There are also corresponding output formats, referred to simply as "formats," to output dates in calendar form.)

```
data days;
input day ddmmyy8.;
cards;
020160
01/02/60
31 12 59
231019
231020
;
run;
proc print data=days;
run;
proc print data=days;
  format day ddmmyy10.;
run;
```

As the example illustrates, if the year is given by only its last two digits, values of 20 or above are assumed to be in the twentieth century.

This data step is also an example of instream data. The data are contained between a cards statement (datalines is a synonym for cards) and a line with a single semicolon on it. The data must always be at the end of the data step.

Occasionally, data values will contain commas separating the thousands. These can be read with the comma format as follows:

```
data commas;
  input bignum comma6.;
cards;
1,860
;
```

Another instance where formatted input may be needed is when numeric data contain an implied decimal point. In this case, the informat has a second number after the period to indicate the number of digits to the right of the decimal point. For example, an informat of 5.2 would read five columns of numeric data and, in effect, move the decimal point two places to the left. Where the data contain an explicit decimal point, this takes precedence over the informat.

```
data decimals;
  input realnum 5.2;
cards;
1234
 4567
123.4
  6789
;
proc print;
run;
```

Leading or trailing spaces, within the field width, as in lines 1 and 2, will not prevent the number from being read correctly. In the case of the last line, the final digit is outside the field width, that is, in column 6, and so is not read as part of the number.

Formatted input can be much more concise than column input, particularly when consecutive data values have the same format. If the first 20 columns of the data line contain the single digit responses to 20 questions, the data could be read as follows:

```
input (q1-q20) (20*1.);
```

In this case, using a numbered variable list makes the statement even more concise. The informats in the format list can be repeated by prefixing them with n*, where n is the number of times the format is to be repeated; 20 in this case. If the format list has fewer informats than the number of variables in the variable list, the whole format list is re-used. So the above input statement could be re-written as:

```
input (q1-q20) (1.);
```

This feature is useful where the data contain repeating groups. If the answers to the 20 questions occupied 1 and 2 columns alternately, they could be read with

```
input (q1-q20) (1. 2.);
```

The different forms of input may be mixed on the same input statement for maximum flexibility.

1.4.4.4 Multiple Lines per Observation

Where the data for an observation occupy several lines, the slash character (/), used as part of the input statement, indicates where to start reading data from the next line. Alternatively, a separate input statement could be written for each line of data, since SAS automatically goes on to the next line of data at the completion of each input statement.

1.4.4.5 Multiple Observations per Line

In some circumstances, it is useful to be able to prevent SAS from automatically going on to the next line, and this is done by adding an @ character to the end of the input statement. The usual reason for doing this is that there are data for more than one observation on the same line. These features of data input will be illustrated in later chapters.

1.4.4.6 Delimited Data

There are two commonly occurring forms of raw data that are worth commenting on specifically: tab- and comma-separated data. Although list input is most commonly used for data separated by spaces, it can also be used to read data with other separators, referred to as "delimiters." One question that arises when delimiters other than spaces are used is how to treat two consecutive delimiters. With spaces as delimiters, list input by default treats consecutive spaces as a single delimiter. This is why spaces cannot be used for missing values. With comma-separated data, it is more likely that two consecutive commas are intended to indicate that the value which would have been between them is missing. Tabs are more commonly treated like spaces but could be intended to be read either way. To change the default, so that two consecutive delimiters are treated as having a missing value between them, use the dsd (delimiter-sensitive data) option on the infile statement.

1.4.4.7 Tab-Separated Data

The simplest way to read tab-delimited data is to use list input with the expandtabs option on the infile statement. This substitutes a number of spaces for the tab character. If consecutive tabs indicate missing values the delimiter= and dsd options are needed, as follows:

```
Infile 'filename' delimiter='09'x dsd;
```

The value 09 is the hexadecimal code for the tab character in the ASCII character set.

1.4.4.8 Comma-Separated Data

Comma-delimited data files may also be referred to as comma separate value (CSV) files, with a file extension .csv, and many PC programs can produce files in this format. For most of these the dsd option on the infile statement will suffice, as it assumes the delimiter is a comma. Some comma-delimited files will have data values enclosed in quotes to avoid problems where data values include commas. The dsd option deals with this too, by ignoring commas within quotes and removing the quotes from the data values. The missover option is also recommended for CSV files to prevent SAS going to a new line where the last value on a data line is missing. There is an example in Chapter 14. CSV files may also contain the names of the variables as the first line of the file. To skip this line when reading the data, use the firstobs = 2 option. So the recommended form of infile statement is

```
infile 'filename' dsd missover;
```

or

```
infile 'filename' dsd missover firstobs = 2;
```

where the variable names are on the first line.

1.4.5 Proc Import

For tab- and comma-delimited data, particularly where the first line contains the variable names, proc import is a useful alternative to reading in the data with a data step. For example, to read a tab-delimited file with the variable names in the first line, use

```
Proc import datafile = 'filename' out = sasdataset dbms = tab replace;
  getnames = yes;
run;
```

For comma-separated value files, substitute dbms = csv. Proc import does a good job of determining whether variables are numeric or character and their format, but the results need to be checked. If a numeric variable has any erroneous, non-numeric values, proc import may make the variable a character variable. An alternative way of using proc import is via the import wizard. From the file menu, select Import data . . .

1.4.6 Reading Data from Other Programs and Databases

SAS has a comprehensive set of modules enabling data held in proprietary databases to be read directly into SAS. This needs the appropriate SAS/ACCESS module to be licensed and is beyond the scope of this book. In the PC context, however, it is worth mentioning that the SAS/ACCESS module for PC file formats will enable proc import to read data from Access, Excel, Dbase and Lotus spreadsheets. The first screen of the import wizard will show which data sources have been licensed.

1.4.7 Temporary and Permanent SAS Data Sets – SAS Libraries

So far, all the examples have shown temporary SAS data sets. They are temporary in the sense that they will be deleted when SAS is exited. To store SAS data sets permanently on disc, and to access such data sets, the libname statement is used and the SAS data set is referred to slightly differently.

```
libname db 'c:\sasbook\sasdata';
data db.SlimmingClub;
   set SlimmingClub;
run;
```

The libname statement specifies that the *libref* db refers to the directory 'c:\sasbook\sasdata'. Thereafter, a SAS data set name prefixed with db. refers to a SAS data set stored in that directory. When used on a data statement, the effect is to create a SAS data set in that directory. The data step above reads data from the temporary SAS data set SlimmingClub and stores it in a permanent data set of the same name.

Because the libname statement is a global statement, the link between the libref db and the directory 'c:\sasbook\sasdata' remains throughout the SAS session, or until reset. If SAS has been exited and restarted, the libname statement will need to be submitted again.

In Table 1.2 we saw that the temporary data set bodyfat was referred to in the log notes as 'WORK.BODYFAT'. This is because work is the libref pointing to the directory where temporary SAS data sets are stored. SAS automatically sets up this directory and deletes the data sets in it when SAS is closed.

To use the SAS explorer window to examine the contents of a temporary data set, or its variables, double click on libraries in the explorer window, then double click on work. To do the same for permanently stored data sets, after opening the libraries folder, double click on the libref (e.g., db in the above example).

1.4.8 Reading Data from an Existing SAS Data Set

To read data from a SAS data set, rather than from a raw data file, the set statement is used in place of the infile and input statements. For example, to retrieve a previously stored data set and continue working with a temporary copy,

```
libname db 'c:\sasbook\sasdata';
data SlimmingClub;
  set db.SlimmingClub;
run;
```

creates a new, temporary SAS data set SlimmingClub reading in the data from the stored version of SlimmingClub.

1.5 Modifying SAS Data

As well as creating a SAS data set, the data step may also be used to modify the data in a variety of ways.

1.5.1 Creating and Modifying Variables

The assignment statement can be used both to create new variables and modify existing ones. The statement

```
weightloss=startweight-weightnow;
```

creates a new variable weightloss and sets its value to the starting weight minus the current weight.

```
startweight=startweight * 0.4536;
```

will convert the starting weight from pounds to kilograms.

SAS has the normal set of arithmetic operators: +, −, / (divide), * (multiply) and ** (exponentiate), plus various arithmetic, mathematical and statistical functions, some of which will be illustrated in later chapters.

1.5.2 Missing Values in Arithmetic Expressions

The result of an arithmetic operation performed on a missing value is itself a missing value. When this happens, a warning message is printed in the log. Missing values for numeric variables are represented by a period (.) and a numeric variable can be set to a missing value by an assignment statement such as:

```
age=. ;
```

With any arithmetical operation, it is worth considering what the effect of missing values will be. Say we want to calculate the mean of five variables, x1-x5. An assignment of the form xmean = (x1 + x2 + x3 + x4 + x5)/5; will result in a missing value if any of x1-x5 are missing. On the other hand, using the mean function, as in xmean = mean(x1,x2,x3,x4,x5); will result in a missing value only if all of them are missing.

To assign a value to a character variable, the text string must be enclosed in quotes, for example,

```
team = 'green';
```

A missing value may be assigned to a character variable as follows:

```
team = '';
```

To modify the value of a variable for some observations and not others, or to make different modifications for different groups of observations, the assignment statement may be used within an if then statement.

```
reward = 0;
if weightloss > 10 then reward = 1;
```

If the condition weightloss > 10 is true then the assignment statement reward = 1 is executed, otherwise the variable reward keeps its previously assigned value of 0. In cases such as these, an else statement could be used in conjunction with the if then statement.

```
if weightloss > 10 then reward = 1;
  else reward = 0;
```

The condition in the if then statement may be a simple comparison of two values. The form of comparison may be one of the following:

Operator		*Meaning*	*Example*
EQ	=	Equal to	a = b
NE	~=	Not equal to	a ne b
LT	<	Less than	a < b
GT	>	Greater than	a gt b
GE	>=	Greater than or equal to	a >= b
LE	<=	Less than or equal to	a le b

Comparisons can be combined into a more complex condition using and (&), or (|) and not.

```
if team='blue' and weightloss gt 10 then reward=1;
```

In more complex cases, it may be advisable to make the logic explicit by grouping conditions together with parentheses.

Some conditions involving a single variable can be simplified. For example, the following two statements are equivalent:

```
if age>18 and age<40 then agegroup=1;
if 18<age<40 then agegroup=1;
```

and conditions of the form

```
x=1 or x=3 or x=5
```

may be abbreviated to

```
x in(1, 3, 5)
```

using the in operator.

If the data contain missing values, it is important to allow for this when recoding. In numeric comparisons, missing values are treated as smaller than any number.

For instance,

```
if age >=18 then adult=1;
  else adult=0;
```

would assign the value 0 to adult if age was missing, whereas it may be more appropriate to assign a missing value. The missing function could be used to do this, by following the else statement with

```
if missing(age) then adult=.;
```

Care needs to be exercised when making comparisons involving character variables because these are case sensitive and sensitive to leading blanks.

A group of statements may be executed conditionally by placing them between a do statement and an end statement:

```
If weightloss > 10 and weightnow < 140 then do;
   target = 1;
   reward = 1;
   team = 'blue';
end;
```

Every observation that satisfies the condition will have the values of target, reward and team set as indicated. Otherwise, they will remain at their previous values.

When the same operation is to be carried out on several variables, it is often convenient to use an array and an iterative do loop in combination. This is best illustrated with a simple example. Suppose we have 20 variables, q1 to q20, for which "not applicable" has been coded -1 and we wish to set those to missing values, we might do it as follows:

```
array qall {20} q1-q20;
do i = 1 to 20;
   if qall{i} = -1 then qall{i} = .;
end;
```

The array statement defines an array by specifying the name of the array, qall here, the number of variables to be included in it in braces and the list of variables to be included. All the variables in the array must be of the same type, that is, all numeric or all character.

The iterative do loop repeats the statements between the do and the end a fixed number of times, with an index variable changing at each repetition. When used to process each of the variables in an array, the do loop should start with the index variable equal to 1 and end when it equals the number of variables in the array.

The array is a shorthand way of referring to a group of variables. In effect, it provides aliases for them so that each variable can be referred to by using the name of the array and its position within the array in braces. For example, q12 could be referred to as qall{12} or when the variable i has the value 12 as qall{i}. However, the array lasts only for the duration of the data step in which it is defined.

1.5.3 Deleting Variables

Variables may be removed from the data set being created by using the drop or keep statements. The drop statement names a list of variables that are to be excluded from the data set, and the keep statement does the converse; that is, it names a list of variables that are to be the only ones retained in the data set, all others being excluded. So the statement drop x y z; in a data step results in a data set that does not contain the variables x y and z, whereas keep x y z; results in a data set that contains *only* those three variables.

1.5.4 Deleting Observations

It may be necessary to delete observations from the data set either because they contain errors or because the analysis is to be carried out on a subset of the data. Deleting erroneous observations is best done by using the if then statement with the delete statement.

```
if weightloss > startweight then delete;
```

In a case such as this, it would also be useful to write out a message giving more information about the observation that contains the error.

```
if weightloss > startweight then do;
  put 'Error in weight data' idno= startweight= weightloss=;
  delete;
end;
```

The put statement writes text (in quotes) and the values of variables to the log.

1.5.5 Subsetting Data Sets

If analysis of a subset of the data is needed, it is often convenient to create a new data set containing only the relevant observations. This can be achieved with either the subsetting if statement or the where statement. The subsetting if statement consists of simply the keyword if followed by a logical condition. Only observations for which the condition is true are included in the data set being created.

```
data women;
  set bodyfat;
  if sex='F';
run;
```

The statement where sex='F'; could be used to the same effect. The difference between the subsetting if statement and the where statement will not concern most users, except that the where statement may also be used with proc steps, as discussed below. More complex conditions may be specified on either statement in the same way as for the if then statement.

1.5.6 Concatenating Data Sets – Adding Observation

Two or more data sets can be combined into one by specifying them on a single set statement.

```
data survey;
  set men women;
run;
```

This is also a simple way of adding new observations to an existing data set. First read the data for the new cases into a SAS data set, then combine this with the existing data set as follows:

```
data survey;
  set survey newcases;
run;
```

1.5.7 Merging Data Sets – Adding Variables

Data for a study may arise from more than one source, or at different times, and need to be combined. For instance, demographic details from a questionnaire may need to be combined with the results of laboratory tests. To deal with this situation, the data are read into separate SAS data sets, and then combined using a merge with a unique subject identifier as a key. Assuming the data have been read into two data sets, demographics and labtests, and that both data sets contain the subject identifier idnumber, they can be combined as follows:

```
proc sort data=demographics;
  by idnumber;
proc sort data=labtests;
  by idnumber;
data combined;
  merge demographics (in=indem) labtest (in=inlab);
  by idnumber;
  if indem and inlab;
run;
```

First both data sets must be sorted by the matching variable, idnumber. This variable should be of the same type, numeric or character, and same length in both data sets. The merge statement in the data step specifies the data sets to be merged. The option in parentheses after the name creates a temporary variable that indicates whether the data set provided an observation for the merged data set. The by statement specifies the matching variable. The subsetting if statement specifies that only observations that have both the demographic data and the lab results should be included in the combined data set. Without this, the combined data set may contain incomplete observations where there are demographic data but no lab results or vice versa. An alternative would be to print messages in the log in such instances as follows:

```
If not indem then put idnumber 'no demographics';
If not inlab then put idnumber 'no lab results';
```

This method of match merging is not confined to situations where there is a one-to-one correspondence between the observations in the data sets; it can be used for one-to-many or many-to-one relationships as well. A common practical application is in the use of lookup tables. For example, the research data set might contain the respondent's post code (or zip code), and another file may contain information on the characteristics of the area. Match merging the two data sets by post code would attach area information to the individual observations. A subsetting if statement would be used so that only observations from the research data were retained.

1.5.8 Operation of the Data Step

In addition to learning the statements that may be used in a data step, it is useful to understand how the data step operates.

The statements that compose the data step form a sequence according to the order in which they occur. The sequence begins with the data statement and finishes at the end of the data step and is executed repeatedly until the source of data runs out. Starting from the data statement a typical data step will read in some data with an input or set statement and use that data to construct an observation. The observation will then be used to execute the statements that follow. The data in the observation may be modified or added to in the process. At the end of the data step, the observation will be written to the data set being created. The sequence will begin again from the data statement, reading the data for the next observation, processing it and writing it to the output data set. This continues until all the data have been read in and processed. The data step will then finish, and the execution of the program will pass on to the next step.

In effect, then, the data step consists of a loop of instructions which is repeated until all the data are processed. The automatic SAS variable, _n_, records the iteration number but is not stored in the data set. Its use will be illustrated in later chapters.

The point at which SAS adds an observation to the data set can be controlled by the use of the output statement. When a data step includes one or more output statements an observation is added to the data set each time an output statement is executed, but not at the end of the data step. In this way the data being read in can be used to construct several observations. This will be illustrated in later chapters.

1.6 Proc Step

Once data have been read into a SAS data set, SAS procedures can be used to analyse the data. Roughly speaking, each SAS procedure performs a specific type of analysis. The proc step is a block of statements that specify the data set to be analysed, the procedure to be used and any further details of the analysis. The step begins with a proc statement and ends with a run statement or when the next data or proc step starts. We recommend including a run statement for every proc step.

1.6.1 Proc Statement

The proc statement names the procedure to be used and may also specify options for the analysis. The most important option is the data = option that names the data set to be analysed. If the option is omitted, the procedure uses the most recently created data set. Although this is usually what is intended, it is safer to explicitly specify the data set.

Many of the statements that follow particular proc statements are specific to individual procedures and will be described in later chapters as they arise. A few, though, are more general and apply to a number of procedures.

1.6.2 Var Statement

The var statement specifies the variables that are to be processed by the proc step. For example,

```
proc print data = SlimmingClub;
  var name team weightloss;
run;
```

restricts the printout to the three variables mentioned, whereas the default would be to print all variables.

1.6.3 Where Statement

The where statement selects the observations to be processed. The keyword where is followed by a logical condition, and only those observations for which the condition is true are included in the analysis.

```
proc print data = SlimmingClub;
  where weightloss > 0;
run;
```

1.6.4 By Statement

The by statement is used to process the data in groups. The observations are grouped according to the values of the variable named in the by statement, and a separate analysis is conducted for each group. In order to do this the data set must first be sorted on the by variable.

```
proc sort data = SlimmingClub;
   by team;
proc means;
   var weightloss;
   by team;
run;
```

1.6.5 Class Statement

The class statement is used with many procedures to name variables that are to be used as classification variables, or factors. The variables named may be character or numeric variables and will typically contain a relatively small range of discreet values. There may be additional options on the class statement depending on the procedure.

1.7 Global Statements

Global statements may occur at any point in a SAS program and remain in effect until reset.

The title statement is a global statement and provides a title that will appear on each page of printed output and each graph until reset. An example would be

```
title 'Analysis of Slimming Club Data';
```

The text of the title must be enclosed in quotes. Multiple lines of titles can be specified with the title2 statement for the second line, title3 for the third line, and so on up to 10. The title statement is synonymous with title1. Titles are reset by a statement of the form:

```
title2;
```

This will reset line two of the titles and all lower lines, that is, title3; etc., and title1; would reset all titles.

Comment statements are global statements in the sense that they can occur anywhere. There are two forms of comment statement. The first form begins with an asterisk and ends with a semicolon, for example,

```
* this is a comment;
```

The second form begins with /* and ends with */.

```
/* this is also a
  comment
*/
```

Comments may appear on the same line as a SAS statement, for example,

```
bmi=weight/height**2; /* Body Mass Index */
```

The enhanced editor colour codes comments green, so it is easier to see if the */ has been omitted from the end or if the semicolon has been omitted in the first form of comment.

The first form of comment is useful for "commenting out" individual statements, whereas the second is useful for commenting out one or more steps since it can include semicolons.

1.7.1 Options

The options statement is used to set SAS system options. Most of these can be safely left at their default values. Some of those controlling the procedure output that may be considered useful are

- Nocenter aligns the output at the left, rather than centring it on the page.
- Nodate suppresses printing of the date and time on the output.
- Pageno = *n* sets the page number for the next page of output, for example, pageno = 1 at the beginning of a program that is to be run repeatedly. Alternatively, nonumber turns page numbering off.

Several options can be set on a single options statement, for example,

```
options nodate nocenter nonumber;
```

1.8 SAS Graphics

When the SAS/GRAPH module has been licensed, there are a number of ways of producing high-quality graphical output. The three main approaches are

- Graphical options within a statistical procedure
- Traditional graphics procedures
- Statistical graphics procedures

The specific graphical options that are available within statistical procedures will be dealt with in later chapters. Where a statistical procedure does not produce the required type of graph, the general purpose graphical procedures within SAS/GRAPH may be used. We refer to the graphics procedures that existed in versions of SAS prior to 9.2 as "traditional" to distinguish them from the new "statistical" graphics procedures introduced in version 9.2: sgplot, sgpanel, sgmatrix and sgrender. (The traditional graphics procedures have names beginning with g: gplot, gchart, etc.) The new procedures, particularly sgplot, can produce a wide range of attractive graphs relatively simply, and for many users, it will be all they need. To avoid repetition we will concentrate on the new procedures in the text, but will illustrate alternatives using the traditional procedures in the downloadable source code.

1.8.1 xy *Plots – Proc sgplot*

An *xy* plot is one in which the data are represented in two dimensions defined by the values of two variables. The simplest such plot is a scatterplot and can be illustrated using the bodyfat data set described earlier.

```
proc sgplot data=bodyfat;
  scatter y=pctfat x=age;
run;
```

The syntax is straightforward: a scatter statement is used and the *x* and *y* variables specified explicitly. For different types of plot, a statement other than scatter is used. Table 1.6 shows some *xy* plots that could be generated by sgplot. Most of these will be illustrated in later chapters.

For line plots and step plots the points will be plotted in the order in which they occur in the data set, so it is usually necessary to sort the data by the *x* axis variable first.

A common variant of the *xy* plot distinguishes separate groups in the data by using different plotting symbols and/or different lines. This is done by the group=*var* option.

Table 1.6 ***xy*** **Plots Using sgplot**

Type of Plot	*Plotting Statement*
Scatter plot – data values are plotted	Scatter
Line plot – data values are joined with lines	Series
Step plot – data values joined with stepped lines	Step
Needle plot – vertical line joins the value to the *x* axis	Needle
Regression plot – a scatter plot with a regression line	Reg
Locally weighted regression	Loess
Penalized Beta splines	Pbspline

```
proc sgplot data=bodyfat;
  scatter y=pctfat x=age/group=sex;
run;
```

It is often useful to combine the information from two or more plots by overlaying them. Sgplot does this automatically when more than one plotting statement is included. For example, a plot to compare the fits from linear and locally weighted regression could be produced as follows (locally weighted regression is explained in Chapter 10):

```
proc sgplot data=bodyfat;
  reg y=pctfat x=age;
  loess y=pctfat x=age/nomarkers;
run;
```

The nomarkers option is specified on the loess statement to prevent the data points being plotted twice as sgplot uses different plotting symbols for each.

1.8.2 Summary Plots

Plots of summary statistics are often useful when comparing groups. Sgplot can produce plots of means, frequencies or sums as a bar plot, line plot or dot plot. The plot statements are vbar/hbar and vline/hline, depending on whether vertical or horizontal orientation is desired, and dot. To illustrate, age in the bodyfat data set is first recoded into 10-year bands.

```
data bodyfat;
  set bodyfat;
  decade=int(age/10);
run;
```

```
proc sgplot data = bodyfat;
  vline decade/response = pctfat stat = mean limitstat = stddev;
run;
```

Another useful summary plot is the boxplot, described in Chapter 2, which can be produced with the vbox/hbox statement as follows:

```
proc sgplot data = bodyfat;
  vbox pctfat/category = sex;
run;
```

There is also a specific procedure, boxplot, which offers more options.

1.8.3 Panel Plots

Proc sgpanel and proc sgscatter both produce multiple plots contained within a grid of related panels. Within sgpanel, the grid is defined by the values of variables in the data set with the result that each plot contains a subset of the data. With sgscatter, each plot contains the full set of data, and the grid is an arrangement of pairs of variables, with or without common axes. There are examples of both types in later chapters (see the index). Proc sgrender is for programming in the graphics template language, which underlies the statistical graphics procedures and is beyond the scope of this book.

1.9 ODS – The Output Delivery System

The Output Delivery System began as a means of generating SAS output in different formats. From this beginning as something of a cosmetic luxury, ODS has become an almost essential part of SAS. There are three reasons for this:

1. Publication quality procedure output
2. Saving output in SAS data sets
3. ODS graphics

1.9.1 ODS Procedure Output

The first of these might appear cosmetic, but the time and effort saved by using ODS should not be underestimated. Whatever the final form in which the results of an analysis are to be published, ODS simplifies the process by saving the output directly in the appropriate format. Html and rtf are probably the most commonly used formats, but there are many others, including xml, latex and pdf. Rtf is specifically designed for incorporating into word processors. Each of these output formats is referred to as an "ODS destination."

The output of one or more procedures can be saved in a particular format by opening the corresponding ODS destination beforehand and closing it afterwards. The rtf destination is opened by the ods rtf; statement and closed by ods rtf close; statement, as in the following example:

```
ods rtf;
proc print data = bodyfat;
proc corr data = bodyfat;
run;
ods rtf close;
```

The output appears in the output window as usual, but the formatted version is also saved in a file named sasrtf.rtf in the current directory. As this file will be overwritten the next time the rtf destination is opened, it is usually better to save the output to an explicitly named file with the file = '*filename*' option on the ods rtf statement.

1.9.2 ODS Styles

ODS output can be formatted according to a number of built-in styles. Each output destination has a default style optimized for that destination. Somewhat confusingly, the default style for html is called default, whereas that for rtf is called rtf. The output in this book has been produced with the theme style, so the ods rtf statement above could be replaced with

```
ods rtf file = 'c:\handbook\rtfexample.rtf' style = theme;
```

The names of other built-in styles can be listed by submitting:

```
proc template;
  list styles;
run;
```

Where the final output is to be in black and white, with greyscale fills or shading, the journal style is a good choice.

The rtf output may also appear in a results viewer window, and this may need to be closed before more rtf output is generated. The results viewer is switched on or off via Tools, options, preferences, results, then view results as they generate. If the output rtf file has been opened with a word processor, it will need to be closed before more output is sent to it.

1.9.3 Saving Output in SAS Data Sets – ODS Output

Another useful feature of ODS is the ability to save procedure output as SAS data sets. Prior to ODS, SAS procedures could save output – parameter estimates, fitted values, residuals, etc. – in SAS data sets via the output statement or other procedure-specific options. As part of the development of ODS, each procedure's output is broken down into a number of tables and any one of these may be saved to a SAS data set by including a statement of the form

```
ods output table=dataset;
```

within the proc step that generates the output.

The names of the tables created by each procedure are given in the "Details" section of the procedure's documentation. To find the variable names use the SAS explorer window or proc contents data = *dataset*.

1.9.4 ODS Graphics

ODS graphics are a recent development of ODS whereby many of the statistical procedures produce a range of useful plots either automatically or by specifying some plot options. As with the ODS tables, information on the ODS graphics that are available for each procedure is given in the details section of the procedure's documentation. ODS graphics are switched on and off with the ods graphics on; and ods graphics off; statements. They also need an ODS destination to be open, so a full example might be

```
ods html;
ods graphics on;
<one or more procedures that produce graphs>
ods graphics off;
ods html close;
```

We could also have used ods rtf; but there is a difference. With rtf, the graphs are included in the rtf document along with the tables. With html, the graphs are each in a separate file. The html output and the graphs are put in the current directory by default, but different and separate directories can be specified by the path = and gpath = options on the ods html statement. The default image file type of the graphs varies according to the ods destination but gif, jpeg and png are alternatives that can be set via the imagefmt = option on the ods graphics statement. In the example, we could use

```
ods html gpath='c:\handbook\graphs';
ods graphics on/imagefmt=jpeg;
```

to store jpeg format graphs in the named directory.

Because the ODS graphics are sent to the currently open ODS destination, they are also formatted with the same style as the tabular output. Any graphs produced independently using the new statistical graphics procedures will also use the same style and, when ODS graphics are on, even those produced by the traditional graphical procedures will.

Note that producing ODS graphics leads to longer processing times. Those with large data sets to analyse may need to be selective in their use.

It is important to bear in mind that the output produced by ODS for the rtf destination is tailored to the current page setup, and this should, therefore, match that of the document for which the output is intended. Although this can easily be done from the page setup menu (File, Page Setup), it can also be done via the options statement as in the following example:

```
options papersize=a4 orientation=portrait
  bottommargin=1in topmargin=1in
  leftmargin=1in rightmargin=1in;
```

The default for papersize is letter. Margins can also be specified in centimetres, for example, bottommargin=2.5cm.

When output is to be incorporated in a word processor document, the following options and settings can be useful:

options nodate nonumber;	Switches off date and page numbers
ods noproctitle;	Omits the procedure name from the output
ods rtf bodytitle;	Titles are placed in document body rather than header
title;	Sets null titles

1.10 Enhancing Output

1.10.1 Variable Labels

Although variable names are limited to 32 characters with no spaces, variable labels can contain spaces and can be much longer – up to 256 characters. If a variable has been given a label, then this can be used in the output. Whether labels are used in the output is controlled by the options label; and options nolabel; statements. The default is on.

The label statement is used to give variables a label and has the form

```
label variable='variable label';
```

For example,

```
label pctfat='Fat as a % of body mass';
```

If the label statement is used in a data step, the label is permanently associated with the variable, whereas if it is used in a proc step, the label is used only for the output from that procedure. To remove a variable's label include a statement of the form label sex=' '; in the data or proc step.

1.10.2 Value Labels – SAS Formats

SAS formats are used to give variable values more meaningful labels. Proc format is used to create the format, and the format statement is used to associate the format with a variable. For example,

```
proc format;
  value $sex 'M'='Male' 'F'='Female';
run;
proc sgplot data=bodyfat;
  scatter y=pctfat x=age/group=sex;
  format sex $sex.;
  label pctfat='Fat as % of body mass';
run;
```

The value statement within proc format has the general form

```
value format-name value1='label1' value2='label2';
```

There may be as many value='label' pairs as required. Where the values are character values, as in the example, they must be in quotes and the format name must begin with $. Character values are case sensitive, so 'm'='Male' would not have worked in the above example as all the values of sex are uppercase.

Format names are like variable names but should not end in a number. Note that in the format statement the format name ends in a period but not in the value statement. If, instead of the variable sex, the bodyfat data set contained a numeric variable called gender, the value and format statements might be

```
value gender 1='Male' 2='Female';
...
format gender gender.;
```

More than one variable can be associated with the same format. If a number of variables were coded 0 and 1, meaning "No" and "Yes," the value and format statements might be

```
value yn10f 0='No' 1='Yes';
...
format q1-q20 yn10f.;
```

1.11 Some Tips for Preventing and Correcting Errors

The following tips can be followed for preventing and correcting errors.

When writing programs:

- Use one statement per line, where possible.
- End each step with a run statement.
- Indent each statement within a step, that is, each statement between the data or proc statement and the run statement, by a couple of spaces. This is automated in the enhanced editor.
- Give the full path name for raw data files on the infile statement.

Before submitting a program:

- Check that each statement ends with a semicolon.
- Check that all opening and closing quotes match.

Use the enhanced editor colour coding to double check.

- Check any statement that does not begin with a keyword (blue or navy blue) or a variable name (black).
- Large blocks of purple may indicate a missing quotation mark.
- Large areas of green may indicate a missing */ from a comment.

"Collapse" the program to check its overall structure. Hold down the Ctrl and Alt keys and press the numeric keypad minus key. Only the data and proc statements and global statements should be visible. To expand the program, press the numeric keypad plus key while holding down Ctrl and Alt.

After running a program:

- Examine the SAS log for warning and error messages.
- Check for the message: "SAS went to a new line when INPUT statement reached past the end of a line" when using list input.

- Verify that the number of observations and variables read in is correct.
- If reading raw data, check the number of lines read and the maximum and minimum line lengths reported.
- Print out small datasets to ensure that they have been read correctly.

If there is an error message for a statement that appears to be correct, check whether the semicolon was omitted from the previous statement.

The message that a variable is "uninitialized" or "not found" usually means it has been misspelled. If not, it might have been included in a drop statement or left out of a keep statement.

The message "ERROR: File is in use, *filename*.rtf" when writing ODS output to a file often means that the Results Viewer window has not been closed or that a word processor has the file open.

To correct a missing quote, submit '; run; or "; run; then correct the program and resubmit it.

Chapter 2

Data Description and Simple Inference: Mortality and Water Hardness in the United Kingdom

2.1 Introduction

The data to be considered in this chapter were collected in an investigation of environmental causes of diseases, and involve the annual mortality rates per 100,000 for the males, averaged over the years 1958–1964, and the calcium concentration (in parts per million) in the drinking water supply for 61 large towns in England and Wales. (The higher the calcium concentration, the harder the water.) The data appear in Table 7 of *SDS*, and have been rearranged for use here as shown by the values for the first five towns given in Table 2.1. Towns at least as far north as Derby are identified in the table by an asterisk.

The main questions of interest about these data are as follows:

1. How are mortality and water hardness related?
2. Is there a geographical factor in the relationship?

Table 2.1 Water Hardness and Mortality in the United Kingdom

Town	Mortality	Hardness
Bath	1247	105
*Birkenhead	1668	17
Birmingham	1466	5
*Blackburn	1800	14
*Blackpool	1609	18

2.2 Methods of Analysis

Initial examination of the data will involve graphical techniques such as histograms and normal probability plots, in order to assess the distributional properties of the two variables, to make general patterns in the data more visible and to detect possible outliers. Scatterplots will be used to explore the relationship between mortality and calcium concentration.

Following this initial graphical exploration, some type of correlation coefficient may be computed for mortality and calcium concentration. Pearson correlation coefficient is generally used, but others, for example, Spearman's rank correlation, may be more appropriate if the data are not considered to have a bivariate normal distribution. The relationship between the two variables will be examined separately for northern and southern towns.

Finally, it will be of interest to compare the mean mortality and mean calcium concentration in the north and south of the country by using either a *t*-test or its non-parametric alternative, the Wilcoxon rank-sum test.

2.3 Analysis Using SAS

Assuming the data are stored in an ASCII file, water.dat, as listed in Table 2.1, including the '*' to identify location of town, and the name of the town, then they can be read using the following instructions:

```
data water;
  infile 'water.dat';
  input flag $ 1 Town $ 2-18 Mortal 19-22 Hardness 25-27;
  if flag = '*' then location = 'north';
    else location = 'south';
run;
```

The input statement uses SAS's column input where the exact columns containing the data for each variable are specified. Column input is simpler than list input would be in this case for three reasons:

- There is no space between the asterisk and the town name.
- Some town names are longer than eight characters – the default length for character variables.
- Some town names contain spaces which would make list input complicated.

The univariate procedure can be used to examine the distributions of numeric variables. The following simple instructions lead to the results shown in Tables 2.2 and 2.3:

```
proc univariate data=water normal;
  var mortal hardness;
  histogram mortal hardness /normal;
  probplot mortal hardness;
run;
```

Table 2.2 SAS Output for Mortality from Proc Univariate Applied to the Water Hardness Data

Variable: Mortal

Moments			
N	61	**Sum Weights**	61
Mean	1524.14754	**Sum Observations**	92973
Std Deviation	187.668754	**Variance**	35219.5612
Skewness	–0.0844436	**Kurtosis**	–0.4879484
Uncorrected SS	143817743	**Corrected SS**	2113173.67
Coeff Variation	12.3130307	**Std Error Mean**	24.0285217

Basic Statistical Measures			
Location		**Variability**	
Mean	1524.148	**Std Deviation**	187.66875
Median	1555.000	**Variance**	35220
Mode	1486.000	**Range**	891.00000
		Interquartile Range	289.00000

NOTE: The mode displayed is the smallest of 3 modes with a count of 2.

Tests for Location: Mu0=0				
Test	**Statistic**		**p Value**	
Student's t	**t**	63.43077	**Pr > |t|**	<.0001
Sign	**M**	30.5	**Pr >= |M|**	<.0001
Signed Rank	**S**	945.5	**Pr >= |S|**	<.0001

Tests for Normality				
Test	**Statistic**		**p Value**	
Shapiro-Wilk	**W**	0.985543	**Pr < W**	0.6884
Kolmogorov-Smirnov	**D**	0.073488	**Pr > D**	>0.1500
Cramer-von Mises	**W-Sq**	0.048688	**Pr > W-Sq**	>0.2500
Anderson-Darling	**A-Sq**	0.337398	**Pr > A-Sq**	>0.2500

Quantiles (Definition 5)	
Quantile	**Estimate**
100% Max	1987
99%	1987
95%	1800
90%	1742
75% Q3	1668
50% Median	1555
25% Q1	1379
10%	1259
5%	1247
1%	1096
0% Min	1096

Extreme Observations			
Lowest		**Highest**	
Value	**Obs**	**Value**	**Obs**
1096	26	1772	29
1175	38	1800	4
1236	42	1807	7
1247	1	1828	30
1254	15	1987	46

Table 2.3 SAS Output for Hardness from Proc Univariate for Water Hardness Data

Variable: Hardness

Moments			
N	61	**Sum Weights**	61
Mean	47.1803279	**Sum Observations**	2878
Std Deviation	38.0939664	**Variance**	1451.15027
Skewness	0.69223461	**Kurtosis**	-0.6657553
Uncorrected SS	222854	**Corrected SS**	87069.0164
Coeff Variation	80.7412074	**Std Error Mean**	4.8774326

Basic Statistical Measures			
Location		**Variability**	
Mean	47.18033	**Std Deviation**	38.09397
Median	39.00000	**Variance**	1451
Mode	14.00000	**Range**	133.00000
		Interquartile Range	61.00000

Tests for Location: Mu0=0				
Test	**Statistic**		**p Value**	
Student's t	**t**	9.673189	**Pr > |t|**	<.0001
Sign	**M**	30.5	**Pr >= |M|**	<.0001
Signed Rank	**S**	945.5	**Pr >= |S|**	<.0001

Tests for Normality				
Test	**Statistic**		**p Value**	
Shapiro-Wilk	**W**	0.887867	**Pr < W**	<0.0001
Kolmogorov-Smirnov	**D**	0.196662	**Pr > D**	<0.0100
Cramer-von Mises	**W-Sq**	0.394005	**Pr > W-Sq**	<0.0050
Anderson-Darling	**A-Sq**	2.399601	**Pr > A-Sq**	<0.0050

Quantiles (Definition 5)	
Quantile	**Estimate**
100% Max	138
99%	138
95%	122
90%	101
75% Q3	75
50% Median	39
25% Q1	14
10%	8
5%	6
1%	5
0% Min	5

Extreme Observations			
Lowest		**Highest**	
Value	**Obs**	**Value**	**Obs**
5	39	107	38
5	3	122	19
6	41	122	59
6	37	133	35
8	46	138	26

The normal option on the proc statement results in a test for the normality of the variables (see below). The var statement specifies which variables are to be included. If the var statement is omitted, the default is all the numeric variables in the data set. The histogram statement produces histograms for both variables and the /normal option requests a normal distribution curve. Curves for various other distributions, including non-parametric kernel density estimates (see Silverman, 1986), may be produced by varying this option. Probability plots are requested with the probplot statement. Normal probability plots are the default. The resulting histograms and plots are shown in Figures 2.1 through 2.4.

Tables 2.2 and 2.3 provide much information about the distributions of the two variables, mortality and hardness. Much of this is self-explanatory, for example, mean, standard deviation, variance and *N*. The meanings of some of the other statistics printed in these displays are as follows:

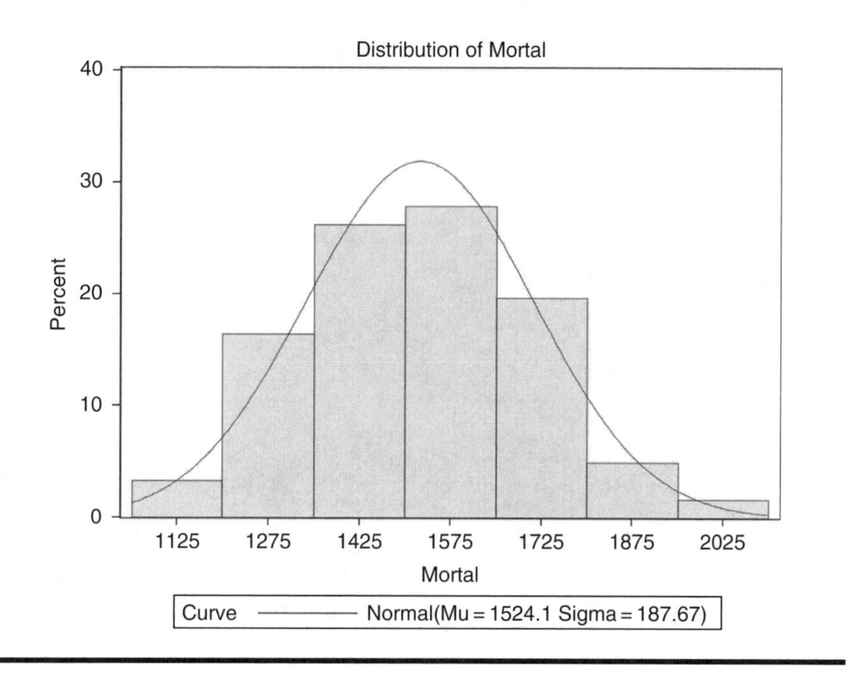

Figure 2.1 Histogram and fitted normal curve for mortality.

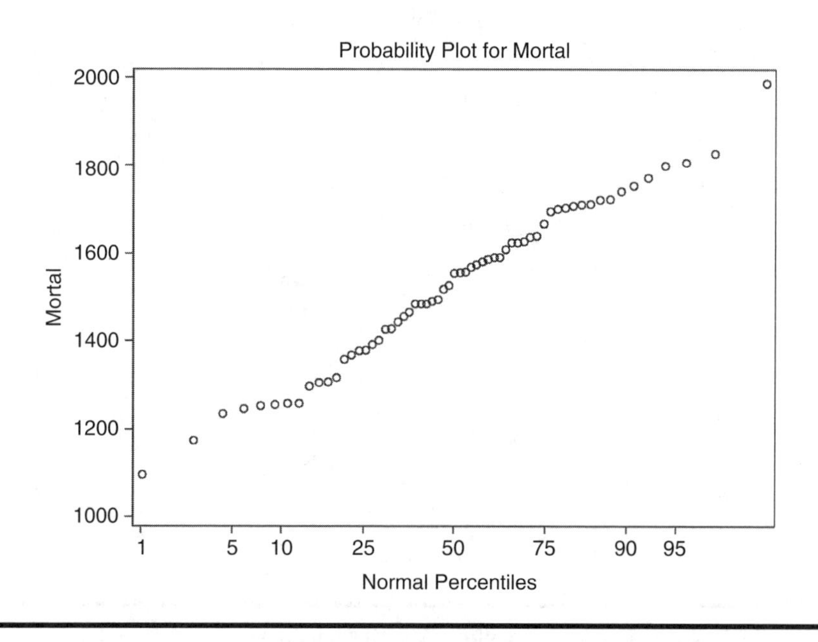

Figure 2.2 Normal probability plot for mortality.

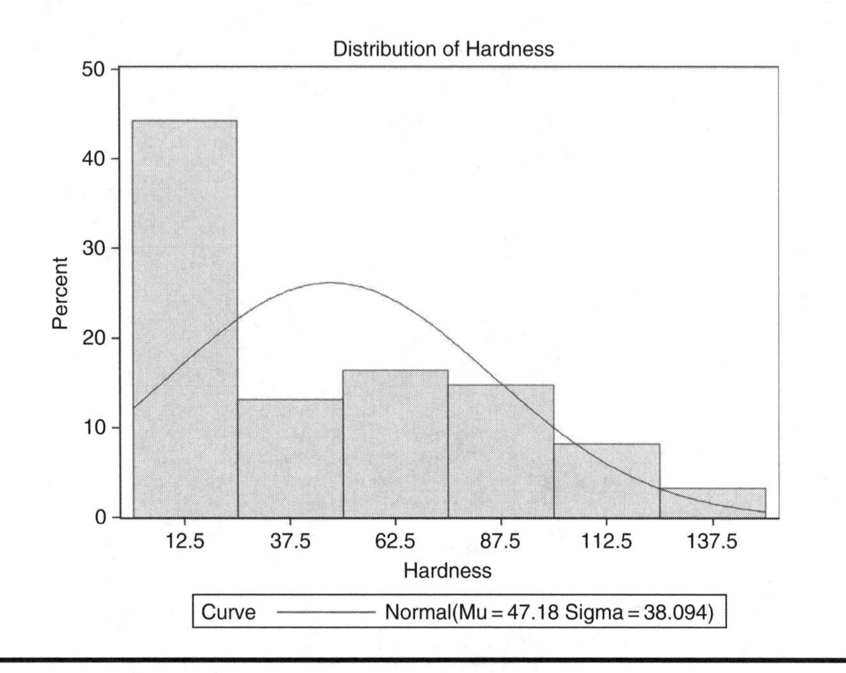

Figure 2.3 Histogram and fitted normal curve for water hardness.

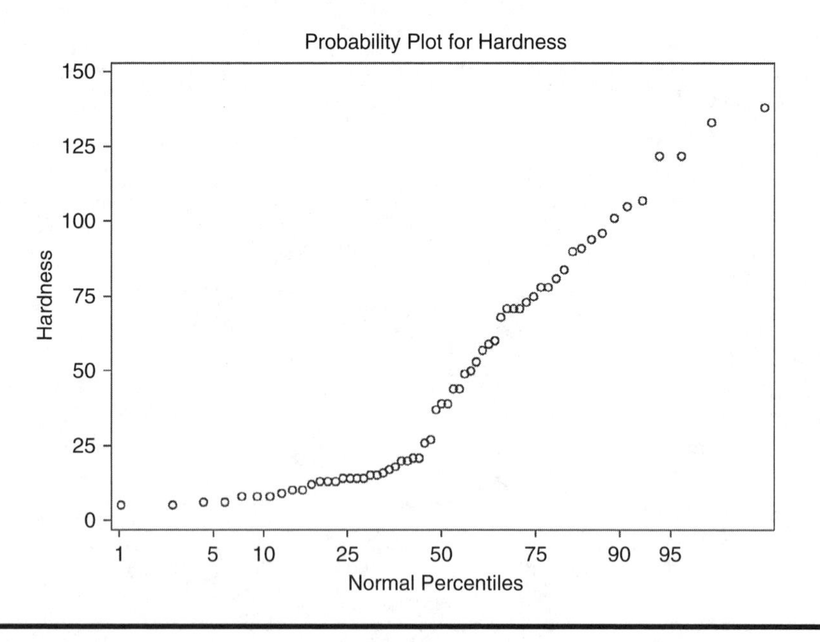

Figure 2.4 Normal probability plot for water hardness.

Uncorrected SS: Uncorrected sum of squares; simply the sum of squares of the observations.

Corrected SS: Corrected sum of squares; simply the sum of squares of deviations of the observations from the sample mean.

Coeff variation: Coefficient of variation; the standard deviation divided by the mean and multiplied by 100.

Std Error Mean: The standard deviation divided by the square root of the number of observations.

Range: Difference between the largest and the smallest observations in the sample.

Interquartile Range: Difference between 25% and 75% quantile (see values of quantiles given later in display to confirm).

Student's t: Student's t-test value for testing that the population mean is zero.

$Pr > |t|$: The probability of a greater absolute value for the t-statistic.

Sign test: A non-parametric test statistic for testing whether the population median is zero.

$Pr > |M|$: An approximation to the probability of a greater absolute value for the Sign test under the hypothesis that the population median is zero.

Signed rank: A non-parametric test statistic for testing whether the population mean is zero.

$Pr >= |S|$: An approximation to the probability of a greater absolute value for the Sign rank statistic under the hypothesis that the population mean is zero.

Shapiro–Wilk W: The Shapiro–Wilk statistic for assessing the normality of the data and the corresponding p value (Shapiro and Wilk, 1965).

Kolmogorov–Smirnov D: The Kolmogorov–Smirnov statistic for assessing the normality of the data and the corresponding p value (Fisher and Van Belle, 1996).

Cramer–von Mises W-Sq: The Cramer–von Mises statistic for assessing the normality of the data and the associated p value (Everitt, 2006).

Anderson–Darling A-Sq: The Anderson–Darling statistic for assessing the normality of the data and the associated p value (Everitt, 2006).

The quantiles give information about the tails of the distribution as well as the five number summaries for each variable. These summaries consist of the minimum value, lower quartile, median, upper quartile and maximum value of a variable. The box plots that can be constructed from these summaries are often very useful in comparing distributions and identifying outliers. Detailed examples will be given in later chapters, but some box plots can also be found later in this chapter (see, for example, Table 2.6).

The listing of extreme values can be useful for identifying outliers, especially when used with an `id` statement.

The numerical information in Table 2.2 and the plots in Figures 2.1 and 2.2 indicate that mortality is symmetrically, approximately normally, distributed. The formal tests of normality result in non-significant values of the test statistic. The results in Table 2.3 and the plots in Figures 2.3 and 2.4, however, strongly suggest that calcium concentration (hardness) has a skew distribution with each of the tests for normality having associated *p* values that are very small.

The first step in examining the relationship between mortality and water hardness is to look at the scatterplot of the two variables. This can be found using proc sgplot as follows:

```
proc sgplot data = water;
  scatter y = mortal x = hardness;
run;
```

The resulting graph is shown in Figure 2.5. The plot shows a clear negative association between the two variables, with high levels of calcium concentration tending to occur with low mortality values and vice versa. The correlation between the two variables is easily found using proc corr, with the following instructions:

```
proc corr data = water pearson spearman;
  var mortal hardness;
run;
```

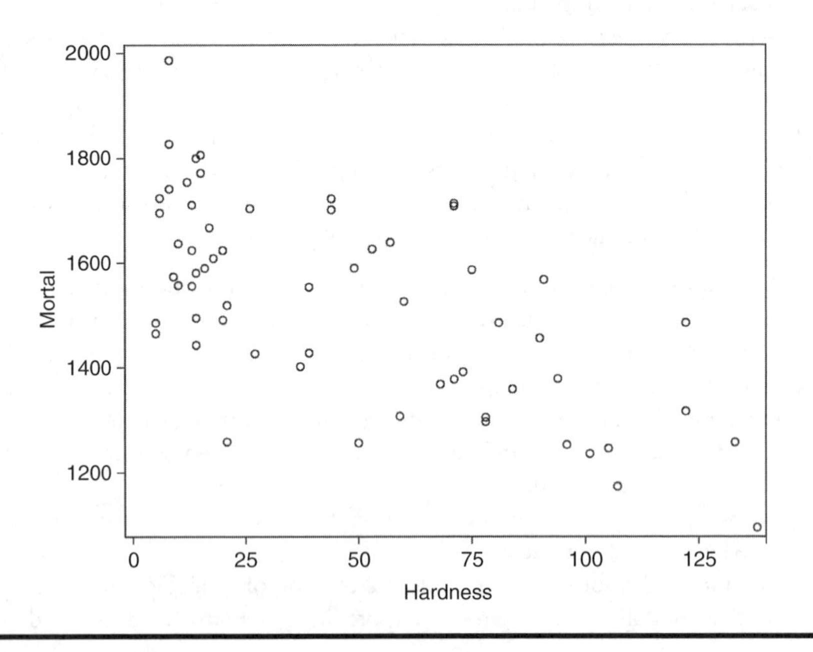

Figure 2.5 Scatterplot of mortality against hardness.

The pearson and spearman options on the proc corr statement request that both types of correlation coefficient be calculated. The default, if neither option is used, is the Pearson coefficient.

The results from these instructions are shown in Table 2.4. The correlation is estimated to be −0.655 using the Pearson coefficient and −0.632 using Spearman's coefficient. In both cases, the test that the population correlation is zero has an associated p value of 0.0001. There is clearly strong evidence of a non-zero correlation between the two variables.

One of the questions of interest about these data is whether there is a geographical factor in the relationship between mortality and water hardness, in particular whether this relationship differs between the towns in the north and those in the south. To examine this question, a useful first step is to replot the scatter diagram in Figure 2.5 with northern and southern towns identified with different symbols using the group as follows.

```
proc sgplot data=water;
  scatter y=mortal x=hardness/group=location;
run;
```

The resulting plot is shown in Figure 2.6. There seems to be no obvious difference in the form of the relationship between mortality and hardness for the two groups of towns.

Separate correlations for northern and southern towns can be produced by using proc corr with a by statement, as follows:

Table 2.4 SAS Output from Applying Proc Corr to the Water Hardness Data

Pearson Correlation Coefficients, N = 61 Prob > |r| under H0: Rho=0		
	Mortal	**Hardness**
Mortal	1.00000	-0.65485 <.0001
Hardness	-0.65485 <.0001	1.00000

Spearman Correlation Coefficients, N = 61 Prob > |r| under H0: Rho=0		
	Mortal	**Hardness**
Mortal	1.00000	-0.63166 <.0001
Hardness	-0.63166 <.0001	1.00000

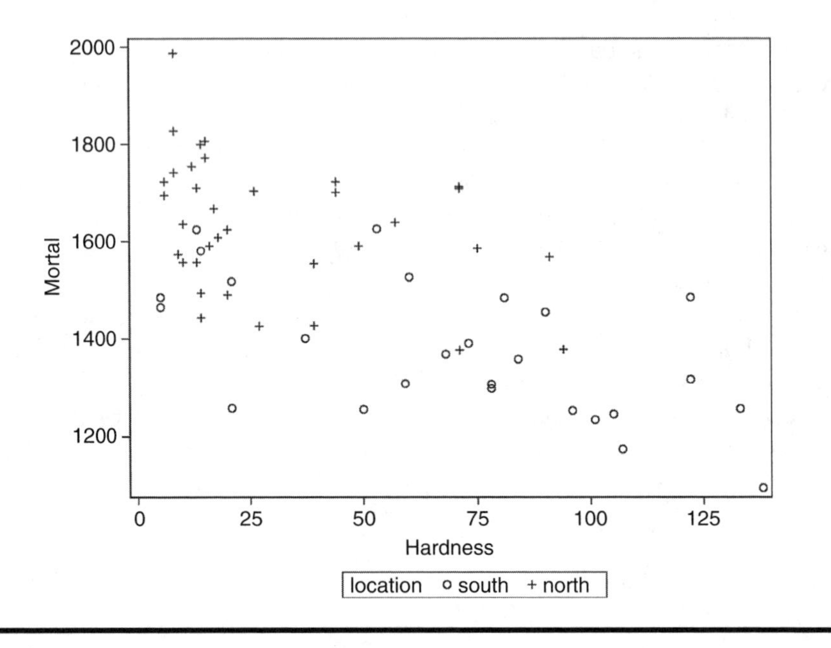

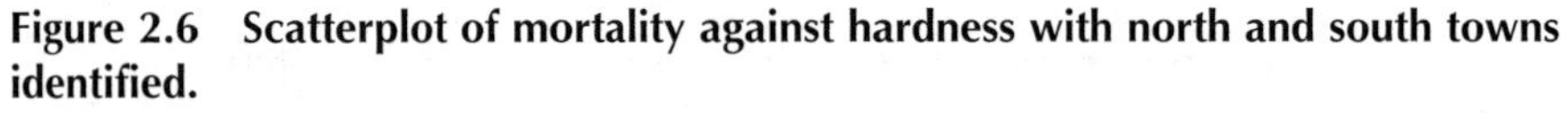

Figure 2.6 Scatterplot of mortality against hardness with north and south towns identified.

```
proc sort;
  by location;
proc corr data=water pearson spearman;
  var mortal hardness;
  by location;
run;
```

The by statement has the effect of producing separate analyses for each sub-group of the data defined by the specified variable, location in this case. However, the data set must first be sorted by that variable.

The results from this series of instructions are shown in Table 2.5. The main items of interest in this display are the correlation coefficients and the results of the tests that the population correlations are zero. The Pearson correlation for towns in the north is −0.369, and for those in the south it is −0.602. Both values are significant beyond the 5% level. The Pearson and Spearman coefficients take very similar values for this example.

Examination of scatterplots often centres on assessing density patterns such as clusters, gaps or outliers. But, humans are not particularly good at visually examining point density, and some type of density estimate added to the scatterplot will frequently be very helpful. Here plotting a bivariate density estimate for mortality and hardness is useful for gaining more insight into the structure of the

Table 2.5 SAS Output from Proc Corr

location=north

2 Variables: Mortal Hardness

Simple Statistics						
Variable	**N**	**Mean**	**Std Dev**	**Median**	**Minimum**	**Maximum**
Mortal	35	1634	136.93691	1637	1378	1987
Hardness	35	30.40000	26.13449	17.00000	6.00000	94.00000

Pearson Correlation Coefficients, N = 35 Prob > |r| under H0: Rho=0		
	Mortal	**Hardness**
Mortal	1.00000	-0.36860 0.0293
Hardness	-0.36860 0.0293	1.00000

Spearman Correlation Coefficients, N = 35 Prob > |r| under H0: Rho=0		
	Mortal	**Hardness**
Mortal	1.00000	-0.40421 0.0160
Hardness	-0.40421 0.0160	1.00000

location=south

2 Variables: Mortal Hardness

Simple Statistics						
Variable	**N**	**Mean**	**Std Dev**	**Median**	**Minimum**	**Maximum**
Mortal	26	1377	140.26918	1364	1096	1627
Hardness	26	69.76923	40.36068	75.50000	5.00000	138.00000

Pearson Correlation Coefficients, N = 26 Prob > |r| under H0: Rho=0		
	Mortal	**Hardness**
Mortal	1.00000	-0.60215 0.0011
Hardness	-0.60215 0.0011	1.00000

Spearman Correlation Coefficients, N = 26 Prob > |r| under H0: Rho=0		
	Mortal	**Hardness**
Mortal	1.00000	-0.59572 0.0013
Hardness	-0.59572 0.0013	1.00000

data. (Details of how to calculate bivariate densities are given in Silverman, 1986.) The following code produces and plots the bivariate density estimate of the two variables:

```
proc kde data=water;
  bivar mortal hardness/plots=surface;
run;
```

The kde procedure produces estimates of a univariate or bivariate probability density function using kernel density estimation. If a single variable is specified in the var statement a univariate density is estimated, and a bivariate density is estimated if two variables are specified. A surface plot is one of the ODS graphics and is shown in Figure 2.7. The two clear modes in the diagram correspond, approximately at least, to northern and southern towns.

The final question to address is whether mortality and calcium concentration differ in northern and southern towns. Because the distribution of mortality appears to be approximately normal, a *t*-test can be applied. Calcium concentration has a relatively high degree of skewness, and so applying a Wilcoxon test or a *t*-test after a log transformation may be more sensible. The relevant SAS instructions are as follows:

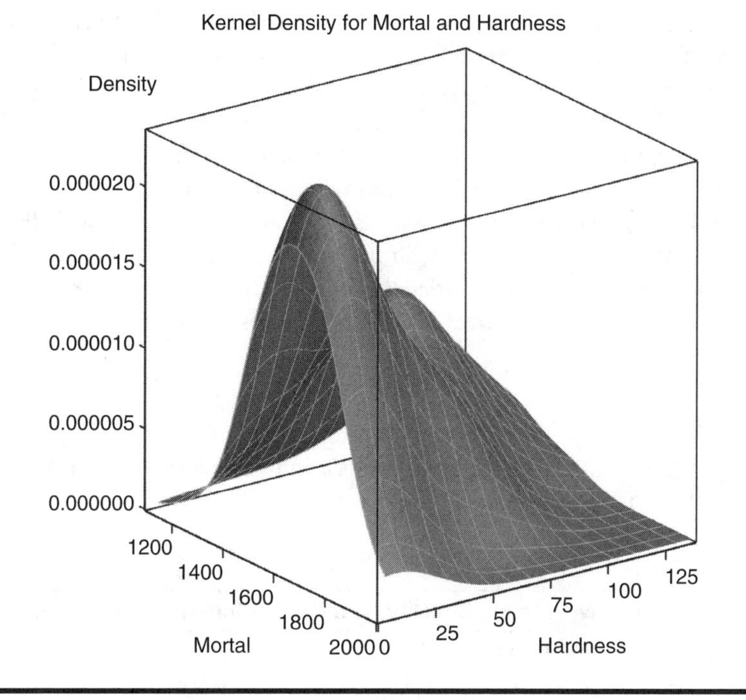

Figure 2.7 Bivariate density estimate for mortality and hardness.

```
data water;
  set water;
  lhardnes = log(hardness);
proc ttest;
  class location;
  var mortal hardness lhardnes;
proc npar1way wilcoxon;
  class location;
  var hardness;
run;
```

The short data step computes the (natural) log of hardness and stores it in the data set as the variable lhardnes. To use proc ttest, the variable that divides the data into two groups is specified in the class statement and the variable, or variables, whose means are to be compared are specified in the var statement. For a Wilcoxon test the npar1way procedure is used with the wilcoxon option.

The results of the *t*-tests are shown in Table 2.6; those for the Wilcoxon tests are shown in Table 2.7. The *t*-test for mortality gives very strong evidence for a difference in mortality in the two regions, with that in the north being considerably larger (the 95% confidence interval for the difference is 185.11, 328.47). Using a test that assumes equal variances in the two populations or one that does not make this assumption (Satterthwaite, 1946) makes little difference in this case. The *t*-test on the untransformed hardness variable also indicates a difference, with the mean hardness in the north being far less than that in the south. Notice here that the test for the equality of population variances (one of the assumptions of the *t*-test) suggests that the variances differ. Examining the results for the log-transformed

Table 2.6 SAS Output (Numerical and Graphical) from Proc *t*-test Applied to Mortality, Hardness and Log(Hardness)

Variable: Mortal

location	N	Mean	Std Dev	Std Err	Minimum	Maximum
north	35	1633.6	136.9	23.1466	1378.0	1987.0
south	26	1376.8	140.3	27.5090	1096.0	1627.0
Diff (1-2)		256.8	138.4	35.8221		

Location	Method	Mean	95% CL Mean		Std Dev	95% CL Std Dev	
North		1633.6	1586.6	1680.6	136.9	110.8	179.4
South		1376.8	1320.2	1433.5	140.3	110.0	193.6
Diff (1-2)	Pooled	256.8	185.1	328.5	138.4	117.3	168.8
Diff (1-2)	Satterthwaite	256.8	184.7	328.9			

Method	Variances	DF	t Value	Pr > |t|
Pooled	Equal	59	7.17	<.0001
Satterthwaite	Unequal	53.29	7.14	<.0001

Equality of Variances				
Method	Num DF	Den DF	F Value	Pr > F
Folded F	25	34	1.05	0.8830

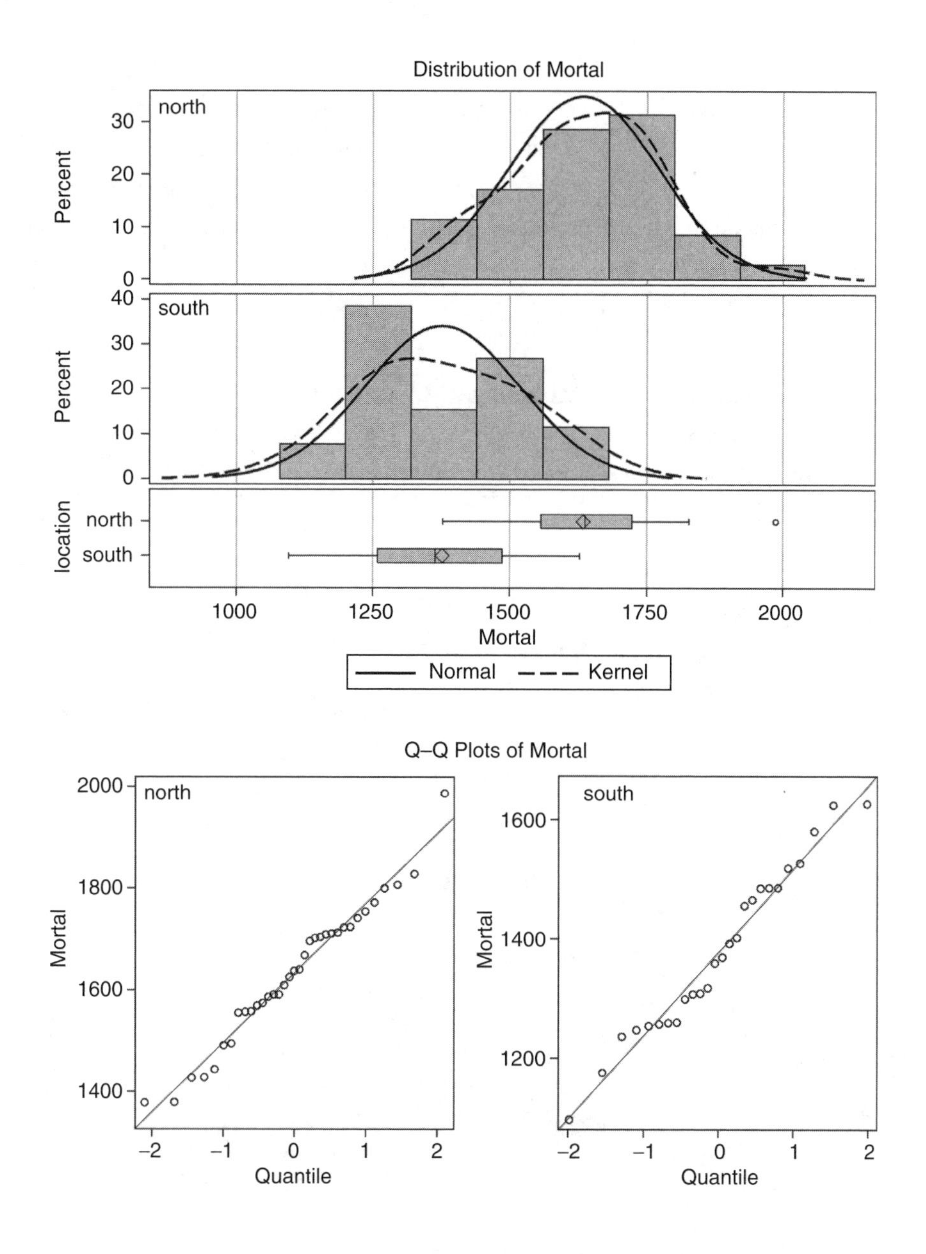

Variable: Hardness

location	N	Mean	Std Dev	Std Err	Minimum	Maximum
north	35	30.4000	26.1345	4.4175	6.0000	94.0000
south	26	69.7692	40.3607	7.9154	5.0000	138.0
Diff (1-2)		-39.3692	32.9218	8.5237		

location	Method	Mean	95% CL Mean		Std Dev	95% CL Std Dev	
north		30.4000	21.4225	39.3775	26.1345	21.1395	34.2415
south		69.7692	53.4672	86.0713	40.3607	31.6532	55.7142
Diff (1-2)	Pooled	-39.3692	-56.4251	-22.3133	32.9218	27.9057	40.1535
Diff (1-2)	Satterthwaite	-39.3692	-57.6876	-21.0508			

Method	Variances	DF	t Value	Pr > |t|
Pooled	Equal	59	-4.62	<.0001
Satterthwaite	Unequal	40.136	-4.34	<.0001

Equality of Variances				
Method	Num DF	Den DF	F Value	Pr > F
Folded F	25	34	2.39	0.0189

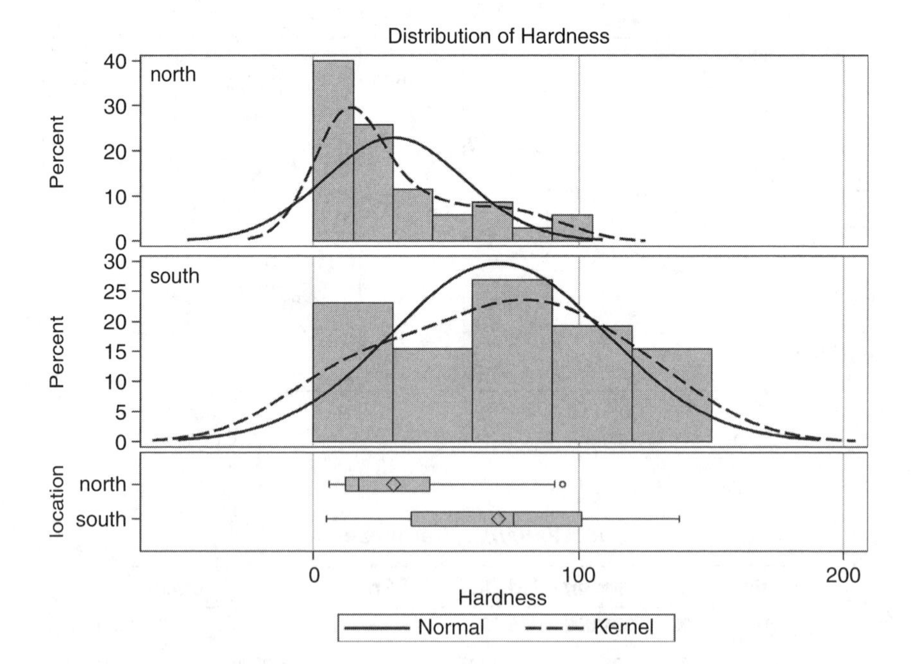

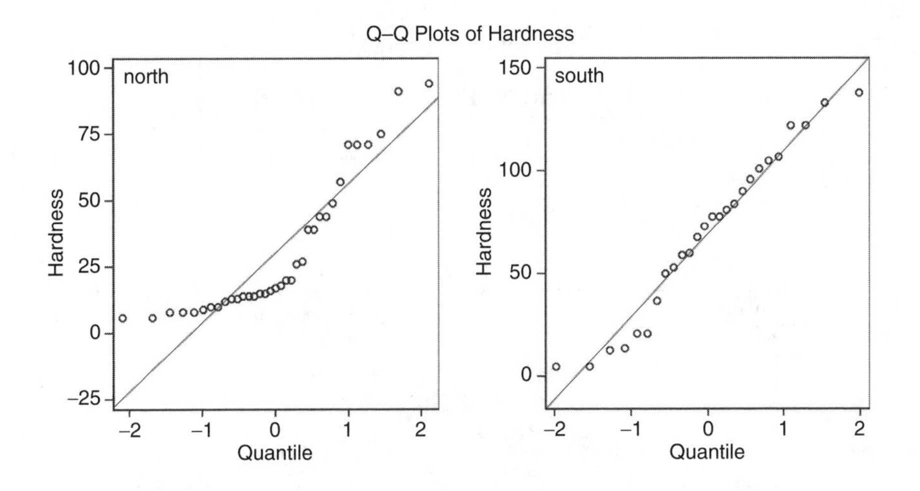

Variable: lhardnes

location	N	Mean	Std Dev	Std Err	Minimum	Maximum
north	35	3.0744	0.8316	0.1406	1.7918	4.5433
south	26	3.9484	0.9544	0.1872	1.6094	4.9273
Diff (1-2)		-0.8740	0.8857	0.2293		

location	Method	Mean	95% CL Mean		Std Dev	95% CL Std Dev	
north		3.0744	2.7887	3.3601	0.8316	0.6727	1.0896
south		3.9484	3.5629	4.3339	0.9544	0.7485	1.3175
Diff (1-2)	**Pooled**	-0.8740	-1.3328	-0.4151	0.8857	0.7508	1.0803
Diff (1-2)	**Satterthwaite**	-0.8740	-1.3442	-0.4037			

Method	Variances	DF	t Value	Pr > |t|
Pooled	Equal	59	-3.81	0.0003
Satterthwaite	Unequal	49.56	-3.73	0.0005

Equality of Variances				
Method	**Num DF**	**Den DF**	**F Value**	**Pr > F**
Folded F	25	34	1.32	0.4496

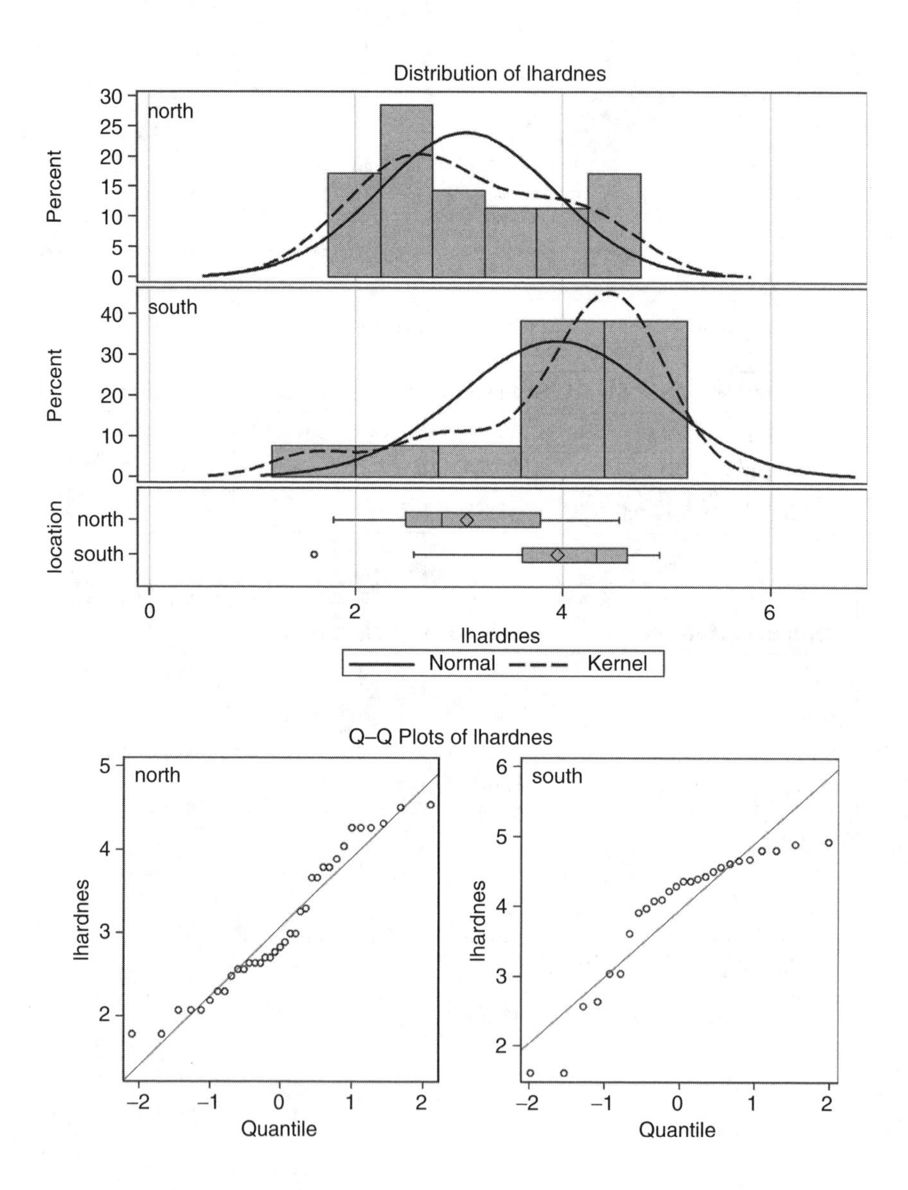

variable, it is seen that the *t*-test still indicates a highly significant difference, but in this case the test for homogeneity is non-significant.

The result from the non-parametric Wilcoxon test (Table 2.7) once again indicates that the mean water hardness of towns in the north differs from that of those in the south. The box plot in Table 2.7 shows the distribution of the Wilcoxon scores on which the test is based. These are simply the rankings of the observations in a joint ranking of the data from both the groups.

Table 2.7 SAS Output (Numerical and Graphical) from Proc Npar1way

The NPAR1WAY Procedure

Wilcoxon Scores (Rank Sums) for Variable Hardness Classified by Variable location					
location	N	Sum of Scores	Expected Under H0	Std Dev Under H0	Mean Score
north	35	832.50	1085.0	68.539686	23.785714
south	26	1058.50	806.0	68.539686	40.711538
Average scores were used for ties.					

Wilcoxon Two-Sample Test	
Statistic	1058.5000
Normal Approximation	
Z	3.6767
One-Sided Pr > Z	0.0001
Two-Sided Pr > |Z|	0.0002
t Approximation	
One-Sided Pr > Z	0.0003
Two-Sided Pr > |Z|	0.0005
Z includes a continuity correction of 0.5.	

Kruskal-Wallis Test	
Chi-Square	13.5718
DF	1
Pr > Chi-Square	0.0002

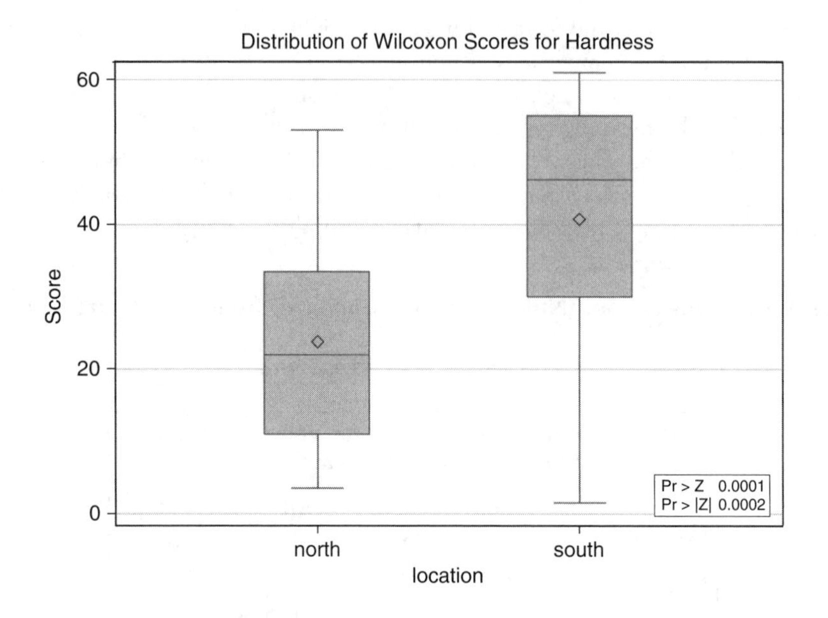

Exercises

1 *Mortality and Water Hardness*

1. Generate box plots of mortality and water hardness by location (use `proc boxplot`).
2. Use `proc univariate` to compare the distribution of water hardness with the log normal and exponential distributions.
3. Produce histograms of both mortality and water hardness with, in each case, a kernel density estimate of the variable's distribution superimposed.
4. Produce separate perspective plots of the estimated bivariate density of northern and southern towns.
5. Reproduce the scatterplot in Figure 2.6 with added linear regression fits of mortality against hardness for both northern and southern towns. Use different line types for the two regions.

2 *Guessing the Width of a Room*

Shortly after metric units of length were officially introduced in Australia in the 1970s, each of a group of 44 students was asked to guess, to the nearest metre, the width of the lecture hall in which they were sitting. Another group of 69 students in the same room was asked to guess the width in feet, to the nearest foot. The true width of the hall was 13.1 m (43.0 ft). The data were collected by professor T. Lewis and are taken from Hand et al. (1994).

The variables in width.dat are

- **ID:** Student identifier
- **Group:** 1 = guesses in metres, 2 = guesses in feet
- **Guess:** the estimate of the room width

1. Construct side-by-side box plots for the guesses made in feet and those made in metres (after conversion to feet). What informal conclusions would you draw from the plot?
2. Carry out appropriate formal tests to assess the hypotheses (a) that the mean of the population of guesses made in metres is 13.1 and (b) that the mean of the population of guesses made in feet is 43.0.
3. Find the 95% confidence interval for the difference in the means of the guesses made in feet and those made in metres (after conversion to feet).
4. What conclusions do you draw from the results of the plots and analyses?

3 Wave Power at Sea

In a design study for a device to generate electricity from wave power at sea, experiments were carried out on scale models in a wave tank to establish how the choice of mooring method for the system affected the bending stress produced in parts of the device. The wave tank could simulate a wide range of sea states and the model system was subjected to the same sample of sea states with each of the two mooring methods, one of which was considerably cheaper than the other. The data are taken from Hand et al. (1994).

The variables in waves.dat are

- **ID:** Sea state identifier
- **Method1:** Bending stress (root mean squared bending moment in Newton metres) for first mooring method
- **Method2:** Bending stress for second mooring method

1. Construct a box plot of the differences in bending stress for the two mooring methods in each sea state.
2. Carry out the appropriate test of whether mean bending stress differs for the two mooring methods.
3. What are your conclusions?

4 Stellar Evolution

The Hertzsprung–Russell (H–R) diagram forms the basis of the theory of stellar evolution. The diagram is essentially a plot of the energy output of stars plotted against their surface temperature. Data from the H–R diagram of Star Cluster CYG OBI, calibrated according to Vanisma and De Greve (1972), are given in Hand et al. (1994).

The variables in stars.dat are

- **ID:** Star identifier
- **Logst:** Log surface temperature
- **Logli:** Log light intensity

1. Construct a scatterplot of the two variables logst and logli and add an estimated bivariate density to the scatterplot. Experiment with different kernels and bandwidths.
2. Estimate the correlation between logst and logli.

Chapter 3

Simple Inference for Categorical Data: From Sandflies to Organic Particulates in the Air

3.1 Introduction

In this chapter, we consider the analysis of categorical data. We begin by looking at tabulating raw data into cross-classifications, that is, *contingency tables,* using the mortality and water hardness data from the previous chapter. We shall then examine six data sets where the data are already tabulated; a description of each of these data sets is given below. The main question of interest in each case involves assessing the relationship between pairs of categorical variables using the chi-square test or some suitable alternative.

The six cross-classified data sets to be examined in this chapter are as follows:

- *Sandflies* (Table 128 in *SDS*): These data are given in Table 3.1 and show the number of male and female sandflies caught in light traps set 3 and 35 ft above the ground at a site in Eastern Panama. The question of interest is, does the proportion of males and females caught at a particular height differ?
- *Acacia ants* (Table 27 in *SDS*): These data, given in Table 3.2, record the results of an experiment with acacia ants. All but 28 trees of two species

Table 3.1 Data on Sandflies Caught in Light Traps

	Sandflies	
	3 ft	*35 ft*
Males	173	125
Females	150	73
Total	323	198

Table 3.2 Data on Acacia Ants

Acacia Species	*Not Invaded*	*Invaded*	*Total*
A	2	13	15
B	10	3	13
Total	12	16	28

Table 3.3 Data on Piston-Ring Failures

Compressor No.	*North*	*Centre*	*South*	*Total*
1	17	17	12	46
2	11	9	13	33
3	11	8	19	38
4	14	7	28	49
Total	53	41	72	166

(15 species A, 13 species B) of acacia were cleared from an area in Central America, and the 28 trees were cleared of ants using insecticide. Sixteen colonies of a particular species of ant were obtained from other trees of species A. The colonies were placed roughly equidistant from the 28 trees and were allowed to invade them. The question of interest is whether the invasion rate differs between the two species of trees?

- *Piston-ring failures* (Table 15 in *SDS*): These data are reproduced in Table 3.3, and show the number of failures of piston rings in each of three legs in each of four steam-driven compressors located in the same building. The compressors have identical design and are orientated in the same way. The question of interest is whether the failure location pattern is different for different compressors?
- *Oral contraceptives*: These data appear in Table 3.4 and arise from a study reported by Sartwell et al. (1969). The study was conducted in a number of

Table 3.4 Data on Oral Contraceptive Use

	Controls	
Oral Contraceptive Use	*Used*	*Not Used*
Cases used	10	57
Cases not used	13	95

Table 3.5 Data or Oral Lesions in Different Regions of India

		Region	
Site of Lesion	*Kerala*	*Gujarat*	*Andhra*
Buccal mucosa	8	1	8
Labial mucosa	0	1	0
Commissure	0	1	0
Gingiva	0	1	0
Hard palate	0	1	0
Soft palate	0	1	0
Tongue	0	1	0
Floor of mouth	1	0	1
Alveolar ridge	1	0	1

hospitals in several large American cities. In those hospitals all the married women identified as suffering from idiopathic thromboembolism (blood clots) over a 3-year period were individually matched with a suitable control, those being female patients discharged alive from the same hospital in the same 6-month time interval as the case. In addition, they were individually matched to cases on age, marital status, race and so on. Patients and controls were then asked about their use of oral contraceptives. The question of interest is whether the rate of oral contraceptive use is different in those women suffering from blood clots from those who do not have the condition?

- *Oral lesions*: These data appear in Table 3.5; they give the location of oral lesions obtained in house-to-house surveys in three geographic regions of rural India, and the question of interest is whether the proportion of oral lesions in different locations differ between the regions?
- *Particulates in the air*: These data are given in Table 3.6; they arise from a study involving cases of bronchitis by level of organic particulates in the air and by age (Somes and O'Brien, 1985). Here the question to be addressed is, does the rate of bronchitis differ by level of organic particulates?

Table 3.6 Data on Bronchitis and Organic Particulates in the Air

Age	Particulates Level	Bronchitis Yes	Bronchitis No	Total
15–24	High	20	382	402
	Low	9	214	223
23–30	High	10	172	182
	Low	7	120	127
40+	High	12	327	339
	Low	6	183	189

3.2 Methods of Analysis

Contingency tables are one of the most common ways to summarize categorical data. Tables 3.1, 3.2 and 3.4 are examples of 2×2 contingency tables (although Table 3.4 has a quite different structure from Tables 3.1 and 3.2, as we shall explain later), Table 3.3 is an example of a 3×4 table, and Table 3.5 an example of a 9×3 table with very sparse data. Table 3.6 is an example of a series of 2×2 tables involving the same two variables. For all such tables interest generally lies in assessing whether or not there is an association between the row variables and the column variables that form the table. Most commonly a *chi-square test of independence* is used to answer this question, although alternatives such as *Fisher's exact test* or *McNemar's test* may be needed when the sample size is small (Fisher's test) or the data consist of matched samples (McNemar's test). In addition, in 2×2 tables it may be required to calculate a confidence interval for the difference in two population proportions. For a series of 2×2 tables the Mantel–Haenszel test may be appropriate (see later). (Details of all the tests mentioned are given in Everitt, 1992.)

3.3 Analysis Using SAS

3.3.1 Cross-Classifying Raw Data

We will first demonstrate how raw data may be put into the form of a cross-classification using the data on mortality and water hardness from the previous chapter.

```
data water;
  infile 'c:\handbook3\datasets\water.dat';
  input flag $ 1 Town $ 2-18 Mortal 19-22 Hardness 25-27;
```

```
  if flag = '*' then location = 'north';
    else location = 'south';
  mortgrp = mortal > 1555;
  hardgrp = hardness > 39;
run;
proc freq data = water;
  tables mortgrp*hardgrp/chisq;
run;
```

The raw data are read into a SAS data set, water, as described in the previous chapter. In this instance, two new variables are computed – mortgrp and hardgrp – which dichotomize mortality and water hardness at their medians. Strictly speaking, the expression mortal > 1555 is a logical expression yielding the result "true" or "false," but in SAS these are represented by the values 1 or 0, respectively.

Proc freq is used both to produce contingency tables and to analyse them. The tables statement defines the table to be produced and specifies the analysis of it. The variables that form the rows and columns are joined with an asterisk. These may be numeric or character variables. One-way frequency distributions are produced where variables are not joined by asterisks. Several tables may be specified on a single tables statement.

The options after the "/" specify the types of analysis. The chisq option requests chi-square tests of independence and measures of association based on chi-square. The output is shown in Table 3.7. We leave commenting on the contents of this type of output until later.

Table 3.7 SAS Output from Applying Proc Freq to the Mortality and Water Hardness Data in Chapter 2

Table of mortgrp by hardgrp

mortgrp Frequency Percent Row Pct Col Pct	hardgrp 0	1	Total
0	11 18.03 35.48 34.38	20 32.79 64.52 68.97	31 50.82
1	21 34.43 70.00 65.63	9 14.75 30.00 31.03	30 49.18
Total	32 52.46	29 47.54	61 100.00

Statistics for Table of mortgrp by hardgrp

Statistic	DF	Value	Prob
Chi-Square	1	7.2830	0.0070
Likelihood Ratio Chi-Square	1	7.4403	0.0064
Continuity Adj. Chi-Square	1	5.9647	0.0146
Mantel-Haenszel Chi-Square	1	7.1636	0.0074
Phi Coefficient		-0.3455	
Contingency Coefficient		0.3266	
Cramer's V		-0.3455	

Fisher's Exact Test	
Cell (1,1) Frequency (F)	11
Left-sided Pr <= F	0.0070
Right-sided Pr >= F	0.9986
Table Probability (P)	0.0056
Two-sided Pr <= P	0.0103

Sample Size = 61

Now we move on to consider the six data sets that actually arise in the form of contingency tables. The freq procedure is again used to analyse such tables and compute tests and measures of association.

3.3.2 Sandflies

The data on sandflies in Table 3.1 can be read into a SAS data set with each cell as a separate observation and the rows and columns identified, as follows:

```
data sandflies;
  input sex $ height n;
cards;
m  3   173
m  35  125
f  3   150
f  35   73
;
```

The rows are identified by a character variable sex with values "m" and "f". The columns are identified by the variable height with values 3 and 35. The variable n contains the cell count. Proc freq can then be used to analyse the table.

```
proc freq data = sandflies;
  tables sex*height /chisq riskdiff;
  weight n;
run;
```

The riskdiff option requests differences in risks (or binomial proportions) and their confidence limits.

The weight statement specifies a variable that contains weights for each observation. It is most commonly used to specify cell counts, as in this example. The default weight is 1, so the weight statement is not required when the data set consists of observations on individuals.

The results are shown in Table 3.8. First the 2×2 table of data is printed augmented with percentages of total frequency, row and column percentages. A number of statistics calculated from the frequencies in the table are then printed, beginning with the well-known chi-square statistic used to test for the independence of the two variables forming the table. Here the p value associated with the chi-square statistic suggests that sex and height are not independent. The *likelihood ratio chi-square* is an alternative statistic for testing independence, which is described in Everitt (1992). Here, the usual chi-square statistic and the likelihood ratio statistic take very similar values. Next, the continuity-adjusted chi-square statistic is printed. This involves what is usually known as Yates's correction, again described

Table 3.8 SAS Output from Proc Freq Applied to the Data on Sandflies

Table of sex by height

sex Frequency Percent Row Pct Col Pct	height 3	 35	 Total
f	150 28.79 67.26 46.44	73 14.01 32.74 36.87	223 42.80
m	173 33.21 58.05 53.56	125 23.99 41.95 63.13	298 57.20
Total	323 62.00	198 38.00	521 100.00

Statistics for Table of sex by height

Statistic	DF	Value	Prob
Chi-Square	1	4.5930	0.0321
Likelihood Ratio Chi-Square	1	4.6231	0.0315
Continuity Adj. Chi-Square	1	4.2104	0.0402
Mantel-Haenszel Chi-Square	1	4.5842	0.0323
Phi Coefficient		0.0939	
Contingency Coefficient		0.0935	
Cramer's V		0.0939	

Fisher's Exact Test	
Cell (1,1) Frequency (F)	150
Left-sided Pr <= F	0.9875
Right-sided Pr >= F	0.0199
Table Probability (P)	0.0073
Two-sided Pr <= P	0.0360

Column 1 Risk Estimates

	Risk	ASE	(Asymptotic) 95% Confidence Limits		(Exact) 95% Confidence Limits	
Row 1	0.6726	0.0314	0.6111	0.7342	0.6068	0.7338
Row 2	0.5805	0.0286	0.5245	0.6366	0.5223	0.6372
Total	0.6200	0.0213	0.5783	0.6616	0.5767	0.6618
Difference	0.0921	0.0425	0.0088	0.1754		

Difference is (Row 1 - Row 2)

Column 2 Risk Estimates

	Risk	ASE	(Asymptotic) 95% Confidence Limits		(Exact) 95% Confidence Limits	
Row 1	0.3274	0.0314	0.2658	0.3889	0.2662	0.3932
Row 2	0.4195	0.0286	0.3634	0.4755	0.3628	0.4777
Total	0.3800	0.0213	0.3384	0.4217	0.3382	0.4233
Difference	-0.0921	0.0425	-0.1754	-0.0088		

Difference is (Row 1 - Row 2)

Sample Size = 521

in Everitt (1992). The correction is usually suggested as a simple way of dealing with what was once perceived as the problem of unacceptably small frequencies in contingency tables. Nowadays there are, as we shall see later, far better ways of dealing with the problem and there is now really no need to ever use Yates's correction. We shall leave explaining the Mantel–Haenszel statistic until later, but the next three statistics printed in Table 3.8, namely the *Phi coefficient*, *Contingency coefficient* and *Cramer's V*, are all essentially attempts to quantify the degree of the relationship between the two variables forming the contingency table. They are all described in detail in Everitt (1992).

Following these statistics in Table 3.8 is information on Fisher's exact test. This is more relevant to the data in Table 3.2 and so we leave its discussion until later. Next are the results of estimating a confidence interval for the difference in proportions in the contingency table. So, for example, the estimated difference in the proportion of female and male sandflies caught in the 3 ft light traps is 0.0921 (0.6726 − 0.5805). The standard error of this difference is calculated as

$$\sqrt{\frac{0.6726(1-0.6726)}{223}+\frac{0.5805(1-0.5805)}{298}}. \tag{3.1}$$

That is, the value of 0.0425 given in Table 3.8 as ASE, asymptotic standard error. The confidence interval for the difference in proportions is therefore

$$0.0921 \pm 1.96 \times 0.0425 = (0.0088{,}0.1754), \tag{3.2}$$

as given in Table 3.8. The proportion of female sandflies caught in the 3 ft traps is larger than the corresponding proportion for male sandflies.

3.3.3 Acacia Ants

The data on Acacia ants are also in the form of a contingency table and are read in as four observations representing cell counts.

```
data ants;
  input species $ invaded $ n;
cards;
A  no    2
A  yes  13
B  no   10
B  yes   3
;
proc freq data=ants;
  tables species*invaded/chisq expected;
  weight n;
run;
```

In this example, the expected option on the tables statement is used to print expected values under the independence hypothesis for each cell.

The results are shown in Table 3.9. Here, due to the small frequencies in the table, Fisher's exact test might be the preferred option, although all the tests of independence have very small associated p values, and so very clearly species and invasion are not independent. A higher proportion of ants invaded species A than species B.

Table 3.9 SAS Output from Proc Freq for Data on Acacia Ants

Table of species by invaded species / invaded Frequency Expected Percent Row Pct Col Pct	no	yes	Total
A	2 6.4286 7.14 13.33 16.67	13 8.5714 46.43 86.67 81.25	15 53.57
B	10 5.5714 35.71 76.92 83.33	3 7.4286 10.71 23.08 18.75	13 46.43
Total	12 42.86	16 57.14	28 100.00

Statistics for Table of species by invaded

Statistic	DF	Value	Prob
Chi-Square	1	11.4991	0.0007
Likelihood Ratio Chi-Square	1	12.4173	0.0004
Continuity Adj. Chi-Square	1	9.0491	0.0026
Mantel-Haenszel Chi-Square	1	11.0885	0.0009
Phi Coefficient		-0.6408	
Contingency Coefficient		0.5396	
Cramer's V		-0.6408	

Fisher's Exact Test	
Cell (1,1) Frequency (F)	2
Left-sided Pr <= F	0.0010
Right-sided Pr >= F	1.0000
Table Probability (P)	9.871E-04
Two-sided Pr <= P	0.0016

Sample Size = 28

3.3.4 Piston Rings

Moving on now to the piston-rings data, they are read in and analysed as follows:

```
data pistons;
  input machine site $ n;
cards;
1  North   17
1  Centre  17
1  South   12
2  North   11
2  Centre   9
2  South   13
3  North   11
3  Centre   8
3  South   19
4  North   14
4  Centre   7
4  South   28
;
proc freq data=pistons order=data;
  tables machine*site/chisq deviation cellchi2 norow nocol nopercent;
  weight n;
run;
```

The order = data option on the proc statement specifies that the rows and columns of the tables follow the order in which they occur in the data. The default is the number order for numeric variables and alphabetical order for character variables.

The deviation option on the tables statement requests the printing of residuals in the cells, and the cellchi2 option requests that each cell's contribution to the overall chi square be printed. To make it easier to view the results, the row, column and overall percentages are suppressed with the norow, nocol and nopercent options, respectively.

Here the chi-square test for independence given in Table 3.10 shows only relatively weak evidence of a departure from independence. (The relevant p value is .069.) But the simple residuals (the differences between an observed frequency and that expected under independence) suggest that failures are fewer than might be expected in the South leg of Machine 1 and more than expected in the South leg of Machine 4. (Other types of residuals may be more useful – see Exercise 3.4.2.)

Table 3.10 SAS Output from Proc Freq Applied to the Data on Piston-Ring Failures

Table of machine by site

machine Frequency Deviation Cell Chi-Square	site North	Centre	South	Total
1	17 2.3133 0.3644	17 5.6386 2.7983	12 -7.952 3.1692	46
2	11 0.4639 0.0204	9 0.8494 0.0885	13 -1.313 0.1205	33
3	11 -1.133 0.1057	8 -1.386 0.2045	19 2.5181 0.3847	38
4	14 -1.645 0.1729	7 -5.102 2.1512	28 6.747 2.1419	49
Total	53	41	72	166

Statistics for Table of machine by site

Statistic	DF	Value	Prob
Chi-Square	6	11.7223	0.0685
Likelihood Ratio Chi-Square	6	12.0587	0.0607
Mantel-Haenszel Chi-Square	1	5.4757	0.0193
Phi Coefficient		0.2657	
Contingency Coefficient		0.2568	
Cramer's V		0.1879	

Sample Size = 166

3.3.5 Oral Contraceptives

The oral contraceptives data involve matched observations. Consequently, they cannot be analysed with the usual chi-square statistic. Instead, they require application of McNemar's test, as described in Everitt (1992). The data may be read in and analysed with the following SAS commands:

```
data the_pill;
  input caseuse $ contruse $ n;
cards;
Y Y 10
Y N 57
N Y 13
N N 95
;
proc freq data=the_pill order=data;
  tables caseuse*contruse/agree;
  weight n;
run;
```

The agree option on the tables statement requests measures of agreement, including the one of most interest here, McNemar's test. The results appear in Table 3.11. The test of no association between using oral contraceptives and suffering from blood clots is rejected. The proportion of matched pairs, in which the case has used

Table 3.11 SAS Output from Proc Freq Applied to the Data on Blood Clots and Oral Contraceptive Use

Table of caseuse by contruse

caseuse Frequency Percent Row Pct Col Pct	contruse Y	N	Total
Y	10 5.71 14.93 43.48	57 32.57 85.07 37.50	67 38.29
N	13 7.43 12.04 56.52	95 54.29 87.96 62.50	108 61.71
Total	23 13.14	152 86.86	175 100.00

Statistics for Table of caseuse by contruse

McNemar's Test	
Statistic (S)	27.6571
DF	1
Pr > S	<.0001

Simple Kappa Coefficient	
Kappa	0.0330
ASE	0.0612
95% Lower Conf Limit	-0.0870
95% Upper Conf Limit	0.1530

Sample Size = 175

oral contraceptives and the control has not, is considerably higher than those pairs where the case is the reverse.

3.3.6 Oral Cancers

The data on the regional distribution of oral cancers are read in using the following data step:

```
data lesions;
  length region $8.;
  input site $ 1-16 n1 n2 n3;
  region='Kerala';
  n=n1;
  output;
  region='Gujarat';
  n=n2;
  output;
  region='Andhra';
  n=n3;
  output;
  drop n1-n3;
cards;
Buccal Mucosa   8  1  8
Labial Mucosa   0  1  0
Commissure      0  1  0
Gingiva         0  1  0
```

```
Hard palate     0  1  0
Soft palate     0  1  0
Tongue          0  1  0
Floor of mouth  1  0  1
Alveolar ridge  1  0  1
;
```

This data step reads in the values for three cell counts from a single line of instream data and then creates three separate observations in the output data set. This is achieved by using three output statements in the data step. The effect of each output statement is to create an observation in the data set with the data values that are current at that point. First, the input statement reads a line of data that contains the three cell counts. It uses column input to read the first 16 columns into the site variable, and then list input to read the three cell counts into variables n1 to n3. When the first output statement is executed the region variable has been assigned the value "Kerala" and the variable n has been set equal to the first of the three cell counts read in. At the second output statement, the value of region is "Gujarat," and n equals n2, the second cell count, and so on for the third output statement. When restructuring data such as these, it is wise to check the results either by viewing the resultant data set interactively or by using proc print. The SAS log also gives the number of variables and observations in any data set created and so can be used to provide a check.

The drop statement excludes the variables mentioned from the lesions data set.

```
proc freq data=lesions order=data;
  tables site*region/exact;
  weight n;
run;
```

For 2×2 tables, Fisher's exact test is calculated and printed by default. For larger tables, exact tests must be explicitly requested with the exact option on the tables statement. Here, due to the very sparse nature of the data it is likely that the exact approach will differ from the usual chi-square procedure. The results given in Table 3.12 confirm this. The chi-square test has an associated p value of .14, indicating that the hypothesis of independence site and region is acceptable. The exact test has an associated p value of .01 indicating that site of lesion and region are associated. Here the chi-square test is unsuitable because of the very sparse nature of the data.

3.3.7 Particulates and Bronchitis

The final data set to be analysed in this chapter, the bronchitis data in Table 3.6, involves 2×2 tables for bronchitis and level of organic particulates for three age

Table 3.12 SAS Output from Proc Freq Applied to the Data on Oral Lesions

Table of site by region

site / Frequency / Percent / Row Pct / Col Pct	region: Kerala	Gujarat	Andhra	Total
Buccal Mucosa	8 29.63 47.06 80.00	1 3.70 5.88 14.29	8 29.63 47.06 80.00	17 62.96
Labial Mucosa	0 0.00 0.00 0.00	1 3.70 100.00 14.29	0 0.00 0.00 0.00	1 3.70
Commissure	0 0.00 0.00 0.00	1 3.70 100.00 14.29	0 0.00 0.00 0.00	1 3.70
Gingiva	0 0.00 0.00 0.00	1 3.70 100.00 14.29	0 0.00 0.00 0.00	1 3.70
Hard palate	0 0.00 0.00 0.00	1 3.70 100.00 14.29	0 0.00 0.00 0.00	1 3.70
Soft palate	0 0.00 0.00 0.00	1 3.70 100.00 14.29	0 0.00 0.00 0.00	1 3.70
Tongue	0 0.00 0.00 0.00	1 3.70 100.00 14.29	0 0.00 0.00 0.00	1 3.70
Floor of mouth	1 3.70 50.00 10.00	0 0.00 0.00 0.00	1 3.70 50.00 10.00	2 7.41
Alveolar ridge	1 3.70 50.00 10.00	0 0.00 0.00 0.00	1 3.70 50.00 10.00	2 7.41
Total	10 37.04	7 25.93	10 37.04	27 100.00

Statistics for Table of site by region

Statistic	DF	Value	Prob
Chi-Square	16	22.0992	0.1400
Likelihood Ratio Chi-Square	16	23.2967	0.1060
Mantel-Haenszel Chi-Square	1	0.0000	1.0000
Phi Coefficient		0.9047	
Contingency Coefficient		0.6709	
Cramer's V		0.6397	

WARNING: 93% of the cells have expected counts less than 5. Chi-Square may not be a valid test.

Fisher's Exact Test	
Table Probability (P)	5.334E-06
Pr <= P	0.0101

Sample Size = 27

groups. The data could be collapsed over age and the aggregate 2×2 table analysed as described previously. But the dangers of this procedure are well documented (see, e.g., Everitt, 1992). In particular, such pooling of contingency tables can generate an association when in separate tables there is none. A more appropriate test in this situation is the Mantel–Haenszel test. For a series of k, 2×2 tables the test statistic for testing the hypothesis of no association is

$$X^2 = \frac{\left[\sum_{i=1}^{k} a_i - \sum_{i=1}^{k} \frac{(a_i+b_i)(a_i+c_i)}{n_i}\right]^2}{\sum_{i=1}^{k} \frac{(a_i+b_i)(c_i+d_i)(a_i+c_i)(b_i+d_i)}{n_i^2(n_i-1)}}, \tag{3.3}$$

where a_i, b_i, c_i, d_i represent the counts in the four cells of the ith table and n_i is the total number of observations in the ith table. Under the null hypothesis of independence in all tables, this statistic has a chi-squared distribution with a single degree of freedom.

The data can be read in and analysed using the following SAS code:

```
data bronchitis;
  input agegrp level $ bronch $ n;
```

```
cards;
1 H Y 20
1 H N 382
1 L Y 9
1 L N 214
2 H Y 10
2 H N 172
2 L Y 7
2 L N 120
3 H Y 12
3 H N 327
3 L Y 6
3 L N 183
;
proc freq data=bronchitis order=data;
  tables agegrp*level*bronch/cmh noprint;
  weight n;
run;
```

The tables statement specifies a three-way tabulation with agegrp defining the strata. The cmh option requests the Cochran–Mantel–Haenszel statistics, and the noprint option suppresses the tables. The results are shown in Table 3.13. There is no evidence of an association between level of organic particulates and suffering from bronchitis. The p value associated with the test statistic is 0.64 and the assumed common odds ratio is calculated as

$$\hat{\Psi}_{\text{pooled}} = \frac{\Sigma(a_i d_i / n_i)}{\Sigma(b_i c_i / n_i)}, \qquad (3.4)$$

which takes the value 1.14 with a confidence interval of (0.67, 1.93).

Table 3.13 SAS Output from Proc Freq Applied to the Data on Bronchitis and Organic Particulates in the Air

Summary Statistics for level by bronch
Controlling for agegrp

Cochran-Mantel-Haenszel Statistics (Based on Table Scores)				
Statistic	**Alternative Hypothesis**	**DF**	**Value**	**Prob**
1	Non-zero Correlation	1	0.2215	0.6379
2	Row Mean Scores Differ	1	0.2215	0.6379
3	General Association	1	0.2215	0.6379

Estimates of the Common Relative Risk (Row1/Row2)				
Type of Study	**Method**	**Value**	**95% Confidence Limits**	
Case-Control (Odds Ratio)	Mantel-Haenszel	1.1355	0.6693	1.9266
	Logit	1.1341	0.6678	1.9260
Cohort (Col1 Risk)	Mantel-Haenszel	1.1291	0.6808	1.8728
	Logit	1.1272	0.6794	1.8704
Cohort (Col2 Risk)	Mantel-Haenszel	0.9945	0.9725	1.0170
	Logit	0.9945	0.9729	1.0166

Breslow-Day Test for Homogeneity of the Odds Ratios	
Chi-Square	0.1173
DF	2
Pr > ChiSq	0.9430

Total Sample Size = 1462

Because the Mantel–Haenszel test will give sensible results only if the association between the two variables is both the same size and same direction in each 2×2 table, it is generally sensible to look at the results of the Breslow–Day test for homogeneity of odds ratios given in Table 3.13. Here there is no evidence against homogeneity. The Breslow–Day test is described in Agresti (2007).

Exercises

1 *Oral Contraceptives*

1. Construct a confidence interval for the difference in the proportion of women who used oral contraceptives suffering from blood clots and the corresponding proportion for women who used contraceptives not suffering from blood clots.
2. What conclusion do you draw from the constructed interval? Do you think the confidence interval or the p value given in the text for these data is more useful when assessing these data?

2 *Piston-Ring Data*

1. For the piston-ring data the "residuals" used in the text were simply observed frequency minus expected under independence. Those are not satisfactory for

a number of reasons discussed in Everitt (1992). More suitable residuals are r and r_{adj} given by

$$r = \frac{\text{Observed} - \text{Expected}}{\sqrt{\text{Expected}}} \quad \text{and} \quad r_{\text{adj}} = \frac{r}{\sqrt{\left(1 - \frac{\text{Row total}}{\text{Sample size}}\right)\left(1 - \frac{\text{Column total}}{\text{Sample size}}\right)}}.$$

Calculate for the piston-ring data and compare what each of the three types has to say about the data.

2. Do the simple residuals used in the text. Do the two types of residual just calculated tell you anything different about the data? Which residual do you think is most appropriate and why?

3 Hodgkin's Disease

In a study of Hodgkin's disease, a cancer of the lymph modes, each of 538 patients with the disease, was classified by histological type and by their response to treatment 3 months after it had begun. (The data are described in more detail in Hancock et al., 1979.) The histological types are LP = lymphocyte predominance, NS = nodular sclerosis, MC = mixed cellularity and LD = lymphocyte depletion.

		Response			
		Positive	*Partial*	*None*	*Total*
Histological type	LP	74	18	12	104
	NS	68	16	12	96
	MC	154	54	58	266
	LD	18	10	44	72
Total		314	98	126	538

1. What, if any, is the relationship between histological type and response to treatment?

4 Church Assembly Vote

In 1967 the Church Assembly of the Church of England voted on a motion that individual women who felt called to exercise "the office and work of a priest in the church" should now be considered, on the same basis as men, as candidates for Holy Orders. The voting patterns are given separately for bishops, clergy and laity. (The original data were given in *The Daily Telegraph* and are also given in Hand et al., 1994.)

	Vote		
	Aye	*No*	*Abstained*
Bishops	1	8	8
Clergy	14	96	20
Laity	45	207	52

How do the voting patterns of the three groups differ?

5 *Mouth Cancer*

In the data given in Table 3.5, the frequencies for the Kerala and Andhra regions are identical. Re-analyse the data after simply summing the frequencies for those two regions and reducing the number of columns of the table by one.

Chapter 4

Analysis of Variance I: Treating Hypertension

4.1 Introduction

Maxwell and Delaney (2003) describe a study in which the effects of three possible treatments for hyptertension were investigated. The details of the treatments are as follows:

Treatment	*Description*	*Levels*
Drug	Medication	Drug X, drug Y, drug Z
Biofeed	Psychological feedback	Present, absent
Diet	Special diet	Present, absent

All 12 combinations of the three treatments were included in a $3 \times 2 \times 2$ design. Seventy-two subjects suffering from hypertension were recruited for the study, with six being randomly allocated to each of twelve treatment combinations. Blood pressure measurements were made on each subject after treatment, leading to the data in Table 4.1.

Questions of interest concern differences in mean blood pressure for the different levels of the three treatments and the possibility of *interactions* between the treatments.

Table 4.1 Blood Pressure Measurements

		Special Diet	
Biofeedback	*Drug*	*No*	*Yes*
Present	X	170 175 165 180 160 158	161 173 157 152 181 190
	Y	186 194 201 215 219 209	164 166 159 182 187 174
	Z	180 187 199 170 204 194	162 184 183 156 180 173
Absent	X	173 194 197 190 176 198	164 190 169 164 176 175
	Y	189 194 217 206 199 195	171 173 196 199 180 203
	Z	202 228 190 206 224 204	205 199 170 160 179 179

4.2 Analysis of Variance Model

A possible model for these data is

$$y_{ijkl} = \mu + \alpha_i + \beta_j + \gamma_k + (\alpha\beta)_{ij} + (\alpha\gamma)_{ik} + (\beta\gamma)_{jk} + (\alpha\beta\gamma)_{ijk} + \varepsilon_{ijkl} \quad (4.1)$$

where y_{ijkl} represents the blood pressure of the *l*th subject for the *i*th drug, the *j*th level of biofeedback and the *k*th level of diet; μ is the overall mean; α_i, β_j and γ_k are the main effects of drugs, biofeedback and diets; $(\alpha\beta)_{ij}$, $(\alpha\gamma)_{ik}$ and $(\beta\gamma)_{jk}$ are the first-order interaction terms between pairs of treatments; $(\alpha\beta\gamma)_{ijk}$ represents the second-order interaction term of the three treatments; and ε_{ijkl} represents the residual or error terms assumed to be normally distributed with zero mean and variance σ^2.

The model as specified is over parameterized and the parameters have to be constrained in some way, commonly by requiring them to sum to zero or setting one parameter at zero (see Everitt, 2001, for details).

Such a model leads to a partition of the variation in the observations into parts due to main effects, first-order interactions between pairs of factors and a second-order interaction between all three factors. This partition leads to a series of F tests for assessing the significance or otherwise of these various components. The assumptions underlying these F tests are

1. Observations are independent of one another.
2. Observations in each cell arise from a population having a normal distribution.
3. Observations in each cell are from populations having the same variance.

4.3 Analysis Using SAS

We assume the 72 blood pressure readings shown in Table 4.1 are in an ASCII file, hypertension.dat. The SAS code used for reading and labelling the data is as follows:

```
data hyper;
  infile 'hypertension.dat';
  input n1-n12;
  if _n_<4 then biofeed='P';
  else biofeed='A';;
  if _n_ in(1,4) then drug='X';
  if _n_ in(2,5) then drug='Y';
  if _n_ in(3,6) then drug='Z';
  array nall {12} n1-n12;
  do i=1 to 12;
    if i>6 then diet='Y';
    else diet='N';
        bp=nall{i};
        cell=drug||biofeed||diet;
        output;
  end;
  drop i n1-n12;
run;
```

The 12 blood pressure readings per row, or line, of data are read into variables n1-n12 and used to create 12 separate observations. The row and column positions in the data are used to determine the values of the factors in the design: drug, biofeedback and diet.

First, the input statement reads the 12 blood pressure values into variables n1 to n12. It uses list input, which assumes the data values to be separated by spaces.

The next group of statements uses the SAS automatic variable, _n_, to determine which row of data is being processed and hence to set the values of drug and biofeed. Because six lines of data will be read, one line per iteration of the data step, _n_ will increment from 1 to 6 corresponding to the line of data read with the input statement.

The key elements in splitting the one line of data into separate observations are the array, the do loop and the output statement.

The array statement defines an array by specifying the name of the array, nall here, the number of variables to be included in it in braces, and the list of variables to be included, n1 to n12 in this case.

In SAS, an array is a shorthand way of referring to a group of variables. In effect, it provides aliases for them so that each variable can be referred to by using the name of the array and its position within the array in braces. For example, in this data step, n12 could be referred to as nall{12} or as nall{i} when the variable i has the value 12. However, the array lasts for only the duration of the data step in which it is defined.

The main purpose of an iterative do loop, like the one used here, is to repeat the statements between the do and the end a fixed number of times, with an index variable changing at each repetition. When used to process each of the variables in an array, the do loop should start with the index variable equal to 1 and end when it equals the number of variables in the array.

Within the do loop, in this example, the index variable, i, is first used to set the appropriate values for diet. Then a variable for the blood pressure reading is assigned 1 of the 12 values input. A character variable, cell, is formed by concatenating the values of the drug, biofeed and diet variables. The double bar operator (||) concatenates character values.

The output statement writes an observation to the output data set with the current value of all variables. An output statement is not normally necessary, since without it an observation is automatically written out at the end of the data step. Putting an output statement within the do loop results in 12 observations being written to the data set.

Finally, the drop statement excludes the index variable i, and n1 to n12 from the output data set as they are no longer needed.

As with any relatively complex data manipulation, it is wise to check the results are as they should be, for example, by using proc print.

To begin the analysis, it will be helpful to look at some summary statistics for each of the cells of the design.

```
proc tabulate data=hyper;
  class drug diet biofeed;
  var bp;
  table drug*diet*biofeed,
  bp*(mean std n);
run;
```

The tabulate procedure is useful for displaying descriptive statistics in a concise tabular format. The variables that are used in the table must first be declared in either a class statement or a var statement. Class variables are those used to divide the observations into groups. Those declared on the var statement (analysis variables) are those for which descriptive statistics are to be calculated. The first part of the table statement, up to the comma, specifies how the rows of the table are to be formed and the remaining part specifies the columns. In this example, the rows comprise a hierarchical grouping of biofeed within diet within drug. The columns comprise the blood pressure mean, standard deviation and cell count for each of the groups. The resulting table is shown in Table 4.2. The differences between the standard deviations seen in this display may have implications for the analysis of variance of these data, since one of the assumptions made is that observations in each cell come from populations with the same variance.

There are various ways in which the homogeneity of variance assumption can be tested; we shall use the hovtest option of the anova procedure to apply Levene's test (Levene, 1960). Here the cell variable calculated above, which has 12 levels corresponding to the 12 cells of the design, is used.

Table 4.2 Summary Statistics for Blood Pressure Data

			bp		
Drug	*Diet*	*Biofeed*	*Mean*	*Std*	*N*
X	N	A	188.00	10.86	6
		P	168.00	8.60	6
	Y	A	173.00	9.80	6
		P	169.00	14.82	6
Y	N	A	200.00	10.08	6
		P	204.00	12.68	6
	Y	A	187.00	14.01	6
		P	172.00	10.94	6
Z	N	A	209.00	14.35	6
		P	189.00	12.62	6
	Y	A	182.00	17.11	6
		P	173.00	11.66	6

```
proc anova data=hyper;
  class cell;
  model bp=cell;
  means cell/hovtest;
run;
```

The results are shown in Table 4.3. Concentrating on the results of Levene's test given in this display, we see that there is no formal evidence of heterogeneity of variance, despite the rather different observed standard deviations noted in Table 4.3.

To apply the model specified in Equation 4.1 to the hypertension data, proc anova is now used as follows:

Table 4.3 Results from Using Proc Anova to Test for Homogeneity of Variance in the Blood Pressure Data

The ANOVA Procedure

Class Level Information		
Class	**Levels**	**Values**
cell	12	XAN XAY XPN XPY YAN YAY YPN YPY ZAN ZAY ZPN ZPY

Number of Observations Read	72
Number of Observations Used	72

The ANOVA Procedure

Dependent Variable: bp

Source	DF	Sum of Squares	Mean Square	F Value	Pr > F
Model	11	13194.00000	1199.45455	7.66	<.0001
Error	60	9400.00000	156.66667		
Corrected Total	71	22594.00000			

R-Square	Coeff Var	Root MSE	bp Mean
0.583960	6.784095	12.51666	184.5000

Source	DF	Anova SS	Mean Square	F Value	Pr > F
cell	11	13194.00000	1199.45455	7.66	<.0001

The ANOVA Procedure

Levene's Test for Homogeneity of bp Variance ANOVA of Squared Deviations from Group Means					
Source	**DF**	**Sum of Squares**	**Mean Square**	**F Value**	**Pr > F**
cell	11	180715	16428.6	1.01	0.4452
Error	60	971799	16196.6		

The ANOVA Procedure

Level of cell	N	bp Mean	bp Std Dev
XAN	6	188.000000	10.8627805
XAY	6	173.000000	9.7979590
XPN	6	168.000000	8.6023253
XPY	6	169.000000	14.8189068
YAN	6	200.000000	10.0796825
YAY	6	187.000000	14.0142784
YPN	6	204.000000	12.6806940
YPY	6	172.000000	10.9361785
ZAN	6	209.000000	14.3527001
ZAY	6	182.000000	17.1113997
ZPN	6	189.000000	12.6174482
ZPY	6	173.000000	11.6619038

```
proc anova data = hyper;
  class diet drug biofeed;
  model bp = diet|drug|biofeed;
  means diet*drug*biofeed;
ods output means = outmeans;
run;
```

The anova procedure is specifically for balanced designs, that is, those with the same number of observations in each cell. (Unbalanced designs should be analysed using proc glm as we shall illustrate in Chapter 5.) The class statement specifies the classification variables or factors. These may be numeric or character variables. The model statement specifies the dependent variable on the left-hand side of the equation and the effects, that is, factors and their interactions, on the right-hand side of the equation. Main effects are specified by including the variable name and interactions by joining the variable names with an asterisk. Joining variable names with a bar (|) is a shorthand way of specifying an interaction and all the lower-order interactions and main effects implied by it. So the model statement above is equivalent to

model blood pressure = diet drug diet*drug biofeed diet*biofeed drug*biofeed diet*drug*biofeed;

The order of the effects is determined by the expansion of the bar operator from left to right.

The means statement generates a table of cell means and the ods output statement specifies that this is to be saved in a SAS data set called outmeans.

The results are shown in Table 4.4. Here it is the analysis of variance table that is of most interest (the box plots shown in the table are not particularly helpful here). The diet, biofeed and drug main effects are all significant beyond the 5% level.

Table 4.4 SAS Output (Numerical and Graphical) from Proc Anova Applied to the Blood Pressure Data

The ANOVA Procedure

Class Level Information		
Class	**Levels**	**Values**
diet	2	N Y
drug	3	X Y Z
biofeed	2	A P

Number of Observations Read	72
Number of Observations Used	72

The ANOVA Procedure

Dependent Variable: bp

Source	DF	Sum of Squares	Mean Square	F Value	Pr > F
Model	11	13194.00000	1199.45455	7.66	<.0001
Error	60	9400.00000	156.66667		
Corrected Total	71	22594.00000			

R-Square	Coeff Var	Root MSE	bp Mean
0.583960	6.784095	12.51666	184.5000

Source	DF	Anova SS	Mean Square	F Value	Pr > F
Diet	1	5202.000000	5202.000000	33.20	<.0001
Drug	2	3675.000000	1837.500000	11.73	<.0001
diet*drug	2	903.000000	451.500000	2.88	0.0638
Biofeed	1	2048.000000	2048.000000	13.07	0.0006
diet*biofeed	1	32.000000	32.000000	0.20	0.6529
Drug*biofeed	2	259.000000	129.500000	0.83	0.4425
diet*drug*biofeed	2	1075.000000	537.500000	3.43	0.0388

The ANOVA Procedure

Level of diet	Level of drug	Level of biofeed	N	bp	
				Mean	Std Dev
N	**X**	**A**	6	188.000000	10.8627805
N	**X**	**P**	6	168.000000	8.6023253
N	**Y**	**A**	6	200.000000	10.0796825
N	**Y**	**P**	6	204.000000	12.6806940
N	**Z**	**A**	6	209.000000	14.3527001
N	**Z**	**P**	6	189.000000	12.6174482
Y	**X**	**A**	6	173.000000	9.7979590
Y	**X**	**P**	6	169.000000	14.8189068
Y	**Y**	**A**	6	187.000000	14.0142784
Y	**Y**	**P**	6	172.000000	10.9361785
Y	**Z**	**A**	6	182.000000	17.1113997
Y	**Z**	**P**	6	173.000000	11.6619038

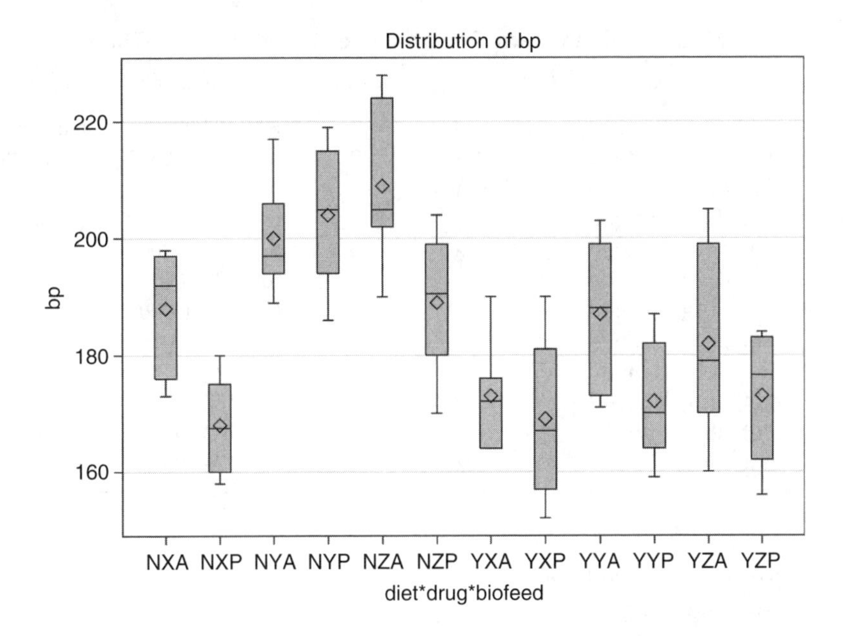

None of the first-order interactions is significant, but the three-way, second-order interaction of diet, drug is biofeedback is significant. Just what does such an effect imply, and what are its implications for interpreting the analysis of variance results?

First, a significant second-order interaction implies that the first-order interaction between two of the variables differs in form or magnitude in the different levels of the remaining variable. And second, the presence of a significant second-order interaction means that there is little point in drawing conclusions about either the non-significant first-order interactions or the significant main effects. The effect of drug, for example, is not consistent for all combinations of diet and biofeedback. It would therefore be potentially misleading to conclude, on the basis of the significant main effect, anything about the specific effects of these three drugs on blood pressure.

Understanding the meaning of the significant second-order interaction is made simpler by plotting some simple graphs. Here the interaction plot of diet and biofeedback separately for each drug will help.

First the outmeans data set is printed. The result is shown in Table 4.5. As well as checking the results, this also shows us the name of the variable containing the means.

To produce separate plots for each drug, we use could the by statement within proc sgplot, having first been sorted by drug. However, the sgpanel procedure will produce a better result with three panels instead of three separate plots.

Table 4.5 Means and Standard Deviations for Blood Pressure Data

Obs	*Effect*	*Diet*	*Drug*	*Bio-feed*	*N*	*Mean_bp*	*SD_bp*
1	diet_drug_biofeed	N	X	A	6	188.000000	10.8627805
2	diet_drug_biofeed	N	X	P	6	168.000000	8.6023253
3	diet_drug_biofeed	N	Y	A	6	200.000000	10.0796825
4	diet_drug_biofeed	N	Y	P	6	204.000000	12.6806940
5	diet_drug_biofeed	N	Z	A	6	209.000000	14.3527001
6	diet_drug_biofeed	N	Z	P	6	189.000000	12.6174482
7	diet_drug_biofeed	Y	X	A	6	173.000000	9.7979590
8	diet_drug_biofeed	Y	X	P	6	169.000000	14.8189068
9	diet_drug_biofeed	Y	Y	A	6	187.000000	14.0142784
10	diet_drug_biofeed	Y	Y	P	6	172.000000	10.9361785
11	diet_drug_biofeed	Y	Z	A	6	182.000000	17.1113997
12	diet_drug_biofeed	Y	Z	P	6	173.000000	11.6619038

```
proc sgpanel data=outmeans;
  panelby drug/rows=3;
  series y=mean_bp x=biofeed/group=diet;
run;
```

The resulting plot is shown in Figure 4.1. For drug X, the diet × biofeedback interaction plot indicates that diet has a negligible effect when biofeedback is given, but substantially reduces blood pressure when biofeedback is absent. For drug Y, the situation is essentially the reverse of that for drug X. For drug Z, the blood pressure difference when the diet is given and when it is not is approximately equal for both levels of biofeedback.

In some cases, a significant high-order interaction may make interpretation of the results from a factorial analysis of variance difficult. In such cases, a transformation of the data may help. For example, we can analyse the log transformed observations, as follows:

```
data hyper;
  set hyper;
  logbp=log(bp);
run;
  proc anova data=hyper;
    class diet drug biofeed;
    model logbp=diet|drug|biofeed;
  run;
```

The data step computes the natural log of bp and stores it in a new variable logbp. The anova results for the transformed variable are given in Table 4.6.

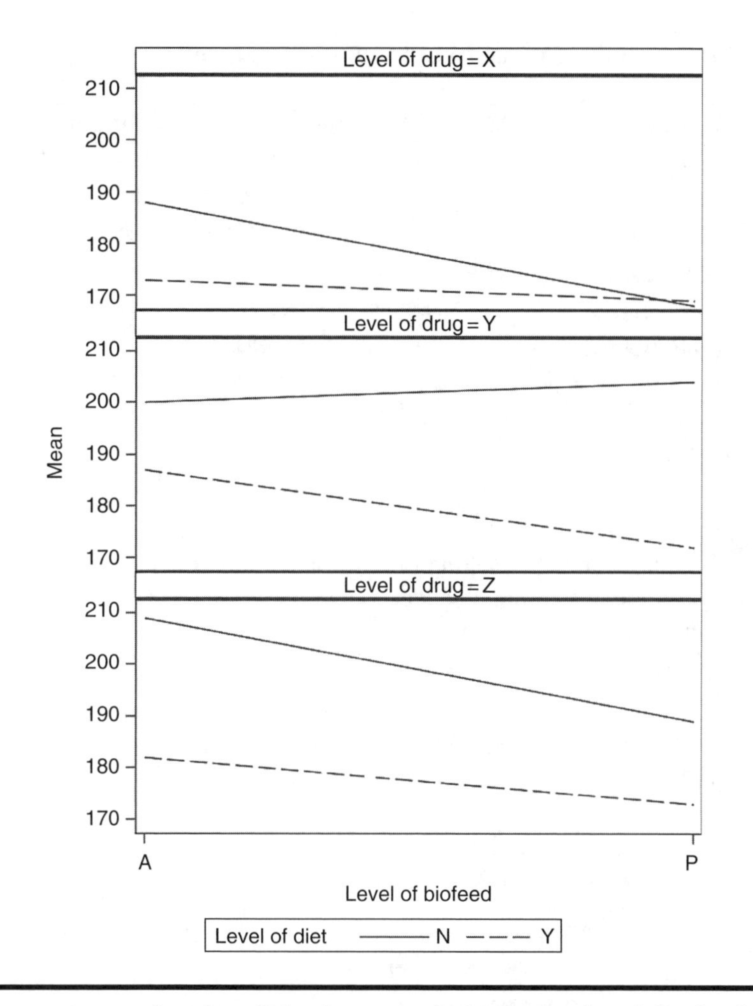

Figure 4.1 Interaction plot of blood pressure for biofeedback and diet for drug X.

Although the results are similar to those for the untransformed observations, the three-way interaction is now only marginally significant. If no substantive explanation of this interaction is forthcoming, it might be preferable now to interpret the results in terms of the very significant main effects and fit a main effects only model to the log transformed blood pressures. In addition, we can use Scheffe's multiple comparison test (Fisher and Van Belle, 1996) to assess which of the three drug means actually differ.

```
proc anova data = hyper;
  class diet drug biofeed;
  model logbp = diet drug biofeed;
  means drug/scheffe;
run;
```

Table 4.6 Analysis of Variance of Log(Blood Pressure)

Source	DF	Anova SS	Mean Square	F Value	Pr > F
Diet	1	0.14956171	0.14956171	32.33	<.0001
Drug	2	0.10706115	0.05353057	11.57	<.0001
diet*drug	2	0.02401168	0.01200584	2.60	0.0830
Biofeed	1	0.06147547	0.06147547	13.29	0.0006
diet*biofeed	1	0.00065769	0.00065769	0.14	0.7075
drug*biofeed	2	0.00646790	0.00323395	0.70	0.5010
diet*drug*biofeed	2	0.03029929	0.01514965	3.28	0.0447

The results are shown in Table 4.7. Each of the main effects is highly significant, and the grouping of means resulting from the application of Scheffe's test indicates that drug X produces lower blood pressures than the other two drugs, whose means do not differ.

Table 4.7 SAS Output from Proc Anova Showing Scheffe's Test for Log(Blood Pressure)

Scheffe's Test for logbp

NOTE: This test controls the Type I experimentwise error rate.

Alpha	0.05
Error Degrees of Freedom	67
Error Mean Square	0.005059
Critical Value of F	3.13376
Minimum Significant Difference	0.0514

Means with the same letter are not significantly different.

Scheffe Grouping	Mean	N	drug
A	5.24709	24	Y
A			
A	5.23298	24	Z
B	5.15915	24	X

Exercises

1 Treating Hypertension

1. Compare the results given by Bonferoni *t*-tests and Ducan's multiple range test for the three drug means with those given by Scheffe's test as reported in the text.
2. Produce box plots of the log-transformed blood pressures for (a) diet present, diet absent, (b) biofeedback present, biofeedback absent and (c) drugs X, Y and Z.

2 Weight Gain in Rats

In an experiment to study the gain in weight of rats fed on four different diets, distinguished by amount of protein (low and high) and by source of protein (beef and cereal), 10 rats were randomized to each of the 4 treatments and the weight gain in grams recorded. The data are given in Hand et al. (1994) and the question of interest is how diet affects weight gain.

The variables in rats.dat are

- **ID:** Observation number
- **Source:** 1 = beef, 2 = cereal
- **Amount:** 1 = low, 2 = high
- **Weightgain:** Gain in weight

1. Plot the mean weight gain for each level of the two factors.
2. Find the variances of weight gain in each of the four cells of the experiment.
3. Carry out the appropriate analysis of variance of the data and interpret the results after constructing any plots that you think will be helpful.

3 Cortisol Levels in Psychotics

Rothschild et al. (1982) give the post-dexamethasone 1600-hour cortisol levels (in micrograms/decilitre) in 31 control patients and four groups of psychotics that are to be compared.

The variables in cortisol.dat are

- **Group:** 1 = control, 2 = major depression, 3 = bipolar depression, 4 = schizophrenia, 5 = atypical
- **Cortisol:** Cortisol level

1. Use box plots to make an initial comparison of the cortisol levels for the different groups of patients. Do these plots indicate any problems with proceeding to a formal analysis of variance of the data?

4 Egyptian Skulls

Hand et al. (1994) give the results of a study in which four measurements were made on Egyptian skulls from five different epochs. The data were collected to try to shed light on whether there was any evidence of differences between the skulls from the five epochs. Non-constant measurements of the skulls over time would indicate interbreeding with immigrant populations.

The variables in skulls.dat are

- **Epoch:** 1 = c 4000 BC, 2 = c 3300 BC, 3 = c 1850 BC, 4 = c 200 BC, 5 = c AD 150
- **Mb:** Maximum breadth of the skull
- **Bh:** Basibregmatic height of the skull
- **Bl:** Basialiveolar length of the skull
- **Nh:** Nasal height of the skull

1. Carry out four one-way analyses of variances to compare each of the four skull measurements between epochs.
2. Carry out a one-way multivariate analysis of variance to simultaneously compare the four skull measurements between epochs.
3. Compare the results from the two analyses. Which do you think is more appropriate for these data and why?

Chapter 5

Analysis of Variance II: School Attendance among Australian Children

5.1 Introduction

The data to be used in this chapter arise from a sociological study of Australian Aboriginal and white children reported by Quine (1975); they are given in Table 5.1. In this study, children of both sexes from four age groups (final grade in primary schools and first, second and third forms in secondary school) and two cultural groups were used. The children in each age group were classified as slow or average learners. The response variable of interest was the number of days absent from school during the school year. (Children who had suffered a serious illness during the year were excluded.)

5.2 Analysis of Variance Model

The basic design of the study is a $4 \times 2 \times 2 \times 2$ factorial. The usual model for y_{ijklm}, the number of days absent for the ith child in the jth sex group, the kth age group, the lth cultural group and the mth learning group, is

Table 5.1 Study of Australian Aboriginal and White Children (Response Variable Is Days Away from School)

Cell	*Origin*	*Sex*	*Grade*	*Type*	*Days*
1	A	M	F0	SL	2, 11, 14
2	A	M	F0	AL	5, 5, 13, 20, 22
3	A	M	F1	SL	6, 6, 15
4	A	M	F1	AL	7, 14
5	A	M	F2	SL	6, 32, 53, 57
			⋮		
30	N	F	F2	AL	1
31	N	F	F3	SL	8
32	N	F	F3	AL	1, 9, 22, 3, 3, 5, 15, 18, 22, 37

Note: A, Aboriginal; N, non-Aboriginal; F, female; M, male; F0, primary; F1, first form; F2, second form; F3, third form; SL, slow learner; AL, average learner.

$$y_{ijklm} = \mu + \alpha_j + \beta_k + \gamma_p + \delta_m + (\alpha\beta)_{jk} + (\alpha\gamma)_{jp} + (\alpha\delta)_{jm} + (\beta\gamma)_{kl} + (\beta\delta)_{km} + (\gamma\delta)_{lm} + (\alpha\beta\gamma)_{jkl} + (\alpha\beta\delta)_{jkm} + (\alpha\gamma\delta)_{jlm} + (\beta\gamma\delta)_{klm} + (\alpha\beta\gamma\delta)_{jklm} + \varepsilon_{ijklm} \quad (5.1)$$

where μ is the overall mean and the terms represent main effects, first-order interactions of pairs of factors, second-order interactions of sets of three factors and a third-order interaction for all four factors. (The parameters have to be constrained in some way to make the model identifiable.) The ε_{ijklm} represents random error terms assumed to be normally distributed with mean zero and variance σ^2.

The unbalanced nature of the data in Table 5.1 (there are different numbers of observations for the different combinations of factors) presents considerably more problems than encountered in the analysis of the balanced factorial data in the previous chapter. The main difficulty is that when the data are unbalanced, there is no unique way of finding a "sum of squares" corresponding to each main effect and each interaction, since these effects are no longer independent of one another. It is no longer possible to partition the total variation in the response variable into non-overlapping or orthogonal sums of squares representing factor main effects and factor interactions. For example, there is a proportion of the variance of the response variable that can be attributed to (explained by) either sex or age group, and so, consequently, sex and age group together explain less of the variation of the response than the sum of which each explains alone. The result of this is that the sum of squares that can be attributed to a factor depends on which factors have already been allocated a sum of squares; in other words, the sums of squares of factors and their interactions depend on the order in which they are considered.

The dependence between the factor variables in an unbalanced factorial design, and the consequent lack of uniqueness in partitioning the variation in the response variable, has led to a great deal of confusion about what is the most appropriate way to analyse such designs. The issues are not straightforward and even statisticians (yes, even statisticians!) have not wholly agreed on the most suitable method of analysis for all situations, as witnessed by the discussion in the papers of Nelder (1977) and Aitkin (1978).

Essentially, the discussion over the analysis of unbalanced factorial designs has involved the question of what type of sums of squares should be used. Basically, there are three possibilities, but here we shall consider only two, which we will illustrate for a design with two factors.

5.2.1 Type I Sums of Squares

These sums of squares represent the effect of adding a term to an existing model, in one particular order. So, for example, a set of Type I sums of squares essentially represents a comparison of the following models:

Source	*Type I SS*
A	SSA
B	SSB|A
AB	SSAB|A,B

1. SSAB|A,B model including an interaction and main effects with one including only main effects.
2. SSB|A model including both main effects, but no interaction, with one including only the main effect of factor A.
3. SSA model containing only the A main effect with one containing only the overall mean.

The use of these sums of squares in a series of tables in which the effects are considered in different orders (see later) will often provide the most satisfactory way of answering the question as to which model is most appropriate for the observations.

5.2.2 Type III Sums of Squares

Type III sums of squares represent the contribution of each term to a model including all other possible terms. So, for a two-factor design the sums of squares represent the following:

Source	*Type III SS*
A	SSA|B,AB
B	SSB|A,AB
AB	SSAB|A,B

SAS also has a Type IV sum of squares which is the same as Type III unless the design contains empty cells.

In a balanced design, Type I and Type III sums of squares are equal, but for an unbalanced design they are not and there have been numerous discussions over which type is more appropriate for the analysis of such designs. Authors such as Maxwell and Delaney (2003) and Howell (2002) strongly recommend the use of Type III sums of squares, and these are the default in SAS. Nelder (1977) and Aitkin (1978), however, are strongly critical of "correcting" main effects sums of squares for an interaction term involving the corresponding main effect; their criticisms are based on both theoretical and pragmatic grounds. The arguments are relatively subtle, but in essence are something like this:

- When fitting models to data, the principle of parsimony is of critical importance. In choosing among possible models, we do not adopt complex models for which there is no empirical evidence.
- So, if there is no convincing evidence of an AB interaction, we do not retain the term in the model. Thus, additivity of A and B is assumed unless there is convincing evidence to the contrary.
- So, the argument proceeds that Type III sums of squares for A in which it is adjusted for AB makes no sense.
- First, if the interaction term is necessary in the model, then the experimenter will usually wish to consider simple effects of A at each level of B separately. A test of the hypothesis of no A main effect would not usually be carried out if the AB interaction is significant.
- If the AB interaction is not significant, then adjusting for it is of no interest and causes a substantial loss of power in testing the A and B main effects.

The issue does not arise so clearly in the balanced case, for there the sum of squares for A, say, is independent of whether interaction is assumed. Thus, in deciding on possible models for the data, the interaction term is not included unless it has been shown to be necessary, in which case tests on main effects involved in the interaction are not carried out or, if carried out, are not interpreted – see biofeedback example in Chapter 4.

The arguments of Nelder (1977) and Aitkin (1978) against the use of Type III sums of squares are powerful and persuasive. Their recommendations to use Type I sums of squares, considering effects in a number of orders, as the most suitable way in which to identify a suitable model for a data set are also convincing and strongly endorsed by the authors of this book.

5.3 Analysis Using SAS

We assume the data are in an ASCII file called ozkids.dat in the current directory and that the values of the factors making up the design are separated by tabs, whereas those recoding days of absence for the subjects within each cell are separated by commas, as in Table 5.1. The data can then be read in as follows:

```
data ozkids;
  infile 'ozkids.dat' dlm=' ,' expandtabs missover;
  input cell origin $ sex $ grade $ type $ days @;
    do until (days=.);
    output;
    input days @;
    end;
    input;
run;
```

The expandtabs option on the infile statement converts tabs to spaces so that list input can be used to read the tab-separated values. In order to read the comma-separated values in the same way, the delimiter option (abbreviated dlm) specifies that both spaces and commas are delimiters. This is done by including a space and a comma in quotes after dlm=. The missover option prevents SAS from reading the next line of data in the event that an input statement requests more data values than contained in the current line. Missing values are assigned to the variables for which there are no corresponding data values. To illustrate this with an example, suppose we have an input statement input x1-x7;. If a line of data only contains five numbers, by default SAS will go to the next line of data to read data values for x6 and x7. This is not usually what is intended, so when it happens there is a warning message in the log: "SAS went to a new line when INPUT statement reached past the end of a line." With the missover option SAS would not go to a new line, but x6 and x7 would have missing values. Here we utilise this to determine when all the values for days of absence from school have been read.

The input statement reads the cell number, the factors in the design and the days absent for the first observation in the cell. The "trailing @" at the end of the statement holds the data line so that more data can be read from it by subsequent input statements. The statements between the do until and the following end are executed repeatedly until the days variable has a missing value. The output statement creates an observation in the output data set. Then another value of days is read, again holding the data line with a trailing @. When all the values from the line have been read and output as observations, the days variable is assigned a missing value and the do until loop finishes. The following input statement then releases the data line, so that the next line of data from the input file can be read.

For unbalanced designs, the glm procedure should be used rather than proc anova. We begin by fitting main effects only models for different orders of main effects.

```
proc glm data=ozkids;
  class origin sex grade type;
  model days=origin sex grade type/ss1 ss3;
proc glm data=ozkids;
  class origin sex grade type;
  model days=grade sex type origin/ss1;
proc glm data=ozkids;
  class origin sex grade type;
  model days=type sex origin grade/ss1;
proc glm data=ozkids;
  class origin sex grade type;
  model days=sex origin type grade/ss1;
run;
```

The class statement specifies the classification variables or factors. These may be numeric or character variables. The model statement specifies the dependent variable on the left-hand side of the equation and the effects, that is, factors and their interactions, on the right-hand side of the equation. Main effects are specified by including the variable name.

The options on the model statement in the first glm step specify that both Type I and Type III sums of squares are to be output. The subsequent proc steps repeat the analysis varying the order of the effects, but because Type III sums of squares are invariant to the order only Type I sums of squares are requested. The relevant output is shown in Table 5.2. Note that when a main effect is ordered last the corresponding Type I sum of squares is the same as the Type III sum of squares for the factor. In fact, when dealing with a main effects only model, the Type III sums of squares can legitimately be used to identify the most important effects. Here it appears that origin and grade have most impact on the number of days a child is absent from school.

Next we fit a full factorial model to the data as follows:

```
proc glm data=ozkids;
  class origin sex grade type;
  model days=origin sex grade type origin|sex|grade|type/
    ss1 ss3;
run;
```

Joining variable names with a bar is a shorthand way of specifying an interaction and all the lower-order interactions and main effects implied by it. This is useful not only to save typing but to ensure that relevant terms in the model are not inadvertently omitted. Here we have specified the main effects explicitly so that they are entered before any interaction terms when calculating Type I sums of squares. The output is shown in Table 5.3. Note first that the only Type I and Type III sums of squares

Table 5.2 SAS Output from Using Proc glm to Fit a Series of Main Effects Only Models to the Data in Table 5.1

Source	DF	Type I SS	Mean Square	F Value	Pr > F
origin	1	2645.652580	2645.652580	11.52	0.0009
sex	1	338.877090	338.877090	1.48	0.2264
grade	3	1837.020006	612.340002	2.67	0.0500
type	1	132.014900	132.014900	0.57	0.4495

Source	DF	Type III SS	Mean Square	F Value	Pr > F
origin	1	2403.606653	2403.606653	10.47	0.0015
sex	1	185.647389	185.647389	0.81	0.3700
grade	3	1917.449682	639.149894	2.78	0.0430
type	1	132.014900	132.014900	0.57	0.4495

Source	DF	Type I SS	Mean Square	F Value	Pr > F
grade	3	2277.172541	759.057514	3.31	0.0220
sex	1	124.896018	124.896018	0.54	0.4620
type	1	147.889364	147.889364	0.64	0.4235
origin	1	2403.606653	2403.606653	10.47	0.0015

Source	DF	Type I SS	Mean Square	F Value	Pr > F
type	1	19.502391	19.502391	0.08	0.7711
sex	1	336.215409	336.215409	1.46	0.2282
origin	1	2680.397094	2680.397094	11.67	0.0008
grade	3	1917.449682	639.149894	2.78	0.0430

Source	DF	Type I SS	Mean Square	F Value	Pr > F
sex	1	308.062554	308.062554	1.34	0.2486
origin	1	2676.467116	2676.467116	11.66	0.0008
type	1	51.585224	51.585224	0.22	0.6362
grade	3	1917.449682	639.149894	2.78	0.0430

that agree are those for the origin × sex × grade × type interaction. Next consider the origin main effect. The Type I sum of squares for origin is "corrected" only for the mean since it appears first in the proc glm statement. The effect is highly significant. But using Type III sums of squares, in which the origin effect is corrected for all other main effects and interactions, the corresponding F value has an associated p value of 0.2736. Now origin is judged non-significant, but this may simply reflect the loss of power after "adjusting" for a lot of relatively unimportant interaction terms.

Arriving at a final model for these data is not straightforward (see Aitkin, 1978, for some suggestions), and we will not pursue the issue here since the data set will be the subject of further analyses in Chapter 9. But some of the exercises encourage readers to try some alternative analyses of variance.

Table 5.3 SAS Output from Using Proc glm to Fit a Model Including all Main Effects and Interactions to the Data in Table 5.1

The GLM Procedure

Class Level Information		
Class	**Levels**	**Values**
origin	2	A N
sex	2	F M
grade	4	F0 F1 F2 F3
type	2	AL SL

Number of Observations Read	154
Number of Observations Used	154

The GLM Procedure

Dependent Variable: days

Source	DF	Sum of Squares	Mean Square	F Value	Pr > F
Model	31	15179.41930	489.65869	2.54	0.0002
Error	122	23526.71706	192.84194		
Corrected Total	153	38706.13636			

R-Square	Coeff Var	Root MSE	days Mean
0.392171	86.05876	13.88675	16.13636

Source	DF	Type I SS	Mean Square	F Value	Pr > F
origin	1	2645.652580	2645.652580	13.72	0.0003
sex	1	338.877090	338.877090	1.76	0.1874
grade	3	1837.020006	612.340002	3.18	0.0266
type	1	132.014900	132.014900	0.68	0.4096
origin*sex	1	142.454554	142.454554	0.74	0.3918
origin*grade	3	3154.799178	1051.599726	5.45	0.0015
sex*grade	3	2009.479644	669.826548	3.47	0.0182
origin*sex*grade	3	226.309848	75.436616	0.39	0.7596
origin*type	1	38.572890	38.572890	0.20	0.6555
sex*type	1	69.671759	69.671759	0.36	0.5489
origin*sex*type	1	601.464327	601.464327	3.12	0.0799
grade*type	3	2367.497717	789.165906	4.09	0.0083
origin*grade*type	3	887.938926	295.979642	1.53	0.2089
sex*grade*type	3	375.828965	125.276322	0.65	0.5847
origi*sex*grade*type	3	351.836918	117.278973	0.61	0.6109

Source	DF	Type III SS	Mean Square	F Value	Pr > F
origin	1	233.201138	233.201138	1.21	0.2736
sex	1	344.037143	344.037143	1.78	0.1841
grade	3	1036.595762	345.531921	1.79	0.1523
type	1	181.049753	181.049753	0.94	0.3345
origin*sex	1	3.261543	3.261543	0.02	0.8967
origin*grade	3	1366.765758	455.588586	2.36	0.0746
sex*grade	3	1629.158563	543.052854	2.82	0.0420
origin*sex*grade	3	32.650971	10.883657	0.06	0.9823
origin*type	1	55.378055	55.378055	0.29	0.5930
sex*type	1	1.158990	1.158990	0.01	0.9383
origin*sex*type	1	337.789437	337.789437	1.75	0.1881
grade*type	3	2037.872725	679.290908	3.52	0.0171
origin*grade*type	3	973.305369	324.435123	1.68	0.1743
sex*grade*type	3	410.577832	136.859277	0.71	0.5480
origi*sex*grade*type	3	351.836918	117.278973	0.61	0.6109

Exercises

1 *School Attendance*

1. The outcome for the school attendance data, number of days absent, is a count variable; consequently assuming normally distributed errors may not be entirely appropriate, as we shall see in Chapter 9. Here, however, we might deal with this potential problem by way of a transformation. One possibility is a log transformation. Investigate this possibility.
2. Find a table of cell means and standard deviations for the school attendance data.
3. Construct a normal probability plot of the residuals from fitting a main effects only model to the school attendance data. Comment on the results.

2 *Foster Feeding of Rats*

In a foster feeding experiment with rat mothers and litters of four different genotypes, A, B, I and J, the litter weight (in grams) after a trial feeding period was found.

The variables in rats.dat are

- **GL:** Genotype of litter, 1 = A, 2 = B, 3 = I and 4 = J
- **GM:** Genotype of mother, 1 = A, 2 = B, 3 = I and 4 = J
- **Weight:** Weight (g) of litter

1. Tabulate number of observations, and mean and standard deviation of litter weight for the cross-classification of genotype of litter and genotype of mother.
2. Construct box plots of litter weights for both genotype of litter and genotype of mother.
3. Carry out an appropriate analysis of variance for the data and report your conclusions.

3 *Slimming Clinics*

Slimming clinics aim to encourage people to lose weight by offering encouragement and support about dieting through regular meetings. In a study to explore the effectiveness of such groups, the question of particular interest was whether adding a technical manual giving advice based on psychological behaviourist theory to the support offered would help clients to control their diets. In addition to comparing "manual" and "no manual" conditions it was also considered important to distinguish between clients who had already been trying to slim and those who had not; consequently two status groups were chosen, the first being those clients who had been trying to slim for more than 1 year, and the second those clients who had been trying for not more than 3 weeks. The response variable was (weight at 3 months – ideal weight)/(initial weight – ideal weight), expressed as a percentage.

The variables in slim.dat are

- **Condition:** 1 = experimental group (those given the manual), 2 = control group (those without the manual)
- **Status:** 1 = those who had been trying to slim for more than 1 year, 2 = those who had been trying to slim for not more than 3 weeks
- **Response:** Value of the response

1. Find the means and standard deviations of the response and number of observations for each of the four combinations of condition and status.
2. Construct side-by-side box plots of the response for the two conditions and for the two status groups.
3. Carry out what you think is an appropriate analysis for these data to answer the question of whether the manual makes any difference in the dieting of these women and how this is affected by the time they have already been trying to slim.

Chapter 6

Simple Linear Regression: Alcohol Consumption and Cirrhosis Deaths and How Old Is the Universe?

6.1 Introduction

Osborne (1979) gives the average alcohol consumption in litres per person per year and the death rate per 100,000 from cirrhosis and alcoholism in 15 countries worldwide. The data are shown in Table 6.1. How is cirrhosis death rate related to a country's alcohol consumption, and can we predict the former from the latter?

Wood (2006) gives the relative velocity and the distance of 24 galaxies, according to measurements made using the Hubble Space Telescope (see Table 6.2). Velocities are assessed by measuring the Doppler red shift in the spectrum of light observed from the galaxies concerned, although some correction for "local" velocity components is required. Distances are measured using the known relationship between the period of Cepheid variable stars and their luminosity. How can these data be used to estimate the age of the universe?

Table 6.1 Average Alcohol Consumption and Death Rate from Cirrhosis and Alcoholism

Country	*Alcohol Consumption (l/Person/Year)*	*Cirrhosis and Alcoholism (Death Rate/100,000)*
France	24.7	46.1
Italy	15.2	23.6
West Germany	12.3	23.7
Austria	10.9	7.0
Belgium	10.8	12.3
United States	9.9	14.2
Canada	8.3	7.4
England and Wales	7.2	3.0
Sweden	6.6	7.2
Japan	5.8	10.6
Netherlands	5.7	3.7
Ireland	5.6	3.4
Norway	4.2	4.3
Finland	3.9	3.6
Israel	3.1	5.4

Table 6.2 Distance and Velocity for 24 Galaxies

Observation	*Galaxy*	*Velocity (km)*	*Distance (mega-parsec)*
1	NGC0300	133	2.00
2	NGC0925	664	9.16
3	NGC1326A	1794	16.14
4	NGC1365	1594	17.95
5	NGC1425	1473	21.88
6	NGC2403	278	3.22
7	NGC2541	714	11.22
8	NGC2090	882	11.75
9	NGC3031	80	3.63
10	NGC3198	772	13.80
11	NGC3351	642	10.00
12	NGC3368	768	10.52
13	NGC3621	609	6.64
14	NGC4321	1433	15.21
15	NGC4414	619	17.70
16	NGC4496A	1424	14.86
17	NGC4548	1384	16.22
18	NGC4535	1444	15.78
19	NGC4536	1423	14.93
20	NGC4639	1403	21.98
21	NGC4725	1103	12.36
22	IC4182	318	4.49
23	NGC5253	232	3.15
24	NGC7331	999	14.72

6.2 Simple Linear Regression

Assume y_i represents the value of what is generally known as the response variable on the ith individual and that x_i represents the individual's values on what is most often called an explanatory variable.

The simple linear regression model is

$$y_i = \beta_0 + \beta_i x_i + \varepsilon_i, \tag{6.1}$$

where β_0 is the intercept, β_i is the slope of the linear relationship assumed between the response and explanatory variables, and ε_i is an error term or residual.

The "simple" here means that the model contains only a single explanatory variable; we shall deal with the situation where there are several explanatory variables in the next chapter.

The residuals are assumed to be independent random variables having a normal distribution with mean zero and constant variance σ^2.

The regression coefficients β_0 and β_1 may be estimated as $\hat{\beta}_0$ and $\hat{\beta}_1$ using least squares estimation in which the sum of squared differences between the observed values of the response variable y_i and the values "predicted" by the regression equation $\hat{y}_i = \hat{\beta}_0 + \hat{\beta}_1 x_i$ is minimized, leading to the estimates

$$\hat{\beta}_0 = \bar{y} - \hat{\beta}_1 \bar{x}, \tag{6.2}$$

$$\hat{\beta}_1 = \frac{\sum (y_i - \bar{y})(x_i - \bar{x})}{\sum (x_i - \bar{x})^2}. \tag{6.3}$$

The predicted values of y from the model are

$$\hat{y}_i = \hat{\beta}_0 + \hat{\beta}_1 x_i. \tag{6.4}$$

The variance σ^2 is estimated as s^2 given by

$$s^2 = \sum_{i=1}^{n} (y_i - \hat{y}_i)^2 / (n - 2). \tag{6.5}$$

The estimated variance of the estimate of the slope parameter is

$$\text{Var}(\hat{\beta}_1) = \frac{s^2}{\sum_{i=1}^{n} (x_i - \bar{x})^2}. \tag{6.6}$$

The estimated variance of a predicted value y_{pred} at a given value of x, say x_0, is

$$\text{Var}(y_{\text{pred}}) = s^2 \sqrt{1 + \frac{1}{n} + \frac{(x_0 - \bar{x})^2}{\sum_{i=1}^{n} (x_i - \bar{x})^2}}. \tag{6.7}$$

In some applications of simple linear regression, a model without an intercept is required (when the data are such that the line must go through the origin), that is, a model of the form

$$y_i = \beta x_i + \varepsilon_i. \tag{6.8}$$

In this case, application of least squares gives the following estimator for β:

$$\hat{\beta} = \sum_{i=1}^{n} x_i y_i \Big/ \sum_{i=1}^{n} x_i^2. \tag{6.9}$$

After a simple linear regression model has been fitted to a data set, some attempt needs to be made to check assumptions such as those of constant variance and normality of the error terms. Violation of these assumptions may invalidate conclusions based on the regression analysis. The estimated residuals $r_i = y_i - \hat{y}_i$ play an essential role in diagnosing a fitted model, although because these do not have the same variance (the precision of $\hat{y}_i$ depends on x_i), they are sometimes standardized before use; see Cook and Weisberg (1982) for details. The following diagnostic plots are generally useful when assessing model assumptions:

- *Residuals versus fitted values*: If the fitted model is appropriate, the plotted points should lie in an approximately horizontal band across the plot. Departures from this appearance may indicate that the functional form of the assumed model is incorrect or, alternatively, that there is non-constant variance.
- *Residuals versus explanatory variables*: Systematic patterns in these plots can indicate violations of the constant variance assumption or an inappropriate model form.
- *Normal probability plot of the residuals*: The plot checks the normal distribution assumptions on which all statistical inference procedures are based.

A further diagnostic that is often very useful is an index plot of the Cook's distances for each observation. This statistic is defined as follows:

$$D_k = \frac{1}{(p+1)s^2} \sum_{i=1}^{n} [\hat{y}_{i(k)} - \hat{y}_i]^2, \tag{6.10}$$

where $\hat{y}_{i(k)}$ is the fitted value of the ith observation when the kth observation is omitted from the model. The values of D_k assess the impact of the kth observation on the estimated regression coefficients. Values of D_k greater than 1 are suggestive that the corresponding observation has undue influence on the estimated regression coefficients (see Cook and Weisberg, 1982).

6.3 Analysis Using SAS

6.3.1 Alcohol and Death from Cirrhosis

To begin we will construct the scatterplot of death rate against average alcohol consumption labelling each point with the appropriate country name.

```
data drinking;
  input country $ 1-12 alcohol cirrhosis;
cards;
France       24.7  46.1
Italy        15.2  23.6
W.Germany    12.3  23.7
Austria      10.9   7.0
Belgium      10.8  12.3
USA           9.9  14.2
Canada        8.3   7.4
E&W           7.2   3.0
Sweden        6.6   7.2
Japan         5.8  10.6
Netherlands   5.7   3.7
Ireland       5.6   3.4
Norway        4.2   4.3
Finland       3.9   3.6
Israel        3.1   5.4
;
run;
proc sgplot data=drinking;
  scatter x=alcohol y=cirrhosis/datalabel=country;
run;
```

Some of the country names are longer than the default of eight for character variables, so column input is used to read them in. The values of the two numeric variables can then be read in with list input. This is an example of mixing different forms of input on one input statement. The datalabel option on the scatter statement specifies a variable whose values are to be used as labels on the plot.

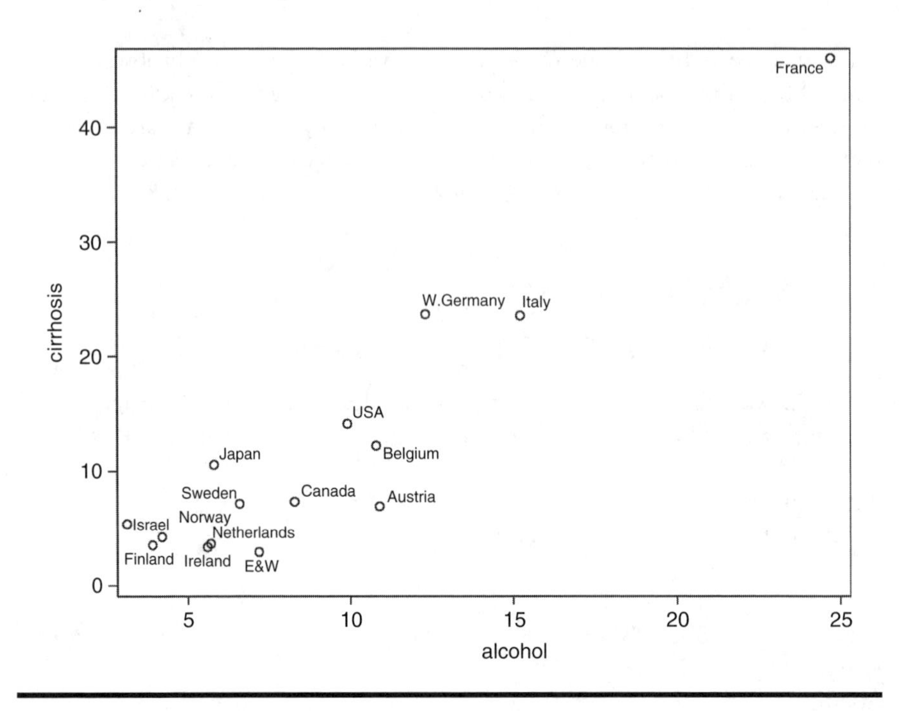

Figure 6.1 Scatterplot of death rate from cirrhosis versus average alcohol consumption.

The resulting scatterplot is shown in Figure 6.1. This diagram indicates that there is a very strong relationship between cirrhosis death rate and a country's average alcohol consumption. One striking feature of the scatterplot is that France lies some way from the other countries on the plot having a very high death rate and a very high average alcohol consumption. We shall return to the potential problems that might be caused by such an "outlier" a little later.

The linear regression model for the data taking death rate as the response variable and average alcohol consumption as the explanatory variable can be fitted using the following SAS code. ODS graphics are used to generate diagnostic plots.

```
ods graphics on;
proc reg data=drinking;
  model cirrhosis=alcohol;
run;
ods graphics off;
```

The numerical results are shown in Table 6.3. The analysis of variance table shown gives a partition of the variation in death rates into a part due to the fitted regression model and another part due to error. The F test can be used to assess the hypothesis that the regression coefficient in the model β_1 takes the value zero. Here the

Table 6.3 Results from Fitting a Simple Linear Regression Model to the Cirrhosis and Alcohol Consumption Data

The REG Procedure
Model: MODEL1
Dependent Variable: cirrhosis

Number of Observations Read	15
Number of Observations Used	15

Analysis of Variance

Source	DF	Sum of Squares	Mean Square	F Value	Pr > F
Model	1	1680.18013	1680.18013	96.61	<.0001
Error	13	226.07987	17.39076		
Corrected Total	14	1906.26000			

Root MSE	4.17022	**R-Square**	0.8814
Dependent Mean	11.70000	**Adj R-Sq**	0.8723
Coeff Var	35.64293		

Parameter Estimates

Variable	DF	Parameter Estimate	Standard Error	t Value	Pr > |t|
Intercept	1	-5.99575	2.09775	-2.86	0.0134
alcohol	1	1.97792	0.20123	9.83	<.0001

associated p value is very small suggesting that the data provide very strong evidence against this hypothesis. The value of R-square gives the value of the square of the multiple correlation coefficient, which in the case of simple linear regression is simply the correlation between the death rate values, predicted by the model $\hat{y}$, where $\hat{y} = \hat{\beta}_0 + \hat{\beta}_1 x$, and alcohol consumption. R-square gives the amount of variation in the response variable accounted for by the explanatory variable, which in this case is 88%. (We shall say more about the multiple correlation coefficient and the Adj R-Sq from Table 6.3 in Chapter 7.) The intercept of the fitted model is estimated to be -6.00 with an estimated standard error of 2.10, and the slope is estimated to be 1.98 with standard error 0.20. Both coefficients are shown in Table 6.3 to be significantly different from zero. The fitted linear regression equation is

$$\text{Death rate} = -6.00 + 1.98 \times \text{Average alcohol consumption.}$$

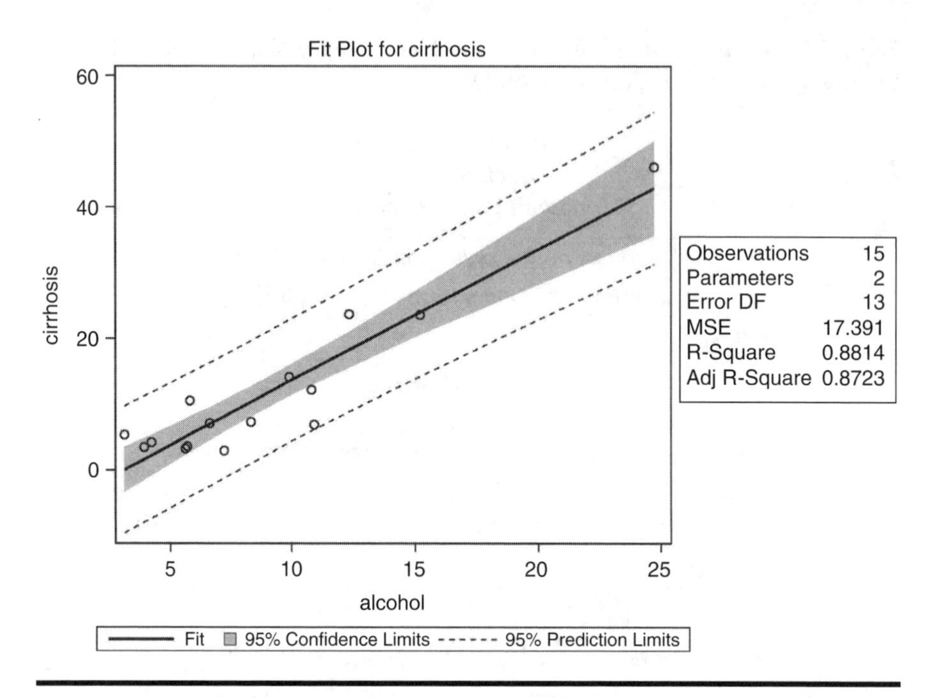

Figure 6.2 Scatterplot of death rate from cirrhosis against alcohol consumption showing fitted linear regression and confidence interval for the fit.

This model leads to an estimated death rate of −6.00 when alcohol consumption is 0. One possible explanation of this clearly unsatisfactory prediction is that the model may not be very sensible for these data; another explanation might be that in making this prediction we are extrapolating outside the range of the observed x values, a potentially dangerous and misleading procedure. A third possible explanation is considered later. The fitted model plus the 95% confidence is generated as one of the ODS graphics and is shown in Figure 6.2.

Now let us return to consider the possible problem of the outlier observation, France. Outliers may distort the value of an estimated regression coefficient and we might consider simply dropping the observation and then re-estimating the regression model. To do this we can re-run the model including a where statement to exclude France.

```
proc reg data=drinking;
  model cirrhosis=alcohol;
  where country ne 'France';
run;
```

The results are shown in Table 6.4 and are similar to those in Table 6.3 except in one important respect, namely that the intercept does not now differ significantly

Table 6.4 Results from Fitting a Simple Linear Regression Model to the Cirrhosis and Alcohol Consumption Data after Removing the Data Point for France

Model: MODEL1
Dependent Variable: cirrhosis

Number of Observations Read	14
Number of Observations Used	14

Analysis of Variance

Source	DF	Sum of Squares	Mean Square	F Value	Pr > F
Model	1	441.84624	441.84624	26.98	0.0002
Error	12	196.52804	16.37734		
Corrected Total	13	638.37429			

Root MSE	4.04689	**R-Square**	0.6921
Dependent Mean	9.24286	**Adj R-Sq**	0.6665
Coeff Var	43.78400		

Parameter Estimates

Variable	DF	Parameter Estimate	Standard Error	t Value	Pr > |t|
Intercept	1	-3.61154	2.70081	-1.34	0.2060
alcohol	1	1.64348	0.31641	5.19	0.0002

from zero. Removing the data point for France has led to a rather more sensible model for the data since we can now assume that the intercept is zero with the consequence that the new model does not lead to a predicted negative death rate when alcohol consumption is zero.

Having fitted a simple linear regression model to the data, we now need to look at some residual plots to assess the validity of the assumptions made, that is, constant variance and normality of error terms. A variety of diagnostic plots will have already been obtained by default from the code used to fit the two models above; those for the fitted model for all the data (including France) are shown in Figures 6.3 and 6.4. The first two plots in the first row of Figure 6.3 show the

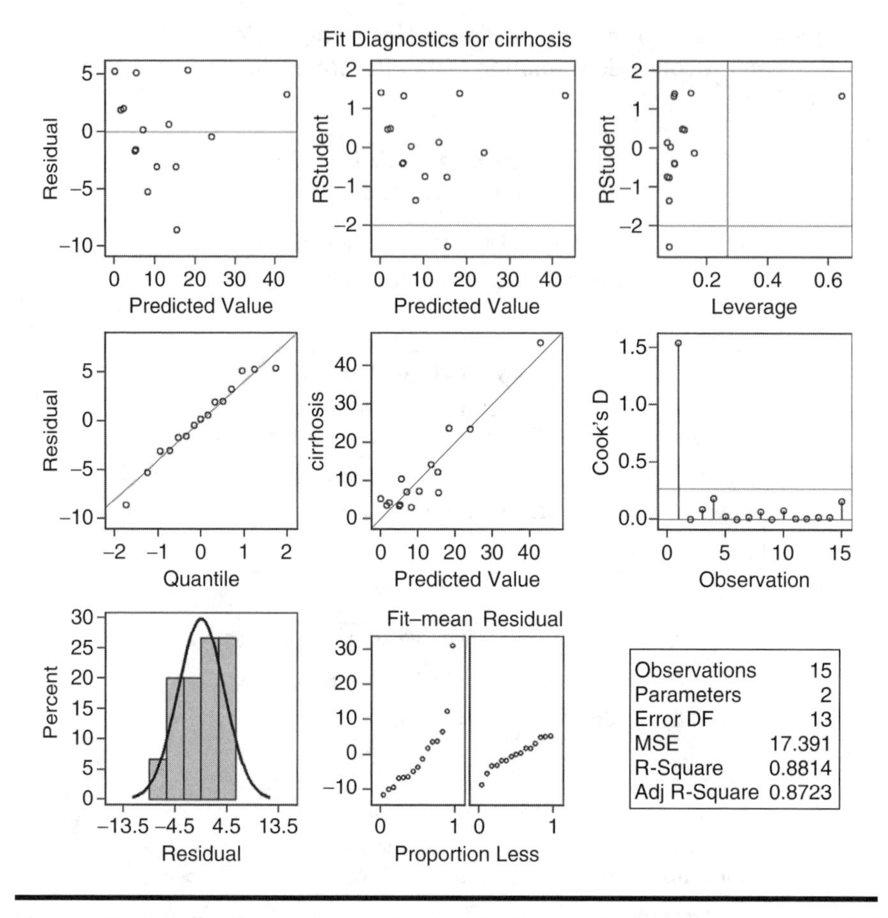

Figure 6.3 A collection of diagnostic plots for the simple linear regression model fitted to all the cirrhosis and alcohol consumption data.

"raw" residuals against predicted values and a standardized residual, known as a studentized residual (see Cook and Weisberg, 1982) also plotted against the predicted values. There is one quite large negative residual on each of these two plots. The third plot in the first row of Figure 6.3 plots the studentized residual against leverage. The later values give an indication of the effect of the x values when fitting the model. Observations with "high" leverage force the fitted model close to the observed value of the response leading to a small residual. The average leverage value is $1/n$ so values much larger than this may give some cause for concern. There is one observation with a very large leverage value. To identify the observations with large residuals or high leverage we need to rerun the procedure adding an output statement to save these quantities.

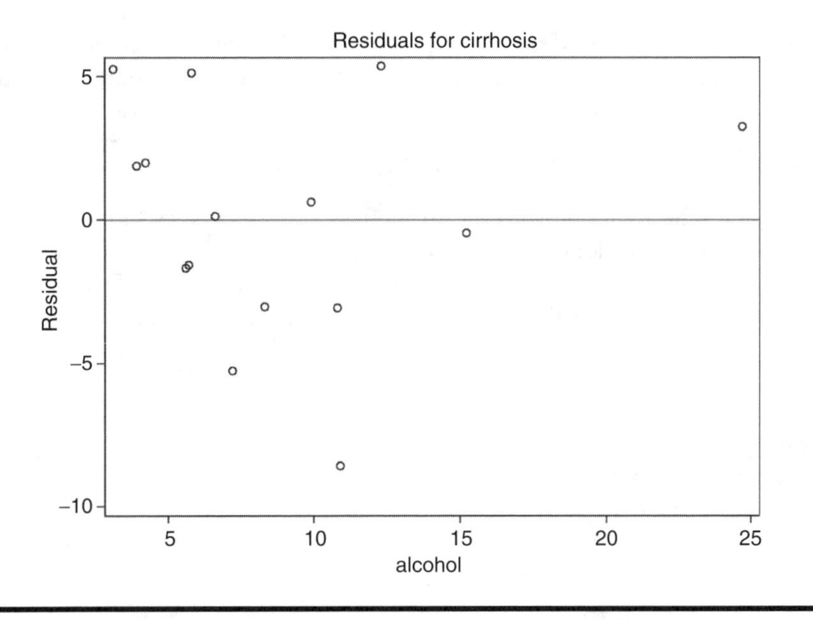

Figure 6.4 Residuals plotted against alcohol consumption for simple linear regression model fitted to cirrhosis and alcohol consumption data.

```
proc reg data=drinking;
  model cirrhosis=alcohol;
  output out=regout p=pred student=zres h=leverage;
run; quit;
proc print data=regout;
  where abs(zres)>2 or leverage>.3;
run;
```

In this example, the output data set, regout, contains the original variables plus the predicted values, studentized residuals and leverage in the variables: pred, zres and leverage, respectively. This data set can then be inspected or printed to identify the relevant observations. The quit statement terminates the reg procedure and so allows the regout data set to be opened for inspection. For larger data sets, it would be preferable to print out the observations with large residuals or leverage values. We use the abs function as we want large negative residuals as well as positive ones. The observation with the large negative residual turns out to be that for Austria and France is the country with high leverage.

The second row of Figure 6.3 begins with a normal probability plot of the residual that is not very informative here since there are only 15 observations. In the next plot, predicted values of the response are compared with observed values, and in the final plot in the second row the values on Cook's distance are given for each of the 15 countries. The largest of these corresponds to the observations for France.

6.3.2 Estimating the Age of the Universe

Again we will begin by constructing a scatterplot of the galaxy velocity and distance data using the same code as in the previous sub-section to give Figure 6.5. The data are separated by spaces and can be read with list input. In the sgplot step yaxis and xaxis statements have been added to give the units of measurement to the axis labels.

```
data universe;
  input id Galaxy $ Velocity Distance;
datalines;
1 NGC0300 133 2.00
2 NGC0925 664 9.16
3 NGC1326A 1794 16.14
...
23 NGC5253 232 3.15
24 NGC7331 999 14.72
;
```

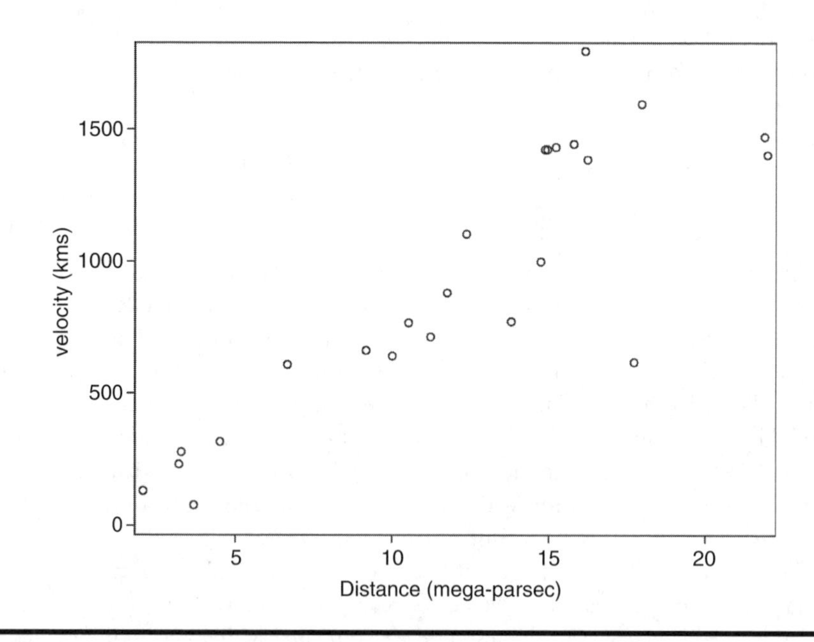

Figure 6.5 Scatterplot of velocity against distance for 24 galaxies.

```
proc sgplot data=universe;
  scatter y=velocity x=Distance;
  yaxis label='Velocity (kms)';
  xaxis label='Distance (mega-parsec)';
run;
```

The diagram shows a clear, strong relationship between velocity and distance. Now, let us fit a simple linear regression model to the data, but in this case the nature of the data requires a model without an intercept since if distance is zero so is relative speed. So the model to be fitted to these data is

$$\text{Velocity} = \beta \text{ Distance} + \text{Error}.$$

This is essentially what astronomers call Hubble's Law, and β is known as Hubble's constant; β^{-1} gives an approximate age of the universe. We can fit the required model as follows, using the noint option on the model statement to omit the intercept.

```
ods graphics on;
proc reg data=universe;
  model velocity= distance/noint;
run;
ods graphics off;
quit;
```

The results of fitting the model are shown in Table 6.5. The slope of the estimated regression line through the origin is 76.58 with an estimated standard error of 3.96.

Table 6.5 Results of Fitting a Linear Regression Model with a Zero Intercept to the Data on Velocities and Distances of Galaxies

Model: MODEL1
Dependent Variable: Velocity

Number of Observations Read	24
Number of Observations Used	24

NOTE: No intercept in model. R-Square is redefined.

Analysis of Variance

Source	DF	Sum of Squares	Mean Square	F Value	Pr > F
Model	1	25013691	25013691	373.08	<.0001
Error	23	1542066	67046		
Uncorrected Total	24	26555757			

Root MSE	258.93306	**R-Square**	0.9419
Dependent Mean	924.37500	**Adj R-Sq**	0.9394
Coeff Var	28.01169		

Parameter Estimates

Variable	DF	Parameter Estimate	Standard Error	t Value	Pr > |t|
Distance	1	76.58117	3.96479	19.32	<.0001

The collection of diagnostic plots, which is also given by using the code above, is shown in Figure 6.6, and the residuals plotted against distance are shown in Figure 6.7. As with the alcohol data, we can identify the observations with large residuals or

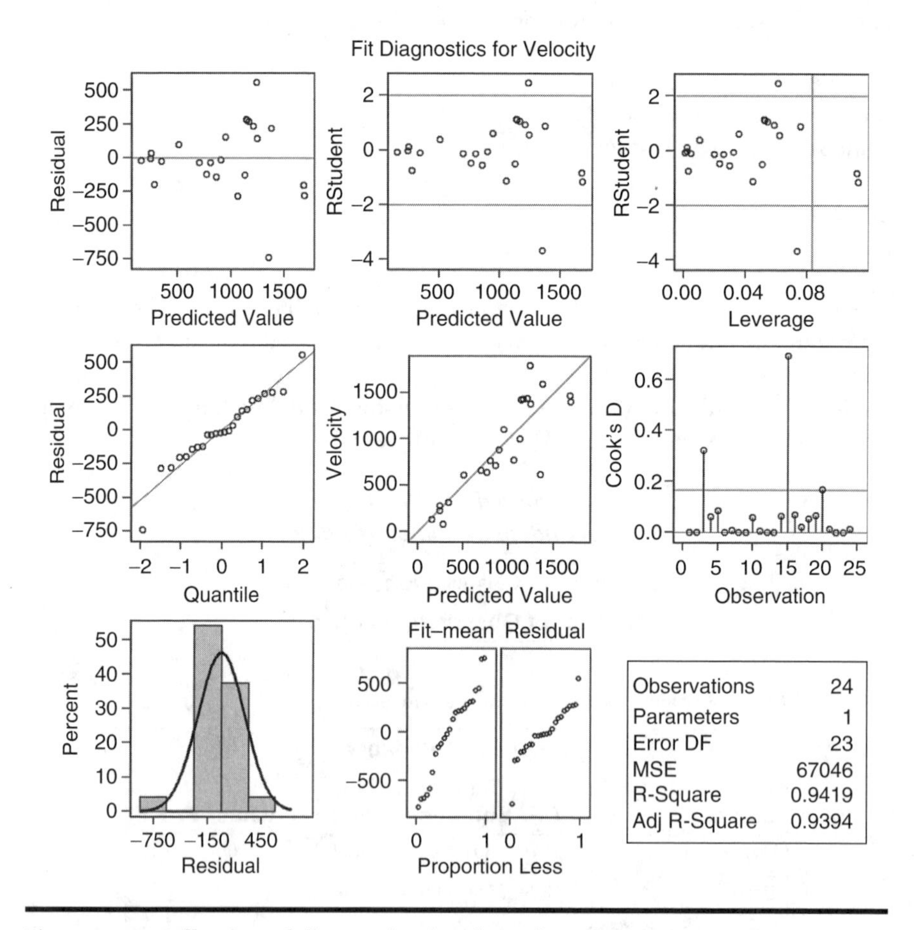

Figure 6.6 Collection of diagnostic plots for galaxy data.

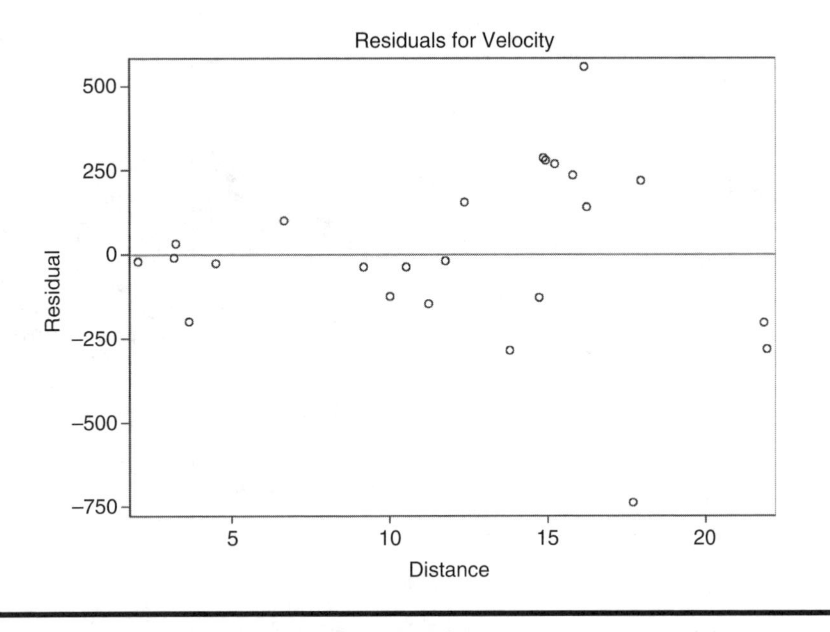

Figure 6.7 Plot of residuals against distance for the galaxy data.

high leverage by saving these values. In this case, we select leverage values greater than 0.08. The results are shown in Table 6.6.

The fit plot given in Figure 6.8 illustrates that the prediction of velocity from distance deteriorates only gradually for larger observed distances.

Now we can use the estimated value of β to find an approximate value for the age of the universe. The Hubble constant itself has unit of (km) s^{-1} $(\text{Mpc})^{-1}$. A mega-parsec is 3.09×10^{19} km, so we need to divide the estimated value of β by this amount in order to obtain Hubble's constant with units of s^{-1}. The approximate of the universe in seconds will then be $1/\beta$. Carrying out the necessary calculations gives an age of 12,794,888,643 years.

Table 6.6 Galaxies with Leverage Values Greater than 0.08 in the Fitted Simple Linear Regression Model for the Data

Obs	id	Galaxy	Velocity	Distance	pred	zres	leverage
3	3	NGC1326A	1794	16.14	1236.02	2.22390	0.06108
5	5	NGC1425	1473	21.88	1675.60	-0.83042	0.11224
15	15	NGC4414	619	17.70	1355.49	-2.95491	0.07345
20	20	NGC4639	1403	21.98	1683.25	-1.14939	0.11327

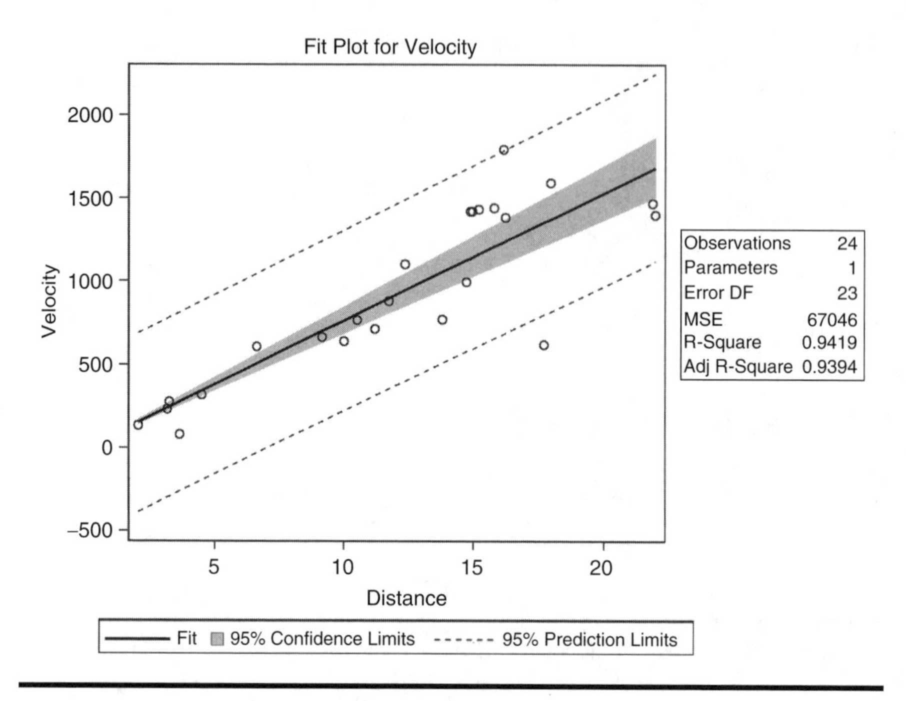

Figure 6.8 Fit plot for galaxy data.

Exercises

1 Alcohol Consumption and Cirrhosis Deaths

1. Refit the simple linear regression model after dropping the observation for France and compare the results with those for the two models considered in the chapter.
2. Investigate the use of a zero intercept model for these data.
3. Fit a simple linear regression model in which it is assumed that the "background" death rate from cirrhosis is 2.5, that is, the death rate when alcohol consumption is zero.

2 Velocity and Distance of Galaxies

1. Remove the observations for galaxies having leverage greater than 0.08 and refit the zero intercept model. What is the estimated age of the universe from this model?
2. Fit a quadratic regression model, that is, a model of the form

$$\text{Velocity} = \beta_1 \text{Distance} + \beta_2 \text{Distance}^2 + \text{Error}$$

and plot the fitted curve and the simple linear regression fit on a scatterplot of the data. Which model do you consider more sensible considering the nature of the data? (The "quadratic model" here is still regarded as a linear regression model since the term *linear* relates to the parameters of the model not to the powers of the explanatory variable.)

3 Vocabulary Size of Children

The average oral vocabulary size of children at various ages was collected to investigate the development of language in children.

The variables in vocab.dat are

- **ID:** Observation number
- **Age:** Age in years
- **Number:** Average number of words in vocabulary

1. Construct a scatterplot of the data; is there any evidence of an outlier?
2. Fit a suitable linear regression model to the data and check the assumptions of the model by suitable residual plots.
3. If the investigator who collected these data asked you to use your model to estimate the vocabulary size of children of age 10, what would be your response?

4 Expenditure on Food

In the United Kingdom Family Expenditure Survey figures were collected to show the average expenditure per household per week in the 11 main regions of the United Kingdom and the average percent of the total that is spent on food.

The variables in expend.dat are

- **Region:** Name of the region in the United Kingdom
- **Expend:** Average expenditure per household per week
- **Percent:** Average percentage spent on food

1. Construct a scatterplot of the data labelling the observations by region name.
2. Fit a simple linear regression model to the data.

Chapter 7

Multiple Regression: Determinants of Crime Rate in States of the United States

7.1 Introduction

The data set of interest in this chapter consists of crime rates for 47 states in the United States along with the values of 13 explanatory variables possibly associated with crime. (The data were originally given in Vandaele, 1978, and also appear in Table 134 of *SDS*.)

A full description of the 14 variables in these data is as follows:

R is the crime rate, that is, the number of offences known to the police per 1,000,000 population.
Age is the age distribution, that is, the number of males aged 14–24 per 1000 of total state population.
S is the binary variable distinguishing southern states ($S = 1$) from the rest.
Ed is the educational level, which is the mean number of years of schooling $\times$ 10 of the population 25 years old and over.
Ex0 is the police expenditure, the per capita expenditure on police protection by state and local government in 1960.
Ex1 is the police expenditure as Ex0 but for 1959.

Table 7.1 U.S. Crime Rates Data

R	*Age*	*S*	*Ed*	*Ex0*	*Ex1*	*LF*	*M*	*N*	*NW*	*U1*	*U2*	*W*	*X*
79.1	151	1	91	58	56	510	950	33	301	108	41	394	261
163.5	143	0	113	103	95	583	1012	13	102	96	36	557	194
57.8	142	1	89	45	44	533	969	18	219	94	33	318	250
196.9	136	0	121	149	141	577	994	157	80	102	39	673	167
123.4	141	0	121	109	101	591	985	18	30	91	20	578	174

LF is the labour force participation rate per 1000 civilian urban males in the age group 14–24.
M is the number of males per 1000 females.
N is the state population size in hundred thousands.
NW is the number of non-whites per 1000.
U1 is the unemployment rate of urban males per 1000 in the age group 14–24.
U2 is the unemployment rate of urban males per 1000 in the age group 35–39.
W is the wealth as measured by the median value of transferable goods and assets or family income (unit 10 dollars).
X is the income inequality, which is the number of families per 1000 earning below one half of the median income.

The data for the first five states are shown in Table 7.1.

The main question of interest about these data is how the crime rate depends on the other variables listed. The central method of analysis will be multiple regression.

7.2 Multiple Regression Model

The multiple regression model has the general form

$$y_i = \beta_0 + \beta_1 x_{1i} + \beta_2 x_{2i} + \cdots + \beta_p x_{pi} + \varepsilon_i, \qquad (7.1)$$

where y_i is the value of a continuous response variable for observation i; $x_{1i}, x_{2i}, \ldots, x_{pi}$ are the values of p explanatory variables for the same observation; and ε_i is the residual or error for individual i and represents the deviation of the observed value of the response for this individual from that expected by the model.

The regression coefficients $\beta_0, \beta_1, \ldots, \beta_p$ are generally estimated by least squares. Significance tests for the regression coefficients can be derived by assuming that the residual terms are normally distributed with zero mean and constant variance σ^2. The estimated regression coefficient corresponding to a particular explanatory variable gives the change in the response variable associated with a unit change in the explanatory variable, conditional on all other explanatory variables remaining constant.

For n observations of the response and explanatory variables, the regression model may be written concisely as

$$\mathbf{y} = \mathbf{X}\boldsymbol{\beta} + \varepsilon, \tag{7.2}$$

where $\boldsymbol{y}$ is the $n \times 1$ vector of responses, $\boldsymbol{X}$ is an $n \times (\mathrm{p}+1)$ matrix of known constraints, the first column containing a series of ones corresponding to the term β_0 in Equation 7.1 and the remaining columns containing values of the explanatory variables.

The elements of the vector $\boldsymbol{\beta}$ are the regression coefficients $\beta_0, \beta_1, \ldots, \beta_p$ and those of the vector ε the residual terms $\varepsilon_1, \varepsilon_2, \ldots, \varepsilon_n$.

The regression coefficients can be estimated by least squares, resulting in the following estimator for $\boldsymbol{\beta}$:

$$\hat{\boldsymbol{\beta}} = (\boldsymbol{X}'\boldsymbol{X})^{-1}\boldsymbol{X}'\boldsymbol{y}. \tag{7.3}$$

The variances and covariances of the resulting estimates can be found from

$$\boldsymbol{S}_{\hat{\boldsymbol{\beta}}} = s^2(\boldsymbol{X}'\boldsymbol{X})^{-1}, \tag{7.4}$$

where s^2 is defined below.

The variation in the response variable can be partitioned into a part due to regression on the explanatory variables and a residual. These can be arranged in an analysis of variance table as follows:

Source	*DF*	*SS*	*MS*	*F*
Regression	P	RGSS	RGMS-RGSS/p	RGMS/RSMS
Residual	*n*−*p*−1	RSS	RSMS-RSS/(*n*−*p*−1)	

Note: DF, degrees of freedom; SS, sum of squares; MS, mean square.

The residual mean square s^2 gives an estimate of σ^2, and the F-statistic is a test that $\beta_1, \beta_2, \ldots, \beta_p$ are all zero.

A measure of the fit of the model is provided by the *multiple correlation coefficient*, R, defined as the correlation between the observed values of the response variable and the values predicted by the model, that is,

$$\hat{y}_i = \hat{\beta}_0 + \hat{\beta}_1 x_{i1} + \cdots + \hat{\beta}_p x_{ip}. \tag{7.5}$$

The value of R-square gives the proportion of the variability of the response variable accounted for by the explanatory variables.

For full details of multiple regression, see, for example, Rawlings et al. (2001).

7.3 Analysis Using SAS

Assuming the data are available as an ASCII file usercrime.dat, they may be read into SAS for analysis using the instructions

```
data uscrime;
  infile 'uscrime.dat' expandtabs;
  input R Age S Ed Ex0 Ex1 LF M N NW U1 U2 W X;
run;
```

Before undertaking a formal regression analysis of these data, it may be helpful to examine them graphically using a scatterplot matrix. This is essentially a grid of scatterplots for each pair of variables. Such a display is often useful in assessing the general relationships between the variables, identifying possible outliers and highlighting potential *multicollinearity* problems among the explanatory variables (i.e., one explanatory variable being essentially predictable from the remainder). The sgscatter procedure is used as follows:

```
proc sgscatter data=uscrime;
  matrix R-X;
run;
```

The variables to be included in the plot are named on the matrix statement. The result is shown in Figure 7.1.

The individual relationships of crime rate to each of the explanatory variables shown in the first column of this plot do not appear to be particularly strong, apart perhaps from Ex0 and Ex1. The scatterplot matrix also clearly highlights the very strong relationship between these two variables. Highly correlated explanatory variables, multicollinearity, can cause several problems when applying the multiple regression model, including the following:

1. It severely limits the size of the multiple correlation coefficient, R, because the explanatory variables are largely attempting to explain much of the same variability in the response variable (see Dizney and Gromen, 1967, for an example).
2. It makes determining the importance of a given explanatory variable difficult because the effects of explanatory variables are confounded due to their intercorrelations.
3. It increases the variances of the regression coefficients, making using the predicted model for prediction less stable. The parameter estimates become unreliable.

Spotting multicollinearity among a set of explanatory variables may not be easy. The obvious course of action is to simply examine the correlations between these variables, but although this is often helpful, it is by no means foolproof – more

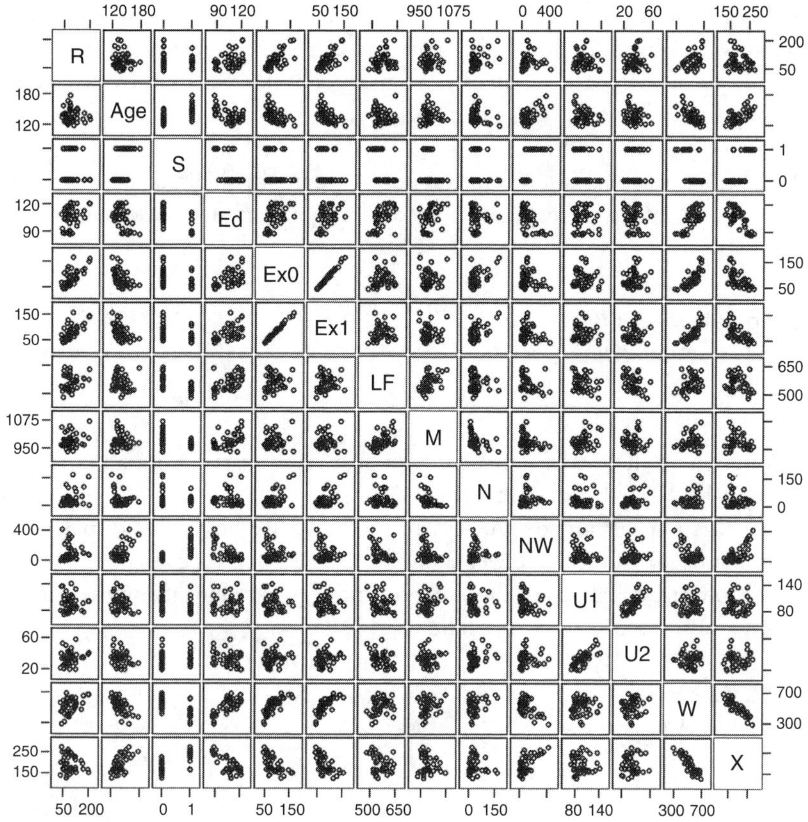

Figure 7.1 Scatterplot matrix of U.S. crime data.

subtle forms of multicollinearity may be missed. An alternative, and generally far more useful, approach is to examine what are known as the variance inflation factors of the explanatory variables. The variance inflation factor VIF_j for the jth variable is given by

$$VIF_j = \frac{1}{1 - R_j^2}, \tag{7.6}$$

where R_j^2 is the square of the multiple correlation coefficient from the regression of the jth explanatory variable on the remaining explanatory variables. The variance inflation factor of an explanatory variable indicates the strength of the linear relationship between the variable and the remaining explanatory variables. A rough rule of thumb is that variance inflation factors greater than 10 give some cause for concern.

How can multi-collinearity be combated? One way is to combine in some way explanatory variables that are highly correlated. An alternative is simply to select one of the set of correlated variables. Two more complex possibilities are regression on principal components and ridge regression, both of which are described in Chatterjee and Price (1999).

We begin our analysis of the crime rate data by looking at the variance inflation factors of the 13 explanatory variables, obtained using the following SAS instructions:

```
proc reg data=uscrime;
  model R=Age--X/vif;
run;
```

The vif option on the model statement requests that the variance inflation factors be included in the output shown in Table 7.2.

Concentrating for now on the variance inflation factors in this display, we see that those for Ex0 and Ex1 are well above the value 10. As a consequence, we will simply drop variable Ex0 from consideration and regress crime rate on the remaining 12 explanatory variables using the following:

```
proc reg data=uscrime;
  model R=Age--Ed Ex1–X/vif;
run;
```

Table 7.2 SAS Output (Numerical Only) from Proc Reg Applied to the U.S. Crime Data Showing Variance Inflation Factors

The REG Procedure

Model: MODEL1

Dependent Variable: R

Number of Observations Read	47
Number of Observations Used	47

Analysis of Variance

Source	DF	Sum of Squares	Mean Square	F Value	Pr > F
Model	13	52931	4071.58276	8.46	<.0001
Error	33	15879	481.17275		
Corrected Total	46	68809			

Root MSE	21.93565	**R-Square**	0.7692
Dependent Mean	90.50851	**Adj R-Sq**	0.6783
Coeff Var	24.23601		

Parameter Estimates

Variable	DF	Parameter Estimate	Standard Error	t Value	Pr > |t|	Variance Inflation
Intercept	1	-691.83759	155.88792	-4.44	<.0001	0
Age	1	1.03981	0.42271	2.46	0.0193	2.69802
S	1	-8.30831	14.91159	-0.56	0.5812	4.87675
Ed	1	1.80160	0.64965	2.77	0.0091	5.04944
Ex0	1	1.60782	1.05867	1.52	0.1384	94.63312
Ex1	1	-0.66726	1.14877	-0.58	0.5653	98.63723
LF	1	-0.04103	0.15348	-0.27	0.7909	3.67756
M	1	0.16479	0.20993	0.78	0.4381	3.65844
N	1	-0.04128	0.12952	-0.32	0.7520	2.32433
NW	1	0.00717	0.06387	0.11	0.9112	4.12327
U1	1	-0.60168	0.43715	-1.38	0.1780	5.93826
U2	1	1.79226	0.85611	2.09	0.0441	4.99762
W	1	0.13736	0.10583	1.30	0.2033	9.96896
X	1	0.79293	0.23509	3.37	0.0019	8.40945

The output is shown in Table 7.3. The square of the multiple correlation coefficient is 0.75 indicating that the 12 explanatory variables account for 75% of the variability in the crime rates of the 47 states. The variance inflation factors are now all less than 10.

The adjusted R^2 statistic given in Table 7.3 is the square of the multiple correlation coefficient adjusted for the number of parameters in the model. The statistic is calculated as

$$\text{adj } R^2 = 1 - \frac{(n-i)(1-R^2)}{n-p}, \tag{7.7}$$

where n is the number of observations used in fitting the model and i is an indicator variable, that is, 1, if the model includes an intercept, and 0 otherwise.

The main features of interest in Table 7.3 are the analysis of variance table and the parameter estimates. In the former, the F test is for the hypothesis that all the regression coefficients in the regression equation are zero. Here, the evidence against this hypothesis is very strong (the relevant p value is 0.0001). In general, however,

Table 7.3 SAS Output (Numerical Only) from Proc Reg Applied to U.S. Crime Data after Dropping Variable Ex0

The REG Procedure

Model: MODEL1

Dependent Variable: R

Number of Observations Read	47
Number of Observations Used	47

Analysis of Variance

Source	DF	Sum of Squares	Mean Square	F Value	Pr > F
Model	12	51821	4318.39553	8.64	<.0001
Error	34	16989	499.66265		
Corrected Total	46	68809			

Root MSE	22.35314	R-Square	0.7531
Dependent Mean	90.50851	Adj R-Sq	0.6660
Coeff Var	24.69727		

Parameter Estimates

Variable	DF	Parameter Estimate	Standard Error	t Value	Pr > |t|	Variance Inflation
Intercept	1	-739.89065	155.54826	-4.76	<.0001	0
Age	1	1.08541	0.42967	2.53	0.0164	2.68441
S	1	-8.16412	15.19508	-0.54	0.5946	4.87655
Ed	1	1.62669	0.65153	2.50	0.0175	4.89076
Ex1	1	1.02965	0.27202	3.79	0.0006	5.32594
LF	1	0.00509	0.15331	0.03	0.9737	3.53357
M	1	0.18686	0.21341	0.88	0.3874	3.64093
N	1	-0.01639	0.13092	-0.13	0.9011	2.28711
NW	1	-0.00213	0.06478	-0.03	0.9740	4.08533
U1	1	-0.61879	0.44533	-1.39	0.1737	5.93432
U2	1	1.94296	0.86653	2.24	0.0316	4.93048
W	1	0.14739	0.10763	1.37	0.1799	9.93013
X	1	0.81550	0.23908	3.41	0.0017	8.37584

this overall test is of little real interest because it is most unlikely in general that none of the explanatory variables will be related to the response. The more relevant question is whether a subset of the regression coefficients is zero, implying that not all the explanatory variables are informative in determining the response. It might be thought that the non-essential variables could be identified by simply examining the estimated regression coefficients and their standard errors, as given in Table 7.3, with those regression coefficients significantly different from zero identifying the explanatory variables needed in the derived regression equation, and those not different from zero corresponding to variables that can be left out. Unfortunately, this very straightforward approach is not in general suitable, simply because, in most cases, the explanatory variables are correlated. Consequently, removing a particular explanatory variable from the regression will alter the estimated regression coefficients (and the standard errors) of the remaining variables. The parameter estimates and their standard errors are conditional on the other variables in the model. A more involved procedure is thus necessary for identifying subsets of the explanatory variables most associated with crime rate. A number of methods are available:

- *Forward selection*: Forward selection method starts with a model containing none of the explanatory variables and then considers variables one by one for inclusion. At each step, the variable added is one that results in the biggest increase in the regression sum of squares. An F-type statistic is used to judge when further additions would not represent a significant improvement in the model.
- *Backward elimination*: Backward elimination method starts with a model containing all the explanatory variables and eliminates variables one by one, at each stage choosing the variable for exclusion as the one leading to the smallest decrease in the regression sum of squares. Once again, an F-type statistic is used to judge when further exclusions would represent a significant deterioration in the model.
- *Stepwise regression*: Stepwise regression method is, essentially, a combination of forward selection and backward elimination. Starting with no variables in the model, variables are added as with the forward selection method. Here, however, with each addition of a variable, a backward elimination process is considered to assess whether variables entered earlier might now be removed because they no longer contribute significantly to the model.

In the best of all possible worlds, the final model selected by each of these procedures would be the same. This is often the case, but it is in no way guaranteed. It should also be stressed that none of the automatic procedures for selecting subsets of variables is foolproof. They must be used with care, and warnings such as the following given in Agresti (2007) should be noted:

> Computerized variable selection procedures should be used with caution. When one considers a large number of terms for potential inclusion in a model, one or two of them that are not really important may look impressive simply due to chance. For instance, when all the true effects are weak, the largest sample effect may substantially overestimate its true effect. In addition it often makes sense to include certain variables of special interest in a model and report their estimated effects even if they are not statistically significant at some level.

In addition, the comments given in McKay and Campbell (1982a,b), concerning the validity of the F tests used to judge whether variables should be included or eliminated, should be considered.

Here we shall apply a stepwise procedure using the following SAS code:

```
proc reg data=uscrime;
  model R=Age–Ed Ex1–X/selection=stepwise sle=.05 sls=.05;
run;
```

The significance levels required for variables to enter and stay in the regression are specified with the sle and sls options, respectively. The default for both is $p = 0.15$. The output is shown in Table 7.4.

Table 7.4 shows the variables entered at each stage in the variable selection procedure. At Step 1 variable Ex1 is entered. This variable is the best single predictor of crime rate. The square of the multiple correlation coefficient is seen to be 0.4445. The variable Ex1 explains 44% of the variation in crime rates. The analysis of variance table shows both the regression and residual or error sums of squares. The F-statistic is highly significant confirming the strong relationship between crime rate and Ex1. The estimated regression coefficient is 0.92220, with a standard error of 0.15368. This implies that a unit increase in Ex1 is associated with an estimated increase in crime rate of 0.92.

At Step 2, variable X is entered. The R-squared value increases to 0.5550. The estimated regression coefficient of X is 0.42312, with a standard error of 0.12803.

In this application of the stepwise option, the default significance levels for the F tests used to judge entry of a variable into an existing model and to judge removal of a variable from a model are each set to 0.05. With these values, the stepwise procedure eventually identifies a subset of five explanatory variables as being important in the prediction of crime rate. The final results are summarized at the end of Table 7.4. The selected five variables account for just over 70% of the variation in crime rates compared with the 75% found when using 12 explanatory variables in the previous analysis. (Notice that in this example, the stepwise procedure gives the same results as would have arisen from using forward selection with the

Table 7.4 SAS Output (Numerical and Graphical) from Proc Reg Applied to U.S. Crime Data Showing Results of Stepwise Selection

The SAS System

Model: MODEL1

Dependent Variable: R

Number of Observations Read	47
Number of Observations Used	47

Stepwise Selection: Step 1

Variable Ex1 Entered: R-Square = 0.4445 and C(p) = 33.4977

Analysis of Variance					
Source	**DF**	**Sum of Squares**	**Mean Square**	**F Value**	**Pr > F**
Model	1	30586	30586	36.01	<.0001
Error	45	38223	849.40045		
Corrected Total	46	68809			

Variable	**Parameter Estimate**	**Standard Error**	**Type II SS**	**F Value**	**Pr > F**
Intercept	16.51642	13.04270	1362.10230	1.60	0.2119
Ex1	0.92220	0.15368	30586	36.01	<.0001

Bounds on condition number: 1, 1

Stepwise Selection: Step 2

Variable X Entered: R-Square = 0.5550 and C(p) = 20.2841

Analysis of Variance					
Source	**DF**	**Sum of Squares**	**Mean Square**	**F Value**	**Pr > F**
Model	2	38188	19094	27.44	<.0001
Error	44	30621	695.94053		
Corrected Total	46	68809			

Variable	Parameter Estimate	Standard Error	Type II SS	F Value	Pr > F
Intercept	-96.96590	36.30976	4963.21825	7.13	0.0106
Ex1	1.31351	0.18267	35983	51.70	<.0001
X	0.42312	0.12803	7601.63672	10.92	0.0019

Bounds on condition number: 1.7244, 6.8978

Stepwise Selection: Step 3

Variable Ed Entered: R-Square = 0.6378 and C(p) = 10.8787

Bounds on condition number: 3.1634, 21.996

Stepwise Selection: Step 4

Variable Age Entered: R-Square = 0.6703 and C(p) = 8.4001

Bounds on condition number: 3.5278, 38.058

Stepwise Selection: Step 5

Variable U2 Entered: R-Square = 0.7049 and C(p) = 5.6452

Analysis of Variance

Source	DF	Sum of Squares	Mean Square	F Value	Pr > F
Model	5	48500	9700.08117	19.58	<.0001
Error	41	20309	495.33831		
Corrected Total	46	68809			

Variable	Parameter Estimate	Standard Error	Type II SS	F Value	Pr > F
Intercept	-528.85572	99.61621	13961	28.18	<.0001
Age	1.01840	0.36909	3771.26606	7.61	0.0086
Ed	2.03634	0.49545	8367.48486	16.89	0.0002
Ex1	1.29735	0.15970	32689	65.99	<.0001
U2	0.99014	0.45210	2375.86580	4.80	0.0343
X	0.64633	0.15451	8667.44486	17.50	0.0001

Bounds on condition number: 3.5289, 57.928

All variables left in the model are significant at the 0.0500 level.

No other variable met the 0.0500 significance level for entry into the model.

Summary of Stepwise Selection

Step	Variable Entered	Variable Removed	Number Vars In	Partial R-Square	Model R-Square	C(p)	F Value	Pr > F
1	Ex1		1	0.4445	0.4445	33.4977	36.01	<.0001
2	X		2	0.1105	0.5550	20.2841	10.92	0.0019
3	Ed		3	0.0828	0.6378	10.8787	9.83	0.0031
4	Age		4	0.0325	0.6703	8.4001	4.14	0.0481
5	U2		5	0.0345	0.7049	5.6452	4.80	0.0343

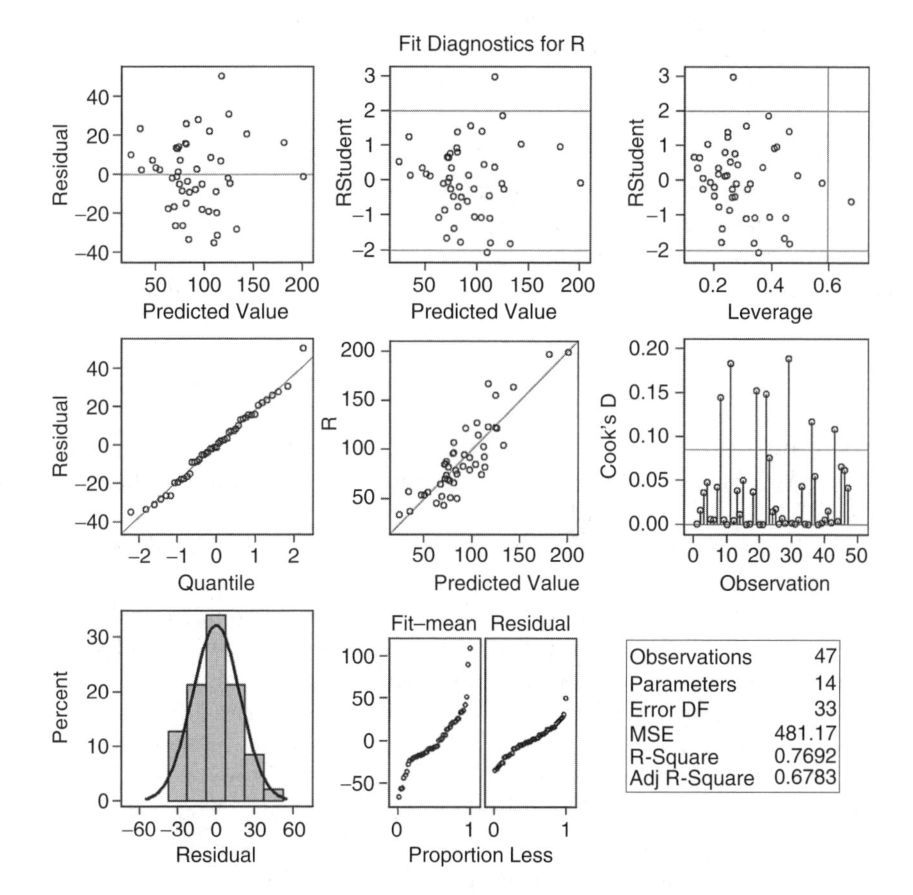

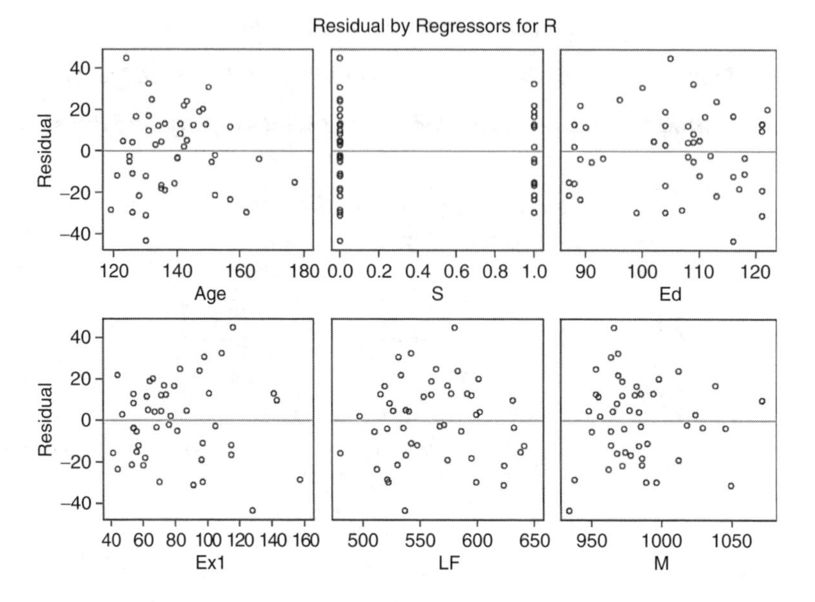

same entry criterion value of 0.05, since none of the variables entered in the "forward" phase is ever removed.)

The statistic C_p was suggested by Mallows (1973) as a possible alternative criterion useful for selecting informative subsets of variables. It is defined as

$$C_p = \frac{SSE_p}{s^2} - (n - 2p), \tag{7.8}$$

where s^2 is the mean square error from the regression including all the explanatory variables available, SSE_p is the error sum of squares for a model which includes just a subset of the explanatory variable.

If C_p is plotted against p, Mallows recommends accepting the model where C_p first approaches p.

The "Bounds on condition number" given in Table 7.4 are explained fully in Berk (1977). Briefly, the condition number is the ratio of the largest and smallest eigenvalues of a matrix and is used as a measure of the numerical stability of the matrix. Very large values are indicative of possible numerical problems.

Having arrived at a final multiple regression model for a data set, it is important to go further and check the assumptions made in the modelling process. Most useful at this stage is an examination of residuals from the fitted model along with other of the many regression diagnostics now available – see Chapter 6. Examining the plots in Table 7.4 that involve residuals plotted against the values of variable Ex1 suggests that the variance of the residuals may increase as Ex1 increases, and the plot of Cook's distances indicates a number of relatively high values although there are no values greater than one, which is the usually accepted threshold for concluding that the corresponding observation has undue influence on the estimated regression coefficients.

Exercises

1 Determinants of Crime Rate in the United States

1. Find the subset of five variables considered by the C_p option to be optimal. How does this subset compare with that chosen by the stepwise option described in the text?
2. Apply the C_p criterion to explore all subsets of the five variables chosen by the stepwise procedure (see Table 7.4). Produce a plot of the number of variables in a subset against the corresponding value of C_p.
3. Examine some of the other regression diagnostics available with `proc reg`.
4. In the text, the problem of high-variance inflation factors associated with variables Ex0 and Ex1 was dealt with by excluding Ex0. An alternative approach might be to use the average of the two variables as an explanatory variable. Investigate this possibility.

2 Aflatoxin in Peanuts

Data (taken from Table 48 of *SDS*) were collected on the average level of aflatoxin (parts per million) for 34 batches of peanuts from a mini-lot of 120 pounds of peanuts and the percentage of non-contaminated peanuts in each batch. The aim is to investigate the relationship between the two variables and to derive an equation for predicting contamination percentage from average level of aflatoxin.

The variables in peanuts.dat are

- **ID:** Batch number
- **Percent:** Percentage of non-contaminated peanuts
- **Level:** Average level of aflatoxin

1. Produce a scatterplot of percent against a level that includes the fitted regression line for the two variables.
2. Use some suitable residual plots to investigate the assumptions made in fitting the regression line.

3 Cloud Seeding

Weather modification, or cloud seeding, is the treatment of individual clouds or storm systems with various inorganic or organic materials in the hope of achieving an increase in rainfall. Introduction of such material into a cloud that contains supercooled water, that is, liquid water colder than 0°C, has the aim of inducing freezing, with the consequent ice particles growing at the expense of liquid droplets and becoming heavy enough to fall as rain from the clouds that otherwise would produce none.

Data were collected in the summer of 1975 from an experiment to investigate the use of massive amounts of silver iodine (100–1000 g/cloud) in cloud seeding to increase rainfall (Woodley et al., 1977). In the experiment, which was conducted in an area of Florida, 24 days were judged suitable for seeding on the basis that a measured suitability criterion was met. This criterion (S-NE), which is defined in detail in Woodley et al., biases the decision for experimentation against naturally rainy days. On thus-defined suitable days, a decision was taken at random as to whether to seed. The aim in analysing these data is to see how rainfall is related to the other variables and, in particular, to determine the effectiveness of seeding.

The variables in cloud.dat are

- **ID:** Experiment number
- **Time:** Number of days after first experiment
- **Seeding:** 0 = no, 1 = yes
- **Sne:** Suitability for seeding criterion
- **Cloud:** Percentage cloud cover
- **Prewet:** Total rainfall in area 1 h before seeding (in $m^3 \times 10^7$)
- **Echo:** Was radio echo moving or stationary? 0 = stationary, 1 = moving
- **Rainfall:** The amount of rain in $m^3 \times 10^7$

1. Construct box plots of the rainfall in each category of the dichotomous explanatory variables and scatterplots of rainfall against each of the continuous explanatory variables. What do you conclude from these plots?
2. In this example, it is sensible to assume that the effect of some of the other explanatory variables is modified by seeding; therefore, consider a model that allows interaction terms for seeding with each of the covariates except time. Fit this model to the data.
3. Interpret the results from the fitted model and construct a suitable diagram for displaying the most important aspect of the results.

Chapter 8

Logistic Regression: Psychiatric Screening, Plasma Proteins, Danish Do-It-Yourself and Lower Back Pain

8.1 Description of Data

In this chapter we shall examine four data sets. The first, shown in Table 8.1, arises from a study of a psychiatric screening questionnaire called the General Health Questionnaire (GHQ) – see Goldberg, 1972. Here the question of interest is how "caseness" is related to gender and GHQ score.

The second data set, part of which is given in Table 8.2, was collected to examine the extent to which erythrocyte sedimentation rate (ESR), that is, the rate at which red blood cells (erythrocytes) settle out of suspension in blood plasma, is related to two plasma proteins, fibrinogen and γ-globulin, both measured in g/L (Collett, 2003a). The ESR for a "healthy" individual should be less than 20 mm/h, and since the absolute value of ESR is relatively unimportant, the response variable used here denotes whether or not is less than 20 mm/h. A response of zero signifies a healthy individual (ESR $<$ 20) while a response of one refers to an unhealthy individual

Table 8.1 Psychiatric Screening Data

GHQ Score	*Sex*	*Number of Cases*	*Number of Non-cases*
0	F	4	80
1	F	4	29
2	F	8	15
3	F	6	3
4	F	4	2
5	F	6	1
6	F	3	1
7	F	2	0
8	F	3	0
9	F	2	0
10	F	1	0
0	M	1	36
1	M	2	25
2	M	2	8
3	M	1	4
4	M	3	1
5	M	3	1
6	M	2	1
7	M	4	2
8	M	3	1
9	M	2	0
10	M	2	0

Note: F, female; M, male.

Table 8.2 Plasma Proteins and ESR Data

Fibrinogen	*Gamma*	*ESR*
2.52	38	0
2.56	31	0
2.19	33	0
2.18	31	0
3.41	37	0
2.46	36	0
3.22	38	0
2.21	37	0
3.15	39	0
2.60	41	0
2.29	36	0
2.35	29	0
5.06	37	1
3.34	32	1
2.38	37	1

Table 8.3 Danish Do-It-Yourself

			Accommodation Type					
			Apartment			House		
Work	*Tenure*	*Response*	*<30*	*31–45*	*46+*	*<30*	*31–45*	*46+*
Skilled	Rent	Yes	18	15	6	34	10	2
		No	15	13	9	28	4	6
	Own	Yes	5	3	1	56	56	35
		No	1	1	1	12	21	8
Unskilled	Rent	Yes	17	10	15	29	3	7
		No	34	17	19	44	13	16
	Own	Yes	2	0	3	23	52	49
		No	3	2	0	9	31	51
Office	Rent	Yes	30	23	21	22	13	11
		No	25	19	40	25	16	12
	Own	Yes	8	5	1	54	191	102
		No	4	2	2	19	76	61

(ESR $\geq$ 20). The aim of the analysis for these data is to determine the strength of any relationship between the ESR level and the levels of the two plasmas.

The third of our data sets (taken from Table 31 of *SDS*) is given in Table 8.3 and results from asking a sample of employed men aged between 18 and 67 whether, in the preceding year, they had carried out work in their home they would have previously employed a craftsman to do. The response variable here is the answer (yes/no) to that question. Here we would like to model the relationship between the response variable and each of the four explanatory variables: work, tenure, accommodation type and age.

The last of the data sets to be considered in this chapter arises from a study reported by Kelsey and Hardy (1975), which was designed to investigate whether driving a car is a risk factor for low back pain resulting from acute herniated lumbar intervertebral discs (AHLID). A case-control study was used with cases selected from people who had recently had x-rays taken of the lower back and had been diagnosed as having AHLID. The controls were taken from patients admitted to the same hospital as a case with a condition unrelated to the spine. Further matching was made on age and sex and a total of 217 matched pairs were recruited, consisting of 89 female pairs and 128 male pairs. Part of the data is shown is Table 8.4.

8.2 Logistic Regression Model

In linear regression (see Chapters 6 and 7), the expected value of a response variable, y, is modelled as a linear function of the explanatory variables:

$$E(y) = \beta_0 + \beta_1 x_1 + \beta_2 x_2 + \cdots + \beta_p x_p. \tag{8.1}$$

Table 8.4 Matched Pairs in the AHLID Study According to Driving and Residence

Pair	*Status*	*Driver*	*Suburban Resident*
1	Case	1	1
1	Control	1	0
2	Case	1	1
2	Control	1	1
3	Case	1	0
3	Control	1	1
⋮			
216	Case	0	0
216	Control	1	1
217	Case	1	1
217	Control	1	0

Note: 0, no; 1, yes.

For a dichotomous response variable coded 0 and 1, the expected value is simply the probability, π, that the variable takes the value 1. This could be modelled as in Equation 8.1, but there are two problems with using linear regression when the response variable is dichotomous:

1. The predicted probability must satisfy $0 \leq \pi \leq 1$, whereas a linear predictor can yield any value from $-\infty$ to $+\infty$.
2. The observed values of y do not follow a normal distribution with mean π but rather a Bernoulli (or binomial $[1, \pi]$) distribution.

In logistic regression the first problem is addressed by replacing the probability π which is the $E(y)$ on the left-hand side of Equation 8.1 by the logit transformation of the probability, $\log \pi/(1-\pi)$. The model now becomes

$$\text{logit}(\pi) = \log \frac{\pi}{1-\pi} = \beta_0 + \beta_1 x_1 + \cdots + \beta_p x_p. \tag{8.2}$$

The logit of the probability is simply the log of the odds of the event of interest. Setting $\boldsymbol{\beta}' = [\beta_0, \beta_1, \ldots, \beta_p]$ and the augmented vector of scores for the ith individual as $\boldsymbol{x}_i' = [1, x_{i1}, x_{i2}, \ldots, x_{ip}]$, the predicted probabilities as a function of the linear predictor are

$$\pi(\boldsymbol{\beta}'\boldsymbol{x}_i) = \frac{\exp(\boldsymbol{\beta}'\boldsymbol{x}_i)}{1 + \exp(\boldsymbol{\beta}'\boldsymbol{x}_i)}. \tag{8.3}$$

Although the logit can take on any real value, this probability always satisfies $0 \leq \pi(\boldsymbol{\beta}'\boldsymbol{x}_i) \leq 1$. In a logistic regression model the parameter β_i associated with explanatory variable x_i is such that $\exp(\beta_i)$ is the odds that $y = 1$ when x_i increases by 1, conditional on the other explanatory variables remaining the same.

Maximum likelihood is used to estimate the parameters of Equation 8.2, the log likelihood function being

$$l(\boldsymbol{\beta}; \boldsymbol{y}) = \sum_i y_i \log[\pi(\boldsymbol{\beta}'\boldsymbol{x}_i)] + (1 - y_i) \log[1 - \pi(\boldsymbol{\beta}'\boldsymbol{x}_i)], \tag{8.4}$$

where $\boldsymbol{y}' = [y_1, y_2, \ldots, y_n]$ are the n observed values of the dichotomous response variable.

This log likelihood is maximized numerically using an iterative algorithm. For full details of logistic regression, see, for example, Collett (2003a).

8.3 Analysis Using SAS

8.3.1 GHQ Data

The data values are separated by tabs and can be read in as follows:

```
data ghq;
  infile 'c:\handbook3\datasets\ghq.dat' expandtabs;
  input ghq sex $ cases noncases;
  total = cases + noncases;
  prcase = cases/total;
run;
```

The variable prcase contains the observed probability of being a case. This can be plotted against ghq score as follows:

```
proc sgplot data = ghq;
  scatter y = prcase x = ghq;
run;
```

The resulting plot is shown in Figure 8.1. Clearly, as GHQ score increases, the probability of being considered a case increases.

It is a useful exercise to compare the results of fitting both a simple linear regression and a logistic regression to these data using the single explanatory variable

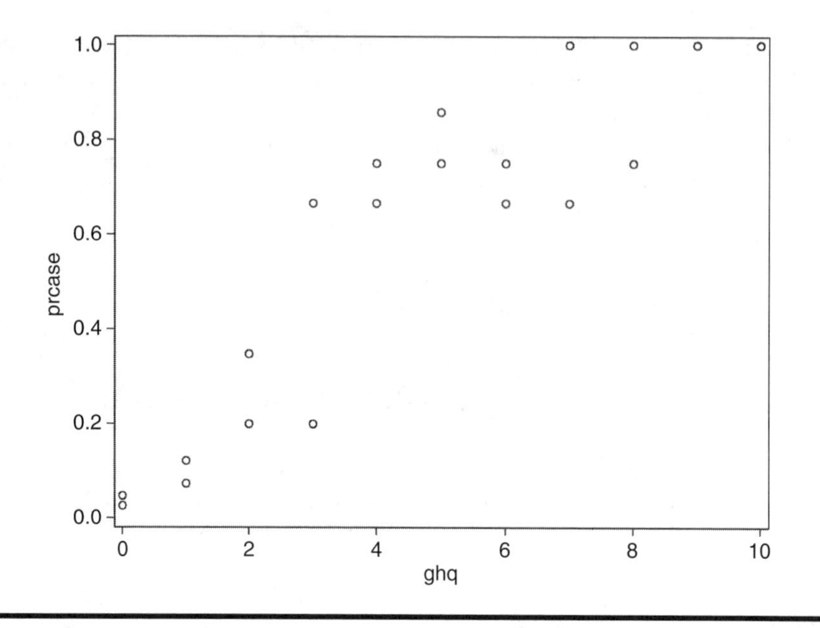

Figure 8.1 Plot of probability of being a case against GHQ score for the psychiatric screening data.

GHQ score. Here we shall make this comparison by plotting graphs of the fitted values from both types of model. We can use sgplot to fit the linear regression. First we perform a logistic regression using proc logistic.

```
proc logistic data=ghq;
  model cases/total=ghq;
  output out=lout p=lpred;
run;
```

There are two forms of model statement within proc logistic. This example shows the events/trials syntax, where two variables are specified separated by a slash. The alternative is to specify a single binary response variable before the equal sign.

The output statement creates an output data set that contains all the original variables plus those created by options. The p=lpred option specifies that the predicted values are included in a variable named lpred. The out=lout option specifies the name of the data set to be created.

Because proc sgplot plots the data in the order in which they occur, for line plots it is usually necessary to sort the data set by the *x*-axis variable. The predicted probabilities from the logistic regression are plotted with a series plot and a linear regression overlaid on the same graph (Figure 8.2). They are labelled

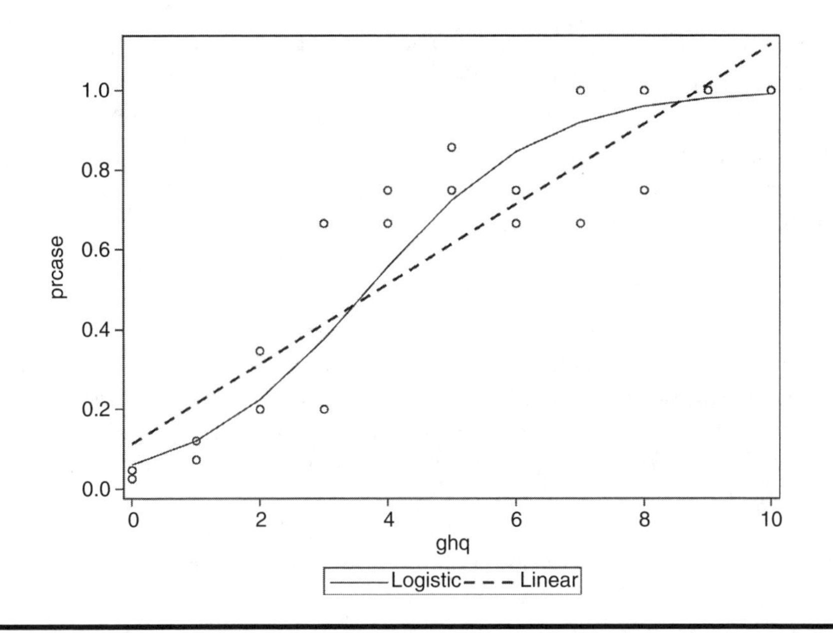

Figure 8.2 Fitted linear and logistic regressions for the probability of being a case with GHQ as the single explanatory variable.

appropriately with the legendlabel option and a dashed line specified for the linear regression.

```
proc sort data = lout;
  by ghq;
run;
proc sgplot data = lout;
  series y = lpred x = ghq/legendlabel = 'Logistic';
  reg y = prcase x = ghq/legendlabel = 'Linear' lineattrs = (pattern = dash);
run;
```

The problems of using the unsuitable linear regression model become apparent on studying Figure 8.2. Using this model, two of the predicted values are greater than one, but the response is a probability constrained to be in the interval (0,1). Additionally, the model provides a very poor fit for the observed data. Using the logistic model, on the other hand, leads to predicted values that are satisfactory in that they all lie between 0 and 1, and the model clearly provides a better description of the observed data.

Next we shall extend the logistic regression model to include both GHQ score and sex as explanatory variables.

```
proc logistic data=ghq;
  class sex;
  model cases/total=sex ghq;
run;
```

The class statement specifies classification variables, or factors, and these may be numeric or character variables. The specification of explanatory effects on the model statement is the same as for proc glm: with main effects specified by variable names and interactions by joining variable names with asterisks. The bar operator may also be used as an abbreviated way of entering interactions if these are to be included in the model (see Chapter 5).

The output is shown in Table 8.5. All three tests of the global hypothesis, in which the regression parameters for both GHQ and sex are zero, lead to rejection of the

Table 8.5 Output from Fitting a Logistic Regression Model to the Psychiatric Screening Data Using GHQ Score and Sex as Explanatory Variables

The LOGISTIC Procedure

Model Information	
Data Set	WORK.GHQ
Response Variable (Events)	Cases
Response Variable (Trials)	Total
Model	binary logit
Optimization Technique	Fisher's scoring

Number of Observations Read	22
Number of Observations Used	22
Sum of Frequencies Read	278
Sum of Frequencies Used	278

Response Profile		
Ordered Value	**Binary Outcome**	**Total Frequency**
1	Event	68
2	Nonevent	210

Class Level Information		
Class	**Value**	**Design Variables**
Sex	**F**	1
	M	-1

Model Convergence Status
Convergence criterion (GCONV=1E-8) satisfied.

Model Fit Statistics		
Criterion	**Intercept Only**	**Intercept and Covariates**
AIC	311.319	196.126
SC	314.947	207.009
-2 Log L	309.319	190.126

Testing Global Null Hypothesis: BETA=0			
Test	**Chi-Square**	**DF**	**Pr > ChiSq**
Likelihood Ratio	119.1929	2	<.0001
Score	120.1327	2	<.0001
Wald	61.9555	2	<.0001

Type 3 Analysis of Effects			
Effect	**DF**	**Wald Chi-Square**	**Pr > ChiSq**
sex	1	4.6446	0.0312
ghq	1	61.8891	<.0001

Analysis of Maximum Likelihood Estimates						
Parameter		**DF**	**Estimate**	**Standard Error**	**Wald Chi-Square**	**Pr > ChiSq**
Intercept		1	-2.9615	0.3155	88.1116	<.0001
sex	**F**	1	0.4680	0.2172	4.6446	0.0312
ghq		1	0.7791	0.0990	61.8891	<.0001

Odds Ratio Estimates			
		95% Wald	
Effect	**Point Estimate**	**Confidence Limits**	
sex F vs M	2.550	1.088	5.974
ghq	2.180	1.795	2.646

Association of Predicted Probabilities and Observed Responses			
Percent Concordant	85.8	**Somers' D**	0.766
Percent Discordant	9.2	**Gamma**	0.806
Percent Tied	5.0	**Tau-a**	0.284
Pairs	14280	**c**	0.883

hypothesis. The results in the "Analysis of Maximum Likelihood Estimates" section of Table 8.5 show that the estimated parameters for both sex and GHQ are significant beyond the 5% level. The parameter estimates are best interpreted if they are converted into odds ratios by exponentiating them. For GHQ, for example, this leads to an odds ratio estimate of exp(0.7791), that is, 2.180, with a 95% confidence interval of (1.795, 2.646). A unit increase in GHQ increases the odds of being a case between about 1.8 and 3 times, conditional on sex.

The same procedure can be applied to the parameter for sex, but more care is needed here, since the Class Level Information in Table 8.5 shows that sex is coded 1 for females and -1 for males. Consequently, the required odds ratio is $\exp(2 \times 0.468)$, that is, 2.55 with a 95% confidence interval of (1.088, 5.974). Being female rather than male increases the odds of being a case between about 1.1 and 6 times, conditional on GHQ. (For a dichotomous explanatory variable, 0/1 coding is usually known as reference cell coding and $-1/1$ coding as deviation from means. Close attention should be paid to the method used to code dichotomous variables although the SAS output in Table 8.5 conveniently provides the correct odds ratios and associated confidence intervals.)

8.3.2 ESR and Plasma Levels

We now move on to examine the ESR data in Table 8.2. The data are first read in for analysis using the following SAS code:

```
data plasma;
  infile 'c:\handbook3\datasets\plasma.dat';
  input fibrinogen gamma esr;
run;
```

We can try to identify which of the two plasma proteins, fibrinogen or γ-globulin, has the stronger relationship with ESR level by fitting a logistic regression model and allowing backward elimination of variables as described in Chapter 7 for multiple regression, although the elimination criterion is now based on a likelihood ratio statistic rather than an F value.

```
proc logistic data=plasma desc;
  model esr=fibrinogen gamma fibrinogen*gamma/selection=backward;
run;
```

Where a binary response variable is used on the model statement, as opposed to the events/trials used for the GHQ data, SAS models the lower of the two response categories as the "event." However, it is common practice for a binary response variable to be coded 0,1, with 1 indicating a response, or an event, and 0 indicating no response, or a non-event. In this case, the seemingly perverse default in SAS will be to model the probability of a non-event. The desc (descending) option on the proc statement reverses this behaviour.

It is worth noting that when the model selection option is forward, backward, or stepwise, SAS preserves the hierarchy of effects by default. For an interaction effect to be allowed in the model, all the lower-order interactions and main effects that it implies must also be included.

The results are given in Table 8.6. We see that both the fibrinogen $\times$ γ-globulin interaction effect and the γ-globulin main effect are eliminated from the initial model. It appears that only fibrinogen level is predictive of ESR level.

Table 8.6 Logistic Regression Models for Plasma Protein Data Using Backward Elimination

The LOGISTIC Procedure

Model Information	
Data Set	WORK.PLASMA
Response Variable	Esr
Number of Response Levels	2
Model	binary logit
Optimization Technique	Fisher's scoring

Number of Observations Read	32
Number of Observations Used	32

Response Profile		
Ordered Value	esr	Total Frequency
1	1	6
2	0	26

Probability modeled is esr=1.

Backward Elimination Procedure

Step 0. The following effects were entered:

Intercept fibrinogen gamma fibrinogen*gamma

Model Convergence Status
Convergence criterion (GCONV=1E-8) satisfied.

Model Fit Statistics		
Criterion	Intercept Only	Intercept and Covariates
AIC	32.885	28.417
SC	34.351	34.280
-2 Log L	30.885	20.417

Testing Global Null Hypothesis: BETA=0			
Test	Chi-Square	DF	Pr > ChiSq
Likelihood Ratio	10.4677	3	0.0150
Score	8.8192	3	0.0318
Wald	4.7403	3	0.1918

Step 1. Effect fibrinogen*gamma is removed:

Model Convergence Status
Convergence criterion (GCONV=1E-8) satisfied.

Model Fit Statistics		
Criterion	Intercept Only	Intercept and Covariates
AIC	32.885	28.971
SC	34.351	33.368
-2 Log L	30.885	22.971

Testing Global Null Hypothesis: BETA=0			
Test	**Chi-Square**	**DF**	**Pr > ChiSq**
Likelihood Ratio	7.9138	2	0.0191
Score	8.2067	2	0.0165
Wald	4.7561	2	0.0927

Residual Chi-Square Test		
Chi-Square	**DF**	**Pr > ChiSq**
2.6913	1	0.1009

Step 2. Effect gamma is removed:

Model Convergence Status
Convergence criterion (GCONV=1E-8) satisfied.

Model Fit Statistics		
Criterion	**Intercept Only**	**Intercept and Covariates**
AIC	32.885	28.840
SC	34.351	31.772
-2 Log L	30.885	24.840

Testing Global Null Hypothesis: BETA=0			
Test	**Chi-Square**	**DF**	**Pr > ChiSq**
Likelihood Ratio	6.0446	1	0.0139
Score	6.7522	1	0.0094
Wald	4.1134	1	0.0425

Residual Chi-Square Test		
Chi-Square	**DF**	**Pr > ChiSq**
4.5421	2	0.1032

NOTE: No (additional) effects met the 0.05 significance level for removal from the model.

Summary of Backward Elimination					
Step	Effect Removed	DF	Number In	Wald Chi-Square	Pr > ChiSq
1	fibrinogen*gamma	1	2	2.2968	0.1296
2	Gamma	1	1	1.6982	0.1925

Analysis of Maximum Likelihood Estimates					
Parameter	DF	Estimate	Standard Error	Wald Chi-Square	Pr > ChiSq
Intercept	1	-6.8451	2.7703	6.1053	0.0135
fibrinogen	1	1.8271	0.9009	4.1134	0.0425

Odds Ratio Estimates			
Effect	Point Estimate	95% Wald Confidence Limits	
fibrinogen	6.216	1.063	36.333

Association of Predicted Probabilities and Observed Responses			
Percent Concordant	71.2	Somers' D	0.429
Percent Discordant	28.2	Gamma	0.432
Percent Tied	0.6	Tau-a	0.135
Pairs	156	C	0.715

It is useful to look at a graphical display of the final model selected. With ODS graphics on, the plots = effect option on the proc statement produces a plot of predicted values from the fibrinogen- only logistic model along with the observed values of ESR (remember these can take only the values 0 or 1) and the 95% confidence interval. The plot is shown in Figure 8.3.

```
proc logistic data = plasma desc plots = effect;
  model esr = fibrinogen;
run;
```

Clearly, increase in fibrinogen level is associated with an increase in the probability of the individual being categorized as unhealthy.

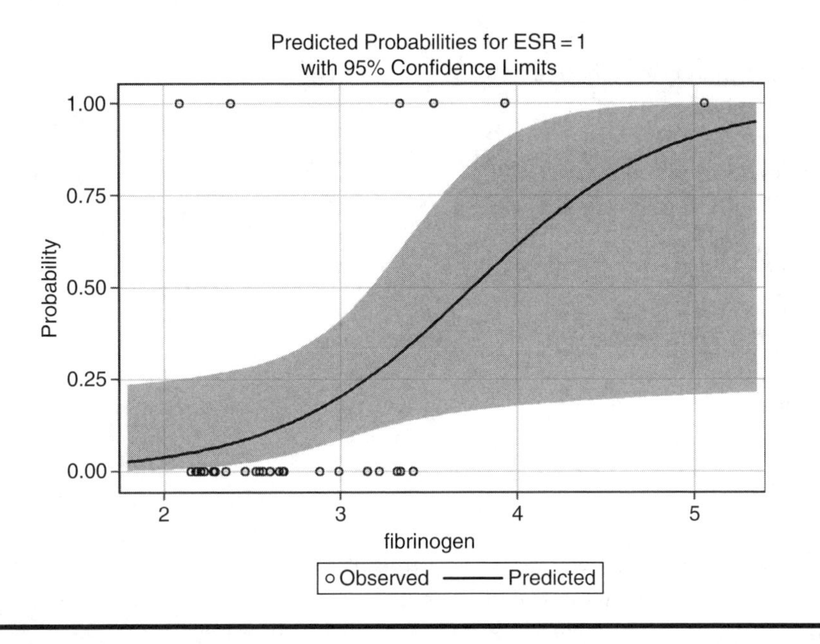

Figure 8.3 Observed values of ESR and values predicted from the logistic regression model chosen by backward elimination.

8.3.3 Danish Do-It-Yourself

Assuming the data shown in Table 8.3 are in a file diy.dat in the current directory and the values are separated by tabs, the following data step may be used to create a SAS data set for analysis. As in previous examples, the values of the grouping variables can be determined from the row and column positions in the data. An additional feature of this data set is that each cell of the design contains two data values: counts of those who answered "yes" and "no" to the question about work in the home. Each observation in the data set needs both of these values so that the events/trials syntax may be used in proc logistic. To do this, two rows of data are input at the same time: six counts of "yes" responses and the corresponding "no" responses.

```
data diy;
  infile 'diy.dat' expandtabs;
  input y1-y6/n1-n6;
  length work $9.;
  work='Skilled';
  if _n_>2 then work='Unskilled';
  if _n_>4 then work='Office';
  if _n_ in(1,3,5) then tenure='rent';
    else tenure='own';
  array yall {6} y1-y6;
  array nall {6} n1-n6;
```

```
  do i=1 to 6;
  if i>3 then type='house';
    else type='flat';
  agegrp=1;
    if i in(2,5) then agegrp=2;
    if i in(3,6) then agegrp=3;
  yes=yall{i};
    no=nall{i};
    total=yes+no;
    prdiy=yes/total;
    output;
  end;
  drop i y1-n6;
run;
```

The expandtabs option on the infile statement allows list input to be used. The input statement reads two lines of data from the file. Six data values are read from the first line into variables y1 to y6. The slash that follows tells SAS to go to the next line and six values from there are read into variables n1 to n6.

There are 12 lines of data in the file, but because each pass through the data step is reading a pair of lines, the automatic variable _n_ will range from 1 to 6. The appropriate values of the work and tenure variables can then be assigned accordingly. Both are character variables and the length statement specifies that work is nine characters. Without this, its length would be determined from its first occurrence in the data step. This would be the statement work='Skilled'; and a length of seven characters would be assigned, insufficient for the value "unskilled."

The variables containing the yes and no responses are declared as arrays and processed in parallel inside a do loop. The values of age group and accommodation type are determined from the index of the do loop, that is, from the column in the data. Counts of yes and corresponding no responses are assigned to the variables yes and no, their sum assigned to total, and the observed probability of a yes to prdiy. The output statement within the do loop writes six observations to the data set. (See Chapter 5 for a fuller explanation.)

As usual with a complicated data step such as this, it is wise to check the results, for example, with proc print.

A useful starting point in examining these data is a tabulation of the observed probabilities using proc tabulate.

```
proc tabulate data=diy order=data f=6.2;
  class work tenure type agegrp;
  var prdiy;
  table work*tenure all,
    (type*agegrp all)*prdiy*mean;
run;
```

Table 8.7 Observed Probabilities for Danish Do-It-Yourself

		Type						
		flat			house			
		agegrp			agegrp			
		1	2	3	1	2	3	All
		prdiy	prdiy	prdiy	prdiy	prdiy	prdiy	prdiy
		Mean	Mean	Mean	Mean	Mean	Mean	Mean
work	tenure							
Skilled	Rent	0.55	0.54	0.40	0.55	0.71	0.25	0.50
	own	0.83	0.75	0.50	0.82	0.73	0.81	0.74
Unskilled	rent	0.33	0.37	0.44	0.40	0.19	0.30	0.34
	own	0.40	0.00	1.00	0.72	0.63	0.49	0.54
Office	rent	0.55	0.55	0.34	0.47	0.45	0.48	0.47
	own	0.67	0.71	0.33	0.74	0.72	0.63	0.63
All		0.55	0.49	0.50	0.62	0.57	0.49	0.54

Basic use of proc tabulate was described in Chapter 5. In this example, the f = option specifies a format for the cell entries of the table, namely six columns with two decimal places. It also illustrates the use of the keyword all for producing totals. The result is shown in Table 8.7. We see that there are considerable differences in the observed probabilities suggesting that some, at least, of the explanatory variables may have an effect.

We continue our analysis of the data with a backwards elimination logistic regression for the main effects of the four explanatory variables only.

```
proc logistic data = diy;
  class work tenure type agegrp/param = ref ref = first;
  model yes/total = work tenure type agegrp/selection = backward;
run;
```

All the predictors are declared as classification variables, or factors, on the class statement. The param option specifies reference coding (more commonly referred to as dummy variable coding) with the ref option setting the first category to be the reference category.

The output is shown in Table 8.8. Work, tenure and age group are all selected in the final model; only type of accommodation is dropped. The estimated conditional odds ratio suggests that skilled workers are more likely to respond “yes” to the

Table 8.8 Results of Using Backward Selection on the Danish Do-It-Yourself Data

The LOGISTIC Procedure

Model Information	
Data Set	WORK.DIY
Response Variable (Events)	Yes
Response Variable (Trials)	Total
Model	binary logit
Optimization Technique	Fisher's scoring

Number of Observations Read	36
Number of Observations Used	36
Sum of Frequencies Read	1591
Sum of Frequencies Used	1591

Response Profile		
Ordered Value	Binary Outcome	Total Frequency
1	Event	932
2	Nonevent	659

Backward Elimination Procedure

Class Level Information			
Class	Value	Design Variables	
work	Office	0	0
	Skilled	1	0
	Unskilled	0	1
tenure	own	0	
	rent	1	
type	flat	0	
	house	1	
agegrp	1	0	0
	2	1	0
	3	0	1

Step 0. The following effects were entered:

Intercept work tenure type agegrp

Step 1. Effect type is removed:

NOTE: No (additional) effects met the 0.05 significance level for removal from the model.

Summary of Backward Elimination

Step	Effect Removed	DF	Number In	Wald Chi-Square	Pr > ChiSq
1	type	1	3	0.0003	0.9865

Type 3 Analysis of Effects

Effect	DF	Wald Chi-Square	Pr > ChiSq
work	2	27.0088	<.0001
tenure	1	78.6133	<.0001
agegrp	2	10.9072	0.0043

Analysis of Maximum Likelihood Estimates

Parameter		DF	Estimate	Standard Error	Wald Chi-Square	Pr > ChiSq
Intercept		1	1.0139	0.1361	55.4872	<.0001
work	Skilled	1	0.3053	0.1408	4.7023	0.0301
work	Unskilled	1	-0.4574	0.1248	13.4377	0.0002
tenure	rent	1	-1.0144	0.1144	78.6133	<.0001
agegrp	2	1	-0.1129	0.1367	0.6824	0.4088
agegrp	3	1	-0.4364	0.1401	9.7087	0.0018

Odds Ratio Estimates

Effect	Point Estimate	95% Wald Confidence Limits	
work Skilled vs Office	1.357	1.030	1.788
work Unskilled vs Office	0.633	0.496	0.808
tenure rent vs own	0.363	0.290	0.454
agegrp 2 vs 1	0.893	0.683	1.168
agegrp 3 vs 1	0.646	0.491	0.851

Association of Predicted Probabilities and Observed Responses			
Percent Concordant	62.8	Somers' D	0.327
Percent Discordant	30.1	Gamma	0.352
Percent Tied	7.1	Tau-a	0.159
Pairs	614188	C	0.663

question asked than office workers (estimated odds ratio 1.357 with 95% confidence interval 1.030, 1.788) and unskilled workers are less likely than office workers to respond "yes" (0.633, 0.496–0.808). People who rent their homes are far less likely to answer "yes" than people who own their homes (0.363, 0.290–0.454). Finally, it appears that people in the two younger age groups are more likely to respond "yes" than the oldest respondents.

8.3.4 Driving and Back Pain

A frequently used design in medicine is the matched case-control study in which each patient suffering from a particular condition of interest included in the study is matched to one or more people without the condition. The most commonly used matching variables are age, ethnic group, mental status, etc. A design with m controls per case is known as a $1{:}m$ matched study. In many cases, m will be 1, and it is the 1–1 matched study that we shall concentrate on here where we analyse the data on low back pain given in Table 8.4. To begin, we shall describe the form of the logistic model appropriate for case-control studies in the simplest case where there is only one binary explanatory variable.

With matched pairs data the form of the logistic model involves the probability, ϕ, that in matched pair i, for a given value of the explanatory variable, the member of the pair is a case. Specifically, the model is

$$\text{logit}(\phi_i) = \alpha_i + \beta x. \tag{8.5}$$

The odds that a subject with $x = 1$ is a case equals $\exp(\beta)$ times the odds that a subject with $x = 0$ is a case.

The model generalizes to the situation where there are p explanatory variables as

$$\text{logit}(\phi_i) = \alpha_i + \beta_1 x_1 + \beta_2 x_2 + \cdots + \beta_p x_p. \tag{8.6}$$

Typically, one x_i is an explanatory variable of real interest, such as past exposure to a risk factor, with the others being used as a form of statistical control in addition to the variables already controlled by virtue of using them to form matched pairs. This is the case in our back pain example where it is the effect of car driving on lower back pain that is of most interest.

The problem with the model above is that the number of parameters increases at the same rate as the sample size with the consequence that maximum likelihood estimation is no longer viable. We can overcome this problem if we regard the parameters α_i as of little interest and so are willing to forgo their estimation. If we do, we can then create a *conditional likelihood function* that will yield maximum likelihood estimators of the coefficients, $\beta_1, \ldots, \beta_p$, that are consistent and asymptotically normally distributed. The mathematics behind this are described in Collett (2003a), who shows that this conditional logistic regression model can be applied using standard logistic regression software as follows:

- Set the sample size to the number of matched pairs.
- Use as the explanatory variables the differences between corresponding covariate values for each case and control.
- Set the value of the response variable to one for all observations.
- Exclude the constant term from the model.

To illustrate this approach, we shall apply it to the back pain data in Table 8.4. The data values are separated by spaces and so can be read in using list input:

```
data backpain;
  infile 'c:\handbook3\datasets\backpain.dat';
  input pair status$ driver suburb;
run;
proc logistic data=backpain;
  model status=driver suburb;
  strata pair;
run;
```

A conditional logistic regression, using the method described above, can be fitted using the strata statement and naming the variable that identifies the matched sets. In this example, the matched sets are pairs, and pair is the variable that identifies them. The output is shown in Table 8.9.

There are several points worth noting: 434 observations are used from 217 strata (pairs). The fact that the response is a character variable poses no problem and the category modelled as the response is the first in alphabetical order. The strata summary shows that there is a single response pattern, that is, each of the 217 strata has one case and one control. (This section would be more informative for a 1:m matched study, where $m > 1$.) As expected, the analysis of maximum likelihood estimates does not contain an intercept term.

The test of the global hypothesis that both regression coefficients in the model are zero is rejected by all three tests. Looking at the estimates of the regression coefficients themselves we see that only that for driver is significant. The estimate of the odds ratio

Table 8.9 Conditional Logistic Regression Model Fitted to Back Pain Data

Conditional Analysis

Model Information	
Data Set	WORK.BACKPAIN
Response Variable	Status
Number of Response Levels	2
Number of Strata	217
Model	Binary logit
Optimization Technique	Newton-Raphson ridge

Number of Observations Read	434
Number of Observations Used	434

Response Profile		
Ordered Value	**status**	**Total Frequency**
1	Case	217
2	control	217

Probability modeled is status='case'.

Strata Summary				
	Status			
Response Pattern	**case**	**control**	**Number of Strata**	**Frequency**
1	1	1	217	434

Newton-Raphson Ridge Optimization

Without Parameter Scaling

Convergence criterion (GCONV=1E-8) satisfied.

Model Fit Statistics		
Criterion	**Without Covariates**	**With Covariates**
AIC	300.826	295.280
SC	300.826	303.426
-2 Log L	300.826	291.280

Testing Global Null Hypothesis: BETA=0			
Test	**Chi-Square**	**DF**	**Pr > ChiSq**
Likelihood Ratio	9.5456	2	0.0085
Score	9.3130	2	0.0095
Wald	8.8536	2	0.0120

Analysis of Maximum Likelihood Estimates					
Parameter	**DF**	**Estimate**	**Standard Error**	**Wald Chi-Square**	**Pr > ChiSq**
driver	1	0.6579	0.2940	5.0079	0.0252
suburb	1	0.2555	0.2258	1.2796	0.2580

Odds Ratio Estimates			
		95% Wald	
Effect	**Point Estimate**	**Confidence Limits**	
driver	1.931	1.085	3.435
suburb	1.291	0.829	2.010

of a herniated disc occurring in a driver relative to a nondriver is 1.93 with a 95% confidence interval of (1.09, 3.43). Conditional on residence we can say that the risk of a herniated disc occurring in a driver is about twice that of a non-driver.

Exercises

1 Psychiatric Caseness

1. In the text, a main effects only logistic regression was fitted to the GHQ data. This assumes that the effect of GHQ on caseness is the same for men and women. Fit a model where this assumption is not made, and construct a graph that displays the fitted model.
2. Investigate which model fits best.

2 ESR and Plasma Proteins

For the ESR and plasma protein data fit a logistic model that includes quadratic effects for both fibrinogen and γ-globulin.

Does the model fit better than the model selected in the text?

3 Car Driving and Herniated Discs

Fit a logistic regression model to the data that allows for a possible interaction between driving and residence.

Assess whether this model provides a better fit to the data than the models considered in the text.

4 Women's Role in Society

In a survey that was concerned with the relationship of education and sex to attitudes toward the role of women in society, each respondent was asked if he or she agreed or disagreed with the statement "Women should take care of running their homes and leave running the country up to men." (The survey was carried out in the United States.)

The variables in role.dat are

- **Years:** Years of education
- **Agreem:** Number of men agreeing with the statement
- **Disagreem:** Number of men disagreeing with the statement
- **Agreef:** Number of women agreeing with the statement
- **Disagreef:** Number of women disagreeing with the statement

1. Plot separately for males and females the observed probabilities of agreement with the statement against years of education.
2. By fitting suitable logistic regression models, investigate how education and sex are related to a person's attitude toward women's role in society.
3. Plot the predicted values, from the model you feel describes the data best, against years of education differentiating between men and women.

Chapter 9

Generalized Linear Models: Polyposis and School Attendance among Australian School Children

9.1 Description of Data

In this chapter, we will re-analyse a number of the data sets from previous chapters, in particular the data on school attendance in Australian school children used in Chapter 5 and a new data set shown in Table 9.1. The latter is taken from Piantadosi (1997) and arises from a study of Familial Adenomatous Polyposis (FAP), an autosomal genetic disease that predisposes those affected to develop large numbers of polyps in the colon which, if untreated, may develop into colon cancer. Patients with FAP were randomly assigned to receive an active drug treatment or a placebo. The response variable was the number of colonic polyps at 3 months after starting treatment. Additional covariates of interest were number of polyps before starting treatment, gender and age. The aim of this chapter is to introduce the concept of generalized linear models and to illustrate how they can be applied in SAS using proc genmod.

Table 9.1 Data from a Study of FAP

Sex	*Treatment*	*Baseline Count of Polyps*	*Age*	*Number of Polyps at 3 Months*
0	1	7	17	6
0	0	77	20	67
1	1	7	16	4
0	0	5	18	5
1	1	23	22	16
		⋮		
0	1	10	23	6
0	1	20	22	5
1	1	12	42	8

9.2 Generalized Linear Models

The analysis of variance models considered in Chapters 4 and 5 and the multiple regression model described in Chapter 7 are, essentially, completely equivalent. Both involve a linear combination of a set of explanatory variables (dummy variables in the case of analysis of variance) as a model for an observed response variable. And both include residual terms assumed to have a normal distribution. (The equivalence of analysis of variance and multiple regression models is discussed in more detail in Everitt, 2001.)

The logistic regression model encountered in Chapter 8 also has similarities to the analysis of variance and multiple regression models. Again, a linear combination of explanatory variables is involved, although here the binary response variable is not modelled directly (for the reasons outlined in Chapter 8), but via a logistic transformation.

Multiple regression, analysis of variance and logistic regression models can all, in fact, be included in a more general class of models known as generalized linear models. The essence of this class of models is a linear predictor of the form

$$\eta = \beta_0 + \beta_1 x_1 + \cdots + \beta_p x_p = \boldsymbol{\beta}'\mathbf{x}, \tag{9.1}$$

where

$$\boldsymbol{\beta}' = [\beta_0, \beta_1, \ldots, \beta_p]$$

$$\mathbf{x}' = [1, x_1, x_2, \ldots, x_p].$$

The linear predictor determines the expectation, μ, of the response variable. In linear regression, where the response is continuous, μ is directly equated with the linear predictor. This is not sensible when the response is dichotomous because in this case the expectation is a probability which must satisfy $0 \leq \mu \leq 1$. Consequently in logistic regression, the linear predictor is equated with the logistic function of μ, $\log \mu/(1-\mu)$.

In the generalized linear model formulation, the linear predictor can be equated with a chosen function of μ, $g(\mu)$, and the model now becomes

$$\eta = g(\mu). \tag{9.2}$$

The function g is referred to as a *link function*.

In linear regression (and analysis of variance), the probability distribution of the response variable is assumed to be normal with mean μ. In logistic regression, a binomial distribution is assumed with probability parameter μ. Both distributions, normal and binomial, come from the same family of distributions, called the *exponential family*, and are given by

$$f(y;\theta,\phi) = \exp\{(y\theta - b(\theta))/a(\phi) + c(y,\phi)\}. \tag{9.3}$$

For example, for the normal distribution,

$$\begin{aligned} f(y;\theta,\phi) &= \frac{1}{\sqrt{(2\pi\sigma^2)}} \exp\{-(y-\mu)^2/2\sigma^2\} \\ &= \exp\left\{(y\mu - \mu^2/2)/\sigma^2 - \frac{1}{2}(y^2/\sigma^2 + \log(2\pi\sigma^2))\right\} \end{aligned} \tag{9.4}$$

so that $\theta = \mu$, $b(\theta) = \theta^2/2$, $\phi = \sigma^2$ and $a(\phi) = \phi$.

The parameter θ, a function of μ, is called the *canonical link*. The canonical link is frequently chosen as the link function, although the canonical link is not necessarily more appropriate than any other link. Table 9.2 lists some of the most common distributions and their canonical link functions used in generalized linear models.

Table 9.2 Canonical Link Functions from Common Distributions Used in Generalized Linear Models

Distribution	*Variance Function*	*Dispersion Parameter*	*Link Function*	$g(\mu) = \theta(\mu)$
Normal	1	σ^2	Identity	μ
Binomial	$\mu(1-\mu)$	1	Logit	$\log(\mu/(1-\mu))$
Poisson	μ	1	Log	$\log(\mu)$
Gamma	μ^2	ν^{-1}	Reciprocal	$1/\mu$

The mean and variance of a random variable Y having the distribution in Equation 9.3 are given by

$$E(Y) = b'(0) = \mu \tag{9.5}$$

and

$$\text{var}(Y) = b''(\theta)a(\phi) = V(\mu)a(\phi), \tag{9.6}$$

where $b'(\theta)$ and $b''(\theta)$ denote the first and second derivatives of $b(\theta)$ with respect to θ and the variance function $V(\mu)$ is obtained by expression $b''(\theta)$ as a function of μ.

It can be seen from Equation 9.4 that the variance for the normal distribution is simply σ^2 regardless of the value of the mean μ, that is, the variance function is 1.

The data on Australian school children will be analysed by assuming a Poisson distribution for the number of days absent from school. The Poisson distribution is the appropriate distribution of the number of events observed if these events occur independently in continuous time at a constant instantaneous probability rate (or incidence rate); see, for example, Clayton and Hills (1993). The Poisson distribution is given by

$$f(y; \mu) = \mu^y e^{-\mu}/y!, \quad y = 0, 1, 2, \ldots. \tag{9.7}$$

Taking the logarithm and summing over observations $y_1, y_2, \ldots, y_n$, the log likelihood is

$$l(\mu; y_1, y_2, \ldots, y_n) = \sum_i \{(y_i \ln \mu - \mu) - \ln(y_i!)\} \tag{9.8}$$

so that $\theta = \ln \mu$, $b(\theta) = \exp(\theta)$, $\phi = 1$ and $\text{var}(y) = \exp(\theta) = \mu$. Therefore, the variance of the Poisson distribution is not constant but equal to the mean. Unlike the normal distribution, the Poisson distribution has no separate parameter for the variance, and the same is true of the binomial distribution. Table 9.2 shows the variance functions and dispersion parameters for some commonly used probability distributions.

9.2.1 Model Selection and Measure of Fit

Lack of fit in a generalized linear model may be expressed by the *deviance*, which is minus twice the difference between the maximized log likelihood of the model and the maximum likelihood achievable, that is, the maximized likelihood of the full or saturated model. For the normal distribution, the deviance is simply the residual sum of squares. Another measure of lack of fit is the generalized Pearson X^2

$$X^2 = \sum_i (y_i - \hat{\mu})^2 / V(\hat{\mu}), \tag{9.9}$$

where, for the Poisson distribution, it is just the familiar statistic for two-way cross-tabulations (since $V(\hat{\mu}) = \hat{\mu}$). Both the deviance and Pearson X^2 have chi-square distributions when the sample size tends to infinity. When the dispersion parameter ϕ is fixed (not estimated), an analysis of variance may be used for testing nested models in the same way as analysis of variance is used for linear models. The difference in deviance between two models is simply compared with the chi-square distribution with degrees of freedom equal to the difference in model degrees of freedom.

The Pearson and deviance residuals are defined as the (signed) square roots of the contributions of the individual observations to the Pearson X^2 and deviance, respectively. These residuals may be used to assess the appropriateness of the link and variance functions. The deviance residuals are more commonly used because their distribution tends to be closer to normal than that of the Pearson residuals.

A relatively common phenomenon with binary and count data is *overdispersion*; that is, the variance is greater than that of the assumed distribution (binomial and Poisson, respectively). This overdispersion may be due to extra variability in the parameter μ, which has not been completely explained by the covariates. One way of addressing the problem is to allow μ to vary randomly according to some (prior) distribution and to assume that conditional on the parameter having a certain value, the response variable follows the binomial (or Poisson) distribution. Such models are called *random effects models* (see Pinheiro and Bates, 2000, and Chapter 11).

A more pragmatic way of accommodating overdispersion in the model is to assume that the variance is proportional to the variance function, but to estimate the dispersion rather than assuming the value 1 appropriate for the distributions. For the Poisson distribution, the variance is modelled as

$$\mathrm{var}(Y) = \phi\mu, \tag{9.10}$$

where ϕ is estimated from the deviance or Pearson X^2. (This is analogous to the estimation of the residual variance in linear regression models from the residual sums of squares.) This parameter is then used to scale the estimated standard errors of the regression coefficients. This approach of assuming a variance function that does not correspond to any probability distribution is an example of *quasi-likelihood*. See McCullagh and Nelder (1989) for more details on generalized linear models.

9.3 Analysis Using SAS

Within SAS, the genmod procedure uses the framework described in the previous section to fit generalized linear models. The distributions covered include those shown in Table 9.2 with, in addition, the inverse Gaussian, negative binomial and multinomial.

To first illustrate the use of proc genmod, we begin by replicating the analysis of U.S. crime rates presented in Chapter 7 using the subset of explanatory variables selected by stepwise regression. We assume the data have been read into a SAS data set, uscrime, as described there.

```
proc genmod data = uscrime;
  model R = ex1 × ed age u2/dist = normal link = identity;
run;
```

The model statement specifies the regression equation in much the same way as proc glm, as described in Chapter 5. For a binomial response, the events/trials syntax described in Chapter 8 for proc logistic may also be used. The distribution and link functions are specified as options on the model statement. Normal and identity may be abbreviated N and id, respectively. The output is shown in Table 9.3. The parameter estimates are equal to those obtained in Chapter 7 using proc reg (see Table 7.4) although the standard errors are not identical. The deviance value of 495.3383 is equal to the error mean square in Table 7.4.

Table 9.3 Output from Proc Genmod Used to Apply Multiple Linear Regression to Data on Crimes in the United States

The GENMOD Procedure

Model Information	
Data Set	WORK.USCRIME
Distribution	Normal
Link Function	Identity
Dependent Variable	R

Number of Observations Read	47
Number of Observations Used	47

Criteria For Assessing Goodness Of Fit			
Criterion	**DF**	**Value**	**Value/DF**
Deviance	41	20308.8707	495.3383
Scaled Deviance	41	47.0000	1.1463
Pearson Chi-Square	41	20308.8707	495.3383
Scaled Pearson X2	41	47.0000	1.1463
Log Likelihood		-209.3037	
Full Log Likelihood		-209.3037	
AIC		432.6075	
AICC		435.4793	

Algorithm converged.

Analysis Of Maximum Likelihood Parameter Estimates							
Parameter	**DF**	**Estimate**	**Standard Error**	**Wald 95% Confidence Limits**		**Chi-Square**	**Pr > ChiSq**
Intercept	1	-528.856	93.0407	-711.212	-346.499	32.31	<.0001
Ex1	1	1.2973	0.1492	1.0050	1.5897	75.65	<.0001
X	1	0.6463	0.1443	0.3635	0.9292	20.06	<.0001
Ed	1	2.0363	0.4628	1.1294	2.9433	19.36	<.0001
Age	1	1.0184	0.3447	0.3428	1.6940	8.73	0.0031
U2	1	0.9901	0.4223	0.1625	1.8178	5.50	0.0190
Scale	1	20.7871	2.1440	16.9824	25.4442		

NOTE: The scale parameter was estimated by maximum likelihood.

9.3.1 School Attendance in Australian Children

Now we can move on to a more interesting application of generalized linear models involving the data on Australian children's school attendance, used previously in Chapter 5 (see Table 5.1). Here, because the response variable, number of days absent, is a count, we will use a Poisson distribution and a log link.

Assuming the data on Australian school attendance have been read into a SAS data set, ozkids, as described in Chapter 5, we fit a main effects model as follows:

```
proc genmod data=ozkids;
  class origin sex grade type;
  model days=sex origin type grade/dist=p link=log type1 type3;
run;
```

The predictors are all categorical variables and so must be declared as such with a class statement. The Poisson probability distribution with a log link is requested with Type 1 and Type 3 analyses. These are analogous to Type I and Type III sums of squares discussed in Chapter 5. The results are shown in Table 9.4. Looking first at the LR statistics for each of the main effects, we see that both Type 1 and Type 3 analyses lead to very similar conclusions, namely that each main effect is significant. For the moment, we shall ignore the Analysis of Parameter Estimates part of the output and examine instead the criteria for assessing goodness of fit. In the absence of overdispersion, the dispersion parameters based on the Pearson X^2 of the deviance should be close to 1. The values of 13.6673 and 12.2147 given in Table 9.4 suggest therefore that there is overdispersion, and as a consequence the p values in this display may be too low.

Table 9.4 Results from Applying Poisson Regression to the Australian School Children Data Using Genmod

The GENMOD Procedure

Model Information	
Data Set	WORK.OZKIDS
Distribution	Poisson
Link Function	Log
Dependent Variable	days

Number of Observations Read	154
Number of Observations Used	154

Class Level Information		
Class	**Levels**	**Values**
origin	2	A N
sex	2	F M
grade	4	F0 F1 F2 F3
type	2	AL SL

Criteria For Assessing Goodness Of Fit			
Criterion	**DF**	**Value**	**Value/DF**
Deviance	147	1795.5665	12.2147
Scaled Deviance	147	1795.5665	12.2147
Pearson Chi-Square	147	2009.0882	13.6673
Scaled Pearson X2	147	2009.0882	13.6673
Log Likelihood		4581.1746	
Full Log Likelihood		-1208.4185	
AIC		2430.8371	
AICC		2431.6042	

Algorithm converged.

Analysis Of Maximum Likelihood Parameter Estimates								
Parameter		DF	Estimate	Standard Error	Wald 95% Confidence Limits		Chi-Square	Pr > ChiSq
Intercept		1	2.7742	0.0628	2.6510	2.8973	1949.78	<.0001
sex	F	1	-0.1405	0.0416	-0.2220	-0.0589	11.39	0.0007
sex	M	0	0.0000	0.0000	0.0000	0.0000	.	.
origin	A	1	0.4951	0.0412	0.4143	0.5758	144.40	<.0001
origin	N	0	0.0000	0.0000	0.0000	0.0000	.	.
type	AL	1	-0.1300	0.0442	-0.2166	-0.0435	8.67	0.0032
type	SL	0	0.0000	0.0000	0.0000	0.0000	.	.
grade	F0	1	-0.2060	0.0629	-0.3293	-0.0826	10.71	0.0011
grade	F1	1	-0.4718	0.0614	-0.5920	-0.3515	59.12	<.0001
grade	F2	1	0.1108	0.0536	0.0057	0.2158	4.27	0.0387
grade	F3	0	0.0000	0.0000	0.0000	0.0000	.	.
Scale		0	1.0000	0.0000	1.0000	1.0000		

NOTE: The scale parameter was held fixed.

LR Statistics For Type 1 Analysis				
Source	Deviance	DF	Chi-Square	Pr > ChiSq
Intercept	2105.9714			
sex	2086.9571	1	19.01	<.0001
origin	1920.3673	1	166.59	<.0001
type	1917.0156	1	3.35	0.0671
grade	1795.5665	3	121.45	<.0001

LR Statistics For Type 3 Analysis			
Source	DF	Chi-Square	Pr > ChiSq
sex	1	11.38	0.0007
origin	1	148.20	<.0001
type	1	8.65	0.0033
grade	3	121.45	<.0001

To rerun the analysis allowing for overdispersion we need an estimate of the dispersion parameter ϕ. One strategy is to fit a model that contains a sufficient number of parameters so that all systematic variation is removed, estimate ϕ from this model as the deviance of Pearson X^2 divided by its degrees of freedom, and then use this estimate in fitting the required model.

So here we shall first fit a model with all first-order interactions included, simply to get an estimate of ϕ. The necessary SAS code is

```
proc genmod data=ozkids;
  class origin sex grade type;
  model days=sex|origin|type|grade@2/dist=p link=log scale=d;
run;
```

The scale=d option on the model statement specifies that the scale parameter is to be estimated from the deviance. The model statement also illustrates a modified use of the bar operator. By appending @2 we limit its expansion to terms involving two effects.

This leads to an estimate of ϕ of 10.1712 (output not shown) with a square root of 3.182, which will now be used as the values of the scale option on the model statement, to allow for overdispersion.

```
proc genmod data=ozkids;
  class origin sex grade type;
  model days=sex origin type grade/dist=p link=log type1 type3
  scale=3.1892;
  output out=genout pred=pr_days stdreschi=resid;
run;
```

The output statement specifies that the predicted values and standardized Pearson (Chi) residuals are to be saved in the variables pr_days and resid, respectively, in the data set genout.

The latest results are shown in Table 9.5. Allowing for overdispersion has had no effect on the regression coefficients, but has had a large effect on the p values and confidence intervals so that sex and type are no longer significant. Interpretation of the significant effect of, say, origin is made in terms of the logs of the predicted mean counts. Here, the estimated coefficient for origin 0.4951 indicates that the log of the predicted mean number of days absent from school for Aboriginal children is 0.4951 higher than that for white children, conditional on the other variables. Exponentiating the coefficient yields count ratios, that is, 1.64 with corresponding 95% confidence interval (1.27, 2.12). Aboriginal children have between about 1.25 and 2 times as many days absent as white children.

Table 9.5 Selected Results after Fitting Model to Australian School Children Allowing for Overdispersion

Analysis Of Maximum Likelihood Parameter Estimates

Parameter		DF	Estimate	Standard Error	Wald 95% Confidence Limits		Chi-Square	Pr > ChiSq
Intercept		1	2.7742	0.2004	2.3815	3.1669	191.70	<.0001
sex	F	1	-0.1405	0.1327	-0.4006	0.1196	1.12	0.2899
sex	M	0	0.0000	0.0000	0.0000	0.0000	.	.
origin	A	1	0.4951	0.1314	0.2376	0.7526	14.20	0.0002
origin	N	0	0.0000	0.0000	0.0000	0.0000	.	.
type	AL	1	-0.1300	0.1408	-0.4061	0.1460	0.85	0.3559
type	SL	0	0.0000	0.0000	0.0000	0.0000	.	.
grade	F0	1	-0.2060	0.2007	-0.5994	0.1874	1.05	0.3048
grade	F1	1	-0.4718	0.1957	-0.8553	-0.0882	5.81	0.0159
grade	F2	1	0.1108	0.1709	-0.2242	0.4457	0.42	0.5169
grade	F3	0	0.0000	0.0000	0.0000	0.0000	.	.
Scale		0	3.1892	0.0000	3.1892	3.1892		

NOTE: The scale parameter was held fixed.

The standardized residuals can be plotted against the predicted values using proc sgplot:

```
proc sgplot data=genout;
  scatter y=resid x=pr_days;
run;
```

The result is shown in Figure 9.1. This plot does not appear to give any cause for concern.

9.3.2 *Analysing the FAP Data*

The FAP data in Table 9.1 also involve counts and so could be modelled using the Poisson regression model used in the previous sub-section. But some of the counts in Table 9.1 are extremely large, indicating that the distribution of counts is very skewed. Consequently, the data might be better modelled by allowing for this with

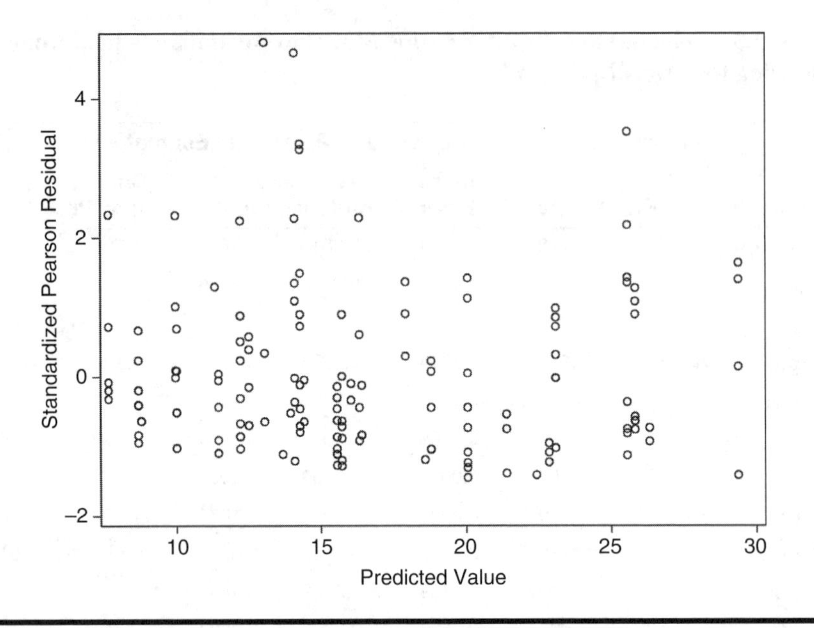

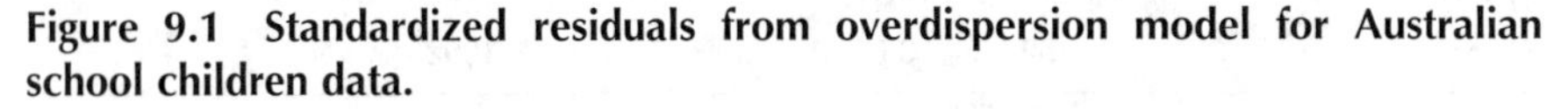

Figure 9.1 Standardized residuals from overdispersion model for Australian school children data.

the use of a gamma distribution (defined in Everitt, 2006) and this we will do here. Because gamma variables are positive, a log-link function will again be used, and the SAS code used to fit the model is

```
data fap;
  infile 'c:\handbook3\datasets\fap.dat';
  input male treat base_n age resp_n;
run;
proc genmod data=fap;
  model resp_n=male treat base_n age/dist=g link=log;
run;
```

The results are shown in Table 9.6. (Note that in this model the scale parameter has been estimated.) Here the covariate of most interest is that for treatment and we find from Table 9.6 that this variable has an estimated regression coefficient of −0.8358 with estimated standard error of 0.2591. The interpretation of the estimated regression coefficient becomes simpler if we exponentiate the estimate (and the upper and lower confidence limits for the coefficient) to give the values 0.4335, [0.2609, 0.7204]. We can conclude that, conditional on the other covariates, patients receiving the active treatment are estimated to have between about 26% and 72% of the number of polyps at 3 months as those receiving the placebo.

Table 9.6 Results from Fitting a Gamma Model to the FAP Data

The GENMOD Procedure

Model Information	
Data Set	WORK.FAP
Distribution	Gamma
Link Function	Log
Dependent Variable	resp_n

Number of Observations Read	22
Number of Observations Used	22

Criteria For Assessing Goodness Of Fit			
Criterion	**DF**	**Value**	**Value/DF**
Deviance	17	7.5870	0.4463
Scaled Deviance	17	23.1875	1.3640
Pearson Chi-Square	17	5.6485	0.3323
Scaled Pearson X2	17	17.2629	1.0155
Log Likelihood		-80.1699	
Full Log Likelihood		-80.1699	
AIC		172.3398	
AICC		177.9398	

Algorithm converged.

Analysis Of Maximum Likelihood Parameter Estimates							
Parameter	**DF**	**Estimate**	**Standard Error**	**Wald 95% Confidence Limits**		**Chi-Square**	**Pr > ChiSq**
Intercept	1	3.0155	0.5048	2.0260	4.0049	35.68	<.0001
male	1	0.5093	0.2940	-0.0668	1.0854	3.00	0.0832
treat	1	-0.8358	0.2591	-1.3437	-0.3280	10.40	0.0013
base_n	1	0.0132	0.0027	0.0079	0.0186	23.46	<.0001
age	1	-0.0223	0.0186	-0.0588	0.0142	1.44	0.2306
Scale	1	3.0562	0.8759	1.7428	5.3595		

NOTE: The scale parameter was estimated by maximum likelihood.

Exercises

1 Australian School Children

1. Dichotomize days absent from school by classifying 14 days or more as frequently absent. Analyse this new response variable using both the logistic and probit link and the binomial family.

2 Polyposis

1. Fit a Poisson regression model to the FAP data allowing for overdispersion. Compare the results with those obtained from the gamma model used in the text.
2. Plot the deviance residuals from the Poisson model fitted in (1) and compare it with the corresponding plot for the gamma model.
3. Which model do you consider preferable for these data?

3 Bladder Cancer

Seeber (1998) describes data arising from a study of 31 male patients who have been treated for superficial bladder cancer and giving the number of recurrent tumours during a particular time period after removal of the primary tumour, as well as the size of the primary tumour (dichotomized as smaller or larger than 3 cm in the variable x). Assuming that the waiting times between the tumours are independent and exponentially distributed with mean $1/\lambda$ (say), then the number of tumours up to time t has a Poisson distribution with mean, $\mu = \lambda t$.

The variables in tumour.dat are

- **ID:** Observation number
- **Time:** Time for which patient was followed up (in months)
- **Tum:** Dichotomous variable coding tumour size, 0 = original tumour smaller than 3 cm, 1 = original tumour larger than 3 cm
- **N:** Number of recurrent tumours in follow-up time

1. Use `proc genmod` to fit the following model for the parameter λ. (*Hint*: You will need to use an offset.)

$$\log(\lambda) = \beta_0 + \beta_1 x$$

2. By exponentiating the parameter estimates in the fitted model, find estimates of λ for smaller and larger tumours.
3. Find estimates of the waiting times between recurrent tumours for larger and smaller tumours.

Chapter 10

Generalized Additive Models: Burning Rubber and Air Pollution in the United States

10.1 Introduction

Table 10.1 shows the first five observations of thirty in a set of data collected in an investigation of wear of tyre rubber. The response variable is a measure of rubber loss, and the two explanatory variables are a measure of the tensile strength and the hardness of the rubber. Here interest focuses on predicting wear from hardness and tensile strength.

The next set of data to be considered in this chapter concerns air pollution in 41 cities in the United States. Observation for the first five cities are shown in Table 10.2. Air pollution is measured by sulphur dioxide concentration and recorded as a binary variable indicating whether the annual mean concentration of sulphur dioxide (over the years 1969–1971) is below 30 $\mu g/m^3$ or above or equal to this value. The values of six explanatory variables are also recorded, two of which relate to human ecology and four to climate. In detail the variables are as follows:

Table 10.1 Tyre Wear Data

ID	*Loss (g/h)*	*Hard (Shore Units)*	*Tens (K)*
1	372	45	162
2	206	55	233
3	175	61	232
4	154	66	231
5	136	71	231

Table 10.2 Air Pollution Data

City	*Hiso2*	*Temperature*	*Factories*	*Population*	*Wind Speed*	*Rain*	*Rainy Days*
Phoenix	0	70.3	213	582	6.0	7.05	36
Little Rock	0	61.0	91	132	8.2	48.52	100
San Francisco	0	56.7	453	716	8.7	20.66	67
Denver	0	51.9	454	515	9.0	12.95	86
Hartford	1	49.1	412	158	9.0	43.37	127

Hiso2: $0 = SO_2 < 30\ \mu g/m^2$
$1 = SO_2 \geq 30\ \mu g/m^2$
Temperature: Average annual temperature in °F
Factories: Number of manufacturing enterprises employing 20 or more workers
Population: Population size in thousands
Wind speed: Average wind speed in miles per hour
Rain: Average annual precipitation in inches
Rainy days: Average number of days with precipitation each year.

In this case, modelling interest is on uncovering the determinants of pollution.

10.2 Scatterplots and Generalized Additive Models

The multiple regression model described in Chapter 7 and the generalized linear model featured in Chapter 9 can accommodate non-linear functions of the explanatory variables, for example, quadratic or cubic terms, if these are thought to be necessary to provide an adequate fit. In this chapter, however, we consider some alternative and generally more flexible statistical methods for modelling non-linear relationships between a response variable and one or more explanatory variables. The main component of these methods, known as *generalized additive models* (GAMs), is the fitting of a "smooth" relationship between the response and each explanatory variable by means of a scatterplot smoother, which we shall describe in detail in the next subsection. GAMs are useful where

- Relationship between the variables is expected to be of complex form not easily fitted by standard linear or non-linear models.
- There is no a priori reason for using a particular model.
- We would like the data themselves to suggest the appropriate functional form for the relationship between an explanatory variable and the response.

Such models should be regarded as philosophically closer to the concepts of exploratory data analysis, in which the form of any functional relationship emerges from a set of data, rather than arising from a theoretical construct.

10.2.1 Scatterplot Smoothers

The scatterplot is an excellent first exploratory graph to study the dependence of two variables. An important second exploratory graph adds a curve to the scatterplot to help us better perceive the pattern of dependence. Most readers will be familiar with adding a parametric curve, such as a simple linear or polynomial regression fit (see Chapter 6); however, there are non-parametric alternatives that are perhaps less familiar, but that can often be more useful, because many bivariate data sets are too complex to be described by a simple parametric family. Perhaps, the simplest of these alternatives is a *locally weighted regression* or *loess* fit, first suggested by Cleveland (1979). In essence, this approach assumes that the explanatory variable x and the response variable y are related by the equation

$$y_i = g(x_i) + \varepsilon_i, \tag{10.1}$$

where g is a "smooth" function and ε_i are random variables with mean zero and constant scale.

Values $\hat{y}_i$, used to "estimate" the y_i at each x_i, are found by fitting polynomials using weighted least squares with large weights for points near to x_i and small weights otherwise. So smoothing takes place essentially by local averaging of the y values of observations having predictor values close to a target value.

Two parameters control the shape of a loess curve; the first is a smoothing parameter, α, with larger values leading to smoother curves – typical values are $1/4$ to 1. The second parameter, λ, is the degree of certain polynomials that are fitted by the method; λ can take values 1 or 2. In any specific application, the choice of the two parameters must be based on a combination of judgement and of trial and error. Residual plots may however be helpful in judging a particular combination of values.

An alternative smoother that can often usefully be applied to bivariate data is some form of spline function. (A spline is a term for a flexible strip of metal or rubber used

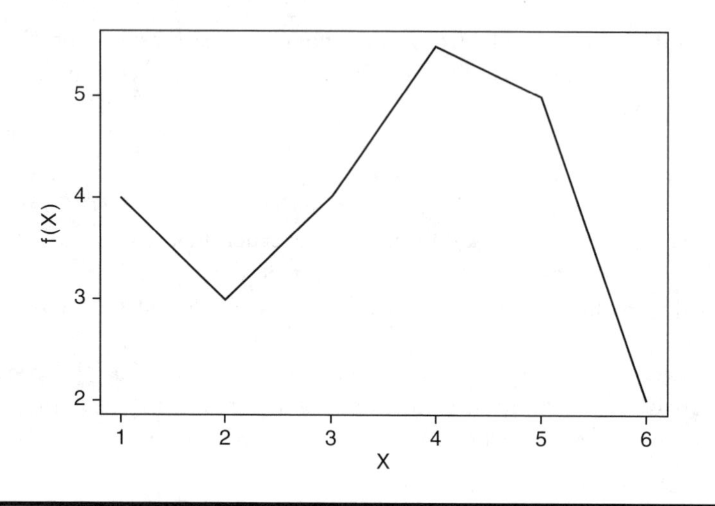

Figure 10.1 Linear spline function with knots at $a = 1$, $b = 3$, $c = 5$.

by a draftsman to draw curves.) Spline functions are polynomials within intervals of the x variable that are connected across different values of x. Figure 10.1, for example, shows a linear spline function, that is, a piecewise linear function, of the form

$$f(x) = \beta_0 + \beta_1 X + \beta_2 (X - a)_+ + \beta_3 (X - b)_+ + \beta_4 (X - c)_+, \tag{10.2}$$

where $(u)_+ = u \quad u > 0,$
$\qquad\qquad = 0 \quad u \leq 0.$

The interval endpoints, a, b and c, are called *knots*. The number of knots can vary according to the amount of data available for fitting the function.

The linear spline is simple and can approximate some relationships, but it is not smooth and so will not fit highly curved functions well. The problem is overcome by using piecewise polynomials – in particular, cubics, which have been found to have nice properties with good ability to fit a variety of complex relationships. The result is a *cubic spline*. Again we wish to fit a smooth curve, $g(x)$, that summarizes the dependence of y on x. A natural first attempt might be to try and determine g by least squares as the curve that minimizes

$$\sum [y_i - g(x_i)]^2. \tag{10.3}$$

But this would simply result in the interpolating curve and would not be smooth at all. Instead of Equation 10.3, the criterion used to determine g is

$$\sum [y_i - g(x_i)]^2 + \lambda \int g''(x)^2 \, dx, \tag{10.4}$$

where $g''(x)$ represents the second derivation of $g(x)$ with respect to x. Although written formally, this criterion looks a little formidable; it is really nothing more than an effort to govern the trade-off between the goodness of fit of the data (as measured by $\Sigma[y_i - g(x_i)]^2$) and the "wiggliness" or departure of linearity of g measured by $\int g''(x)^2\, dx$; for a linear function, this part of Equation 10.4 would be zero. The parameter λ governs the smoothness of g, with larger values resulting in a smoother curve.

The cubic spline that minimizes Equation 10.4 is a series of cubic polynomials joined at the unique observed values of the explanatory variables, x_i (for more details, see Friedman, 1991).

The "effective number of parameters" (analogous to the number of parameters in a parametric fit) or degrees of freedom of a cubic spline smoother is generally used to specify its smoothness rather than λ directly. A numerical search is then used to determine the value of λ corresponding to the required degrees of freedom. Roughly, the complexity of a cubic spline is about the same as a polynomial of degree one less than the degrees of freedom. But the cubic spline smoother "spreads out" its parameters in a more even way and hence is much more flexible than is polynomial regression.

10.2.2 Additive and GAMs

In a linear regression model there is a dependent variable, y, and a set of explanatory variables, $x_1 \ldots, x_p$, and the model assumed is

$$y = \beta_0 + \sum_{j=1}^{p} \beta_j x_j + \varepsilon. \tag{10.5}$$

Additive models replace the linear function $\beta_j x_j$ by a smooth non-parametric function to give

$$y = \beta_0 + \sum_{j=1}^{p} g_j(x_j) + \varepsilon, \tag{10.6}$$

where g_j can be one of the scatterplot smoothers described in the previous subsection, or, if the investigator chooses, a linear function for particular x_j.

A generalized additive model arises from Equation 10.5 in the same way as a generalized linear model arises from a multiple regression model, namely that some function of the expectation of the response variable is now modelled by a sum of non-parametric functions. So, for example, the logistic additive model is

$$\text{logit}[\Pr(y = 1)] = \beta_0 + \sum_{j=1}^{p} g_j(x_j). \tag{10.7}$$

Fitting a GAM involves what is known as a backfitting algorithm. The smooth functions g_j are fitted one at a time by taking the residuals

$$y - \sum_{k \neq j} g_j(x_k) \tag{10.8}$$

and fitting them against x_j using one of the scatterplot smoothers described in Section 10.2.1. The process is repeated until it converges. Linear terms in the model are fitted by least squares. Full details are given in Chambers and Hastie (1993).

Various tests are available to assess the non-linear contributions of the fitted smoothers, and GAMs can be compared with, say, linear models fitted to the same data, by means of an F test on the residual sum of squares of the computing models. In this process, the fitted smooth curve is assigned an estimated equivalent number of degrees of freedom. For full details, again see Chambers and Hastie (1993).

10.3 Analysis Using SAS

10.3.1 Scatterplot Smoothers for the Galaxy Data in Chapter 6

To illustrate the use of scatterplot smoothers we shall return to the data on galaxy speed and distance considered in Chapter 6. To begin we shall fit loess curves to the data. For maximum control over the form of the loess curve, we could use the dedicated procedure, proc loess. For an initial examination of the data proc sgplot is suitable. The clm option specifies confidence bands for the mean. The resulting plot is shown in Figure 10.2.

```
proc sgplot data=universe;
  loess y=velocity x=distance/clm;
run;
```

For more control of the fitted loess regression we could use either proc loess or proc gam within SAS. We illustrate proc gam here because it fits a wider range of GAMs.

```
proc gam data=universe;
  model velocity=loess(distance)/method=gcv;
run;
```

The first point to notice about the syntax of proc gam is that the specification of predictors on the model statement is different from procedures covered in previous chapters. The name of the predictor variable is enclosed in parentheses and prefixed with a keyword indicating the type of smoother to be employed. The keyword

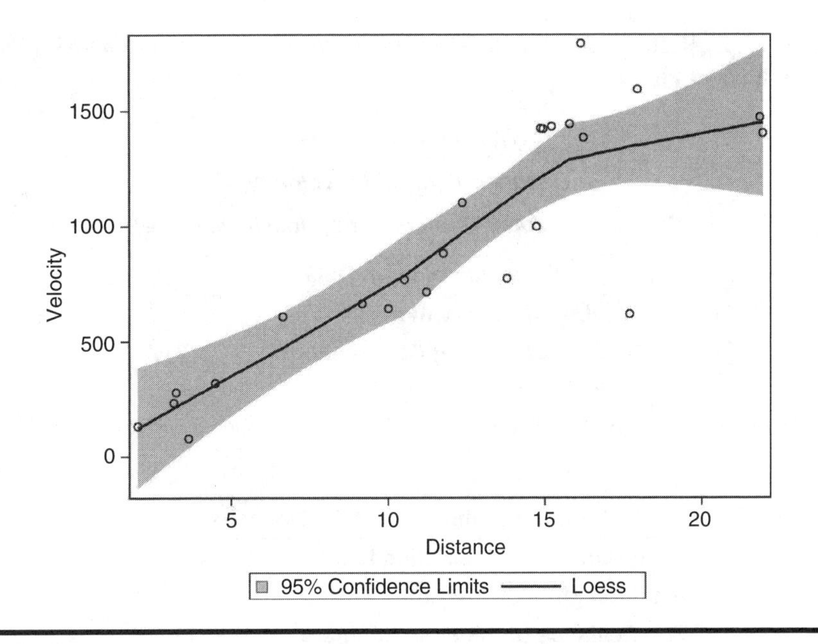

Figure 10.2 Scatterplot of velocity versus distance for galaxy data showing fitted loess curve.

param is used for variables that are not to be smoothed, but entered as parametric linear predictors; loess is used for a locally weighted regression; spline for a cubic smoothing spline (described in the previous section) and spline2 for a thin plate smoothing spline, which is a multivariate version of the cubic spline. Parametric effects must be specified before smoothed effects and must be included in the same set of parentheses when there are more than one. Parametric effects can be categorical, but in that case must also be named on a class statement. For smoothed effects the degree of smoothing can be specified for each in terms of its effective number of parameters, or degrees of freedom. For example,

```
model velocity=loess(distance,df=3).
```

Alternatively, the method=gcv option on the model statement can be used to select a degree of smoothing using generalized cross validation. The dist= option on the model statement specifies the distribution. Gaussian is the default and other possibilities are binomial, binary, gamma, igaussian (inverse Gaussian) or Poisson. The results are shown in Table 10.3. As ODS graphics are on, the partial fit plot is also produced and is shown in Figure 10.3. The results in Table 10.3 appear to confirm the need for a non-linear function for distance since the test for the loess curve is highly significant. The GCV term given in Table 10.3 stands for the Generalized Cross Validation statistic. This is a statistic that can be used to choose

Table 10.3 Results of Fitting a GAM Model to the Velocity of Galaxies Data Using a Loess Fit

The GAM Procedure

Dependent Variable: Velocity

Smoothing Model Component(s): loess(Distance)

Summary of Input Data Set	
Number of Observations	24
Number of Missing Observations	0
Distribution	Gaussian
Link Function	Identity

Iteration Summary and Fit Statistics	
Final Number of Backfitting Iterations	2
Final Backfitting Criterion	0
The Deviance of the Final Estimate	1240398.6897

The local score algorithm converged.

Regression Model Analysis
Parameter Estimates

Parameter	Parameter Estimate	Standard Error	t Value	Pr > |t|
Intercept	924.37500	49.80150	18.56	<.0001

Smoothing Model Analysis
Fit Summary for Smoothing Components

Component	Smoothing Parameter	DF	GCV	Num Unique Obs
Loess(Distance)	0.770833	2.161561	2956.493406	24

Smoothing Model Analysis
Analysis of Deviance

Source	DF	Sum of Squares	Chi-Square	Pr > ChiSq
Loess(Distance)	2.16156	4808099	80.7751	<.0001

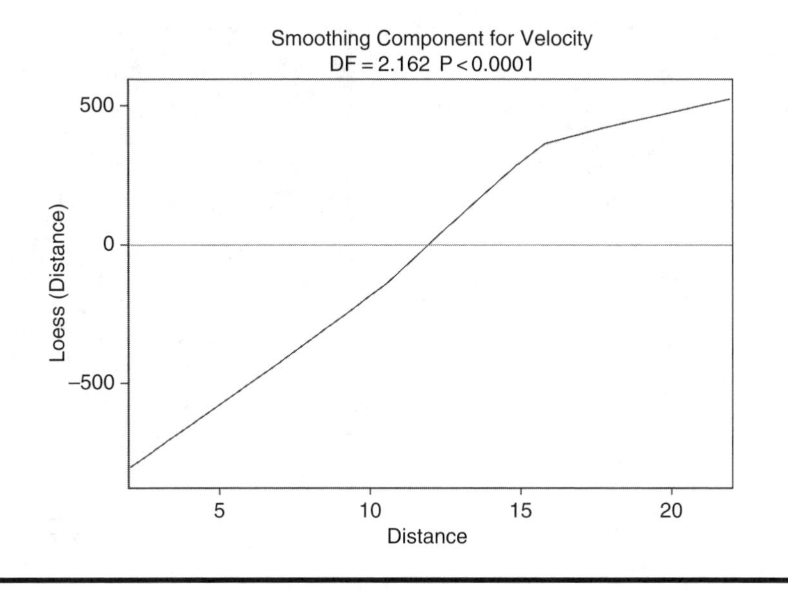

Figure 10.3 Partial fit plot for galaxy data showing loess curve fitted by using proc gam.

the degree of smoothing needed although there is some evidence that its use tends to under-smoothing of the data (see Hastie and Tibshirani, 1990, and Wood, 2006, for more details).

10.3.2 Fitting an Additive Model to the Tyre Wear Data in Table 10.1

Before proceeding to fit a model to the tyre wear data, we should plot them in some way to assess their general characteristics. Here simple scatterplots of tyre wear against hardness and then against tensile strength are a sensible way to begin. We could use proc sgplot as before, but proc sgscatter is useful for comparative scatterplots. Using the compare statement more than one x or y variable can be specified (in parentheses if more than one), and the loess option is used for a loess fit. The result in shown in Figure 10.4.

```
proc sgscatter data=tyres;
  compare y=loss x=(hard tens)/loess;
run;
```

The scatterplot for tyre wear against hardness appears to suggest a simple linear relationship between these two variables, but the one for tyre wear against tensile strength indicates a more complex relationship between these two variables.

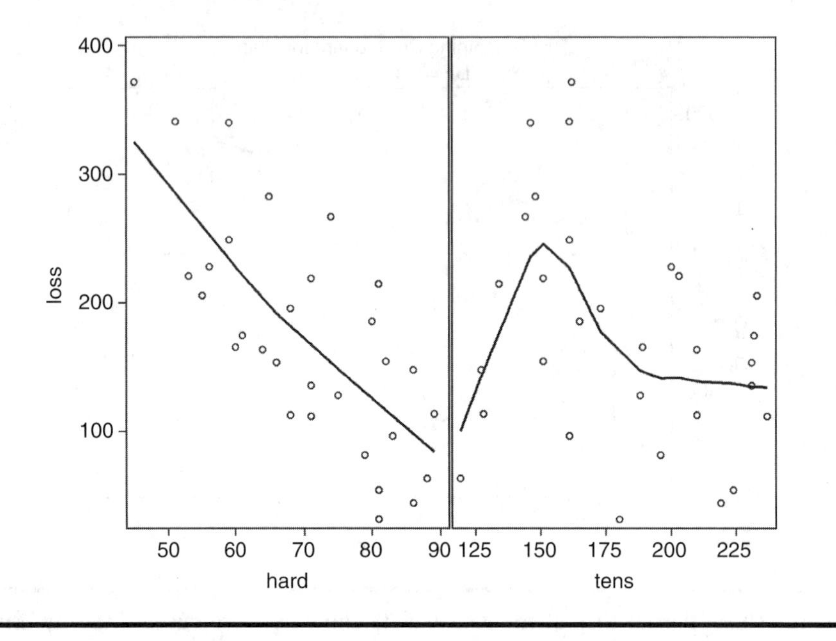

Figure 10.4 Scatterplot of tyre wear versus hardness.

It may also be helpful to look at a three-dimensional plot of tyre wear against the other two variables. For this we need to use one of the "traditional" graphics procedures, proc g3d.

```
proc g3d data=tyres;
  scatter tens*hard=loss /rotate=120;
run;
```

The result is shown in Figure 10.5 and highlights that the relationship between tyre wear and hardness and tensile strength is unlikely to be adequately described by a model in which only the linear effects of the two explanatory variables are included.

We will now fit an additive model with loess fits for both hardness and tensile strength.

```
proc gam data=tyres;
  model loss=loess(hard) loess(tens)/method=gcv;
run;
```

The results are shown in Table 10.4. Turning first to the analysis of deviance table, the results indicate that both hardness and tensile strength contribute to the explanation of tyre wear. But the form of the relationship is not the same.

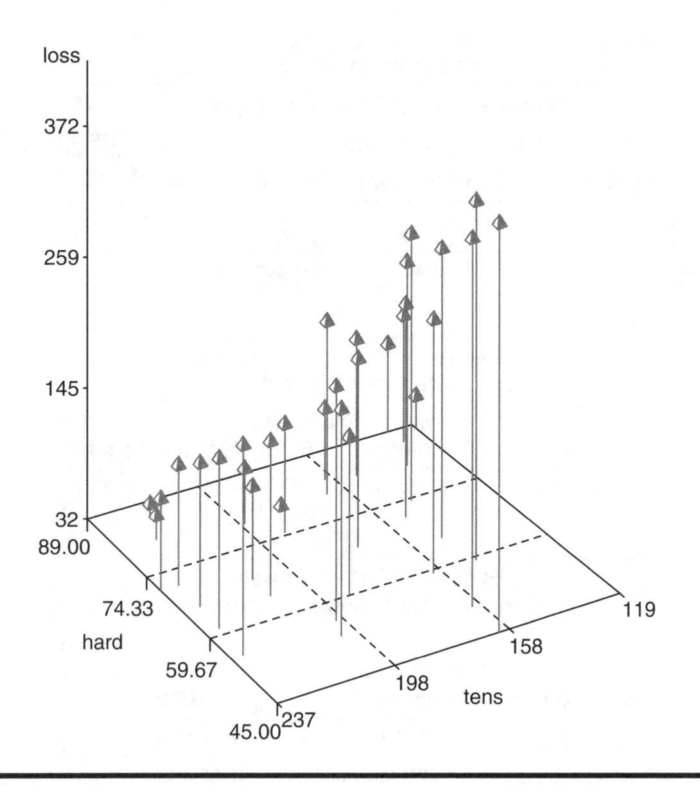

Figure 10.5 Three-dimensional plot of tyre wear against hardness and tensile strength.

Table 10.4 Results of Fitting a GAM Model to the Tyre Wear Data with Loess Fits for Both Explanatory Variables

The GAM Procedure

Dependent Variable: loss

Smoothing Model Component(s): loess(hard) loess(tens)

Summary of Input Data Set	
Number of Observations	30
Number of Missing Observations	0
Distribution	Gaussian
Link Function	Identity

Iteration Summary and Fit Statistics	
Final Number of Backfitting Iterations	20
Final Backfitting Criterion	7.6935731E-9
The Deviance of the Final Estimate	5048.012054

The local score algorithm converged.

Regression Model Analysis
Parameter Estimates

Parameter	Parameter Estimate	Standard Error	t Value	Pr > |t|
Intercept	175.43333	3.29287	53.28	<.0001

Smoothing Model Analysis
Fit Summary for Smoothing Components

Component	Smoothing Parameter	DF	GCV	Num Unique Obs
Loess(hard)	0.450000	3.582844	8.302104	30
Loess(tens)	0.216667	9.898640	15.830918	30

Smoothing Model Analysis
Analysis of Deviance

Source	DF	Sum of Squares	Chi-Square	Pr > ChiSq
Loess(hard)	3.58284	93764	288.2477	<.0001
Loess(tens)	9.89864	94683	291.0733	<.0001

For hardness, the smoothing parameter is much larger, implying a smoother function and the degrees of freedom correspondingly smaller. The partial fit plots shown in Figure 10.6 confirm this. For hardness, the relationship is linear whereas for tensile strength it is not. As the GCV statistic is known to under-smooth in many cases, we might suspect that has happened here.

10.3.3 Fitting a GAM to the Air Pollution Data

We will use the air pollution data to illustrate how GAM models may uncover a relationship that could easily be overlooked if the data were analysed using logistic

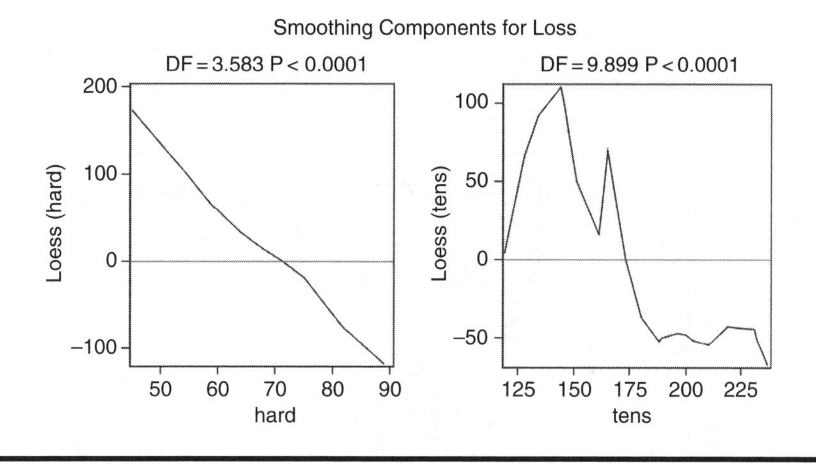

Figure 10.6 Partial fit plots for GAM fitted to tyre wear data.

regression. A naïve approach using logistic regression might conclude that none of the six predictor variables is related to sulphur dioxide concentration. However, some exploratory plots suggest the possibility of non-linear relationships. We concentrate on two of the six variables, population size and average rainfall, and begin by constructing some box plots.

```
proc sgplot data=usair;
  vbox population/category=hiso2 datalabel=city;
run;
proc sgplot data=usair;
  vbox rain/category=hiso2 datalabel=city;
run;
```

The resulting diagrams are shown in Figures 10.7 and 10.8.

The datalabel option identifies the outliers – those observations more than 1.5 times the inter-quartile range from the upper and the lower quartiles. This shows that Chicago is an outlier in population size, and it is dropped from the analysis.

```
data usair;
  set usair;
  if city=:'Chicago' then delete;
run;
```

The spline smoothing plot introduced earlier can also be useful with binary data. We could use sgscatter with the compare statement again, but to ensure that the *y* axis is restricted to the range 0 to 1, sgplot is used with a yaxis statement.

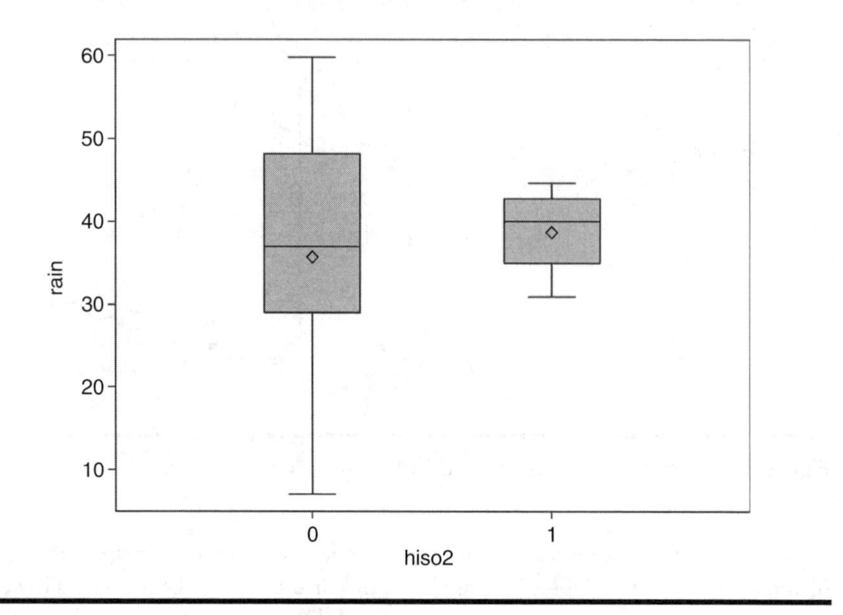

Figure 10.7 Box plots for average rainfall for the air pollution data.

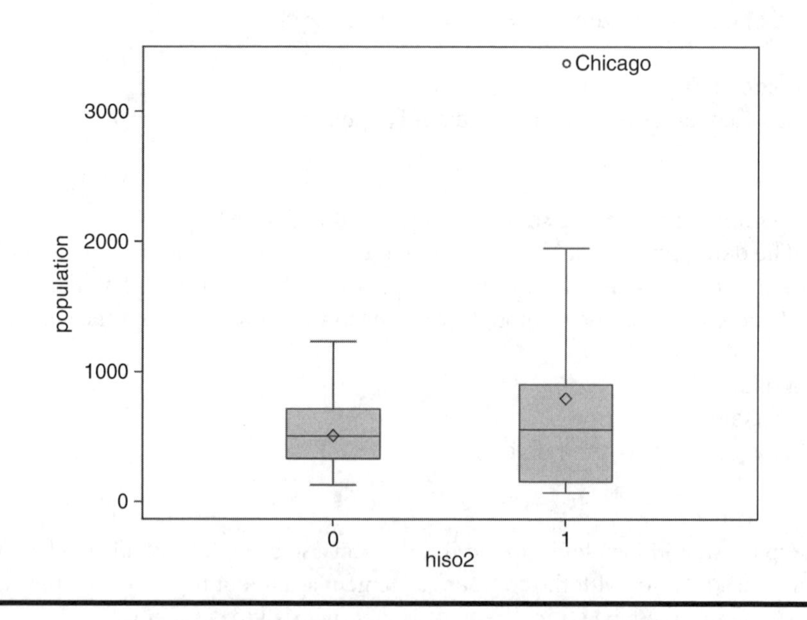

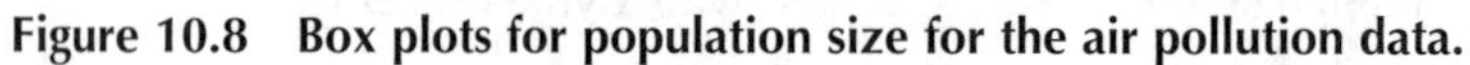

Figure 10.8 Box plots for population size for the air pollution data.

```
proc sgplot data=usair;
  pbspline y=hiso2 x=population/clm;
  yaxis min=0 max=1;
run;
proc sgplot data=usair;
  pbspline y=hiso2 x=rain/clm;
  yaxis min=0 max=1;
run;
```

Figure 10.9 shows the resulting plots, both of which suggest non-linear relationships. We now fit a logistic model with a spline smooth of average rainfall.

```
proc gam data=usair;
  model hiso2(desc)=spline(rain,df=2)/dist=binary;
  output out=gamout p;
run;
```

The dist= option on the model statement specifies that the outcome is binary. For a binary outcome, the default is to model the probability of the lower level (i.e., zero in this case). The desc (descending) option for hiso2 reverses this. The degree of smoothing for spline smooth of rain is set to two degrees of freedom. In the case of

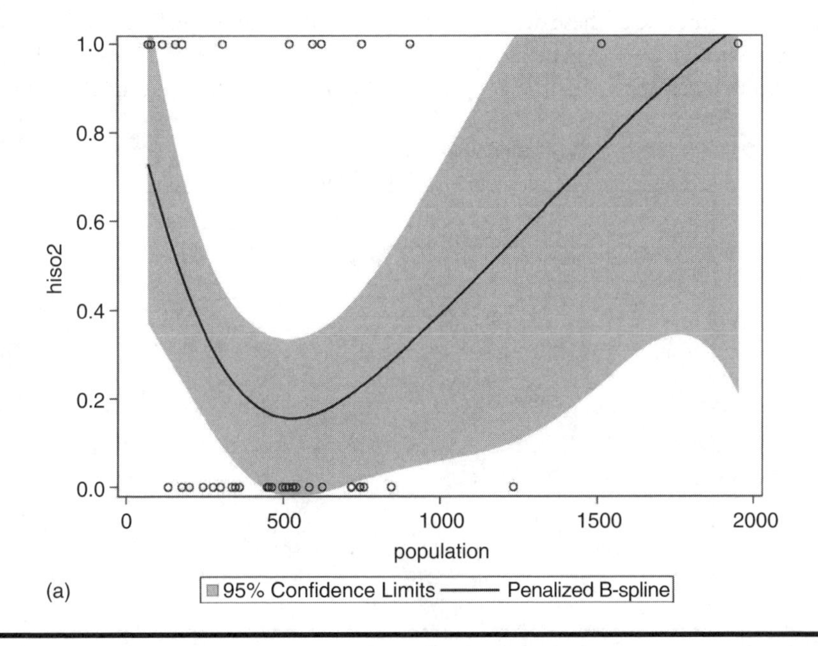

Figure 10.9 Plots of (a) population size against high SO_2 and (b) average rainfall against high SO_2 for the U.S. air pollution data.

(continued)

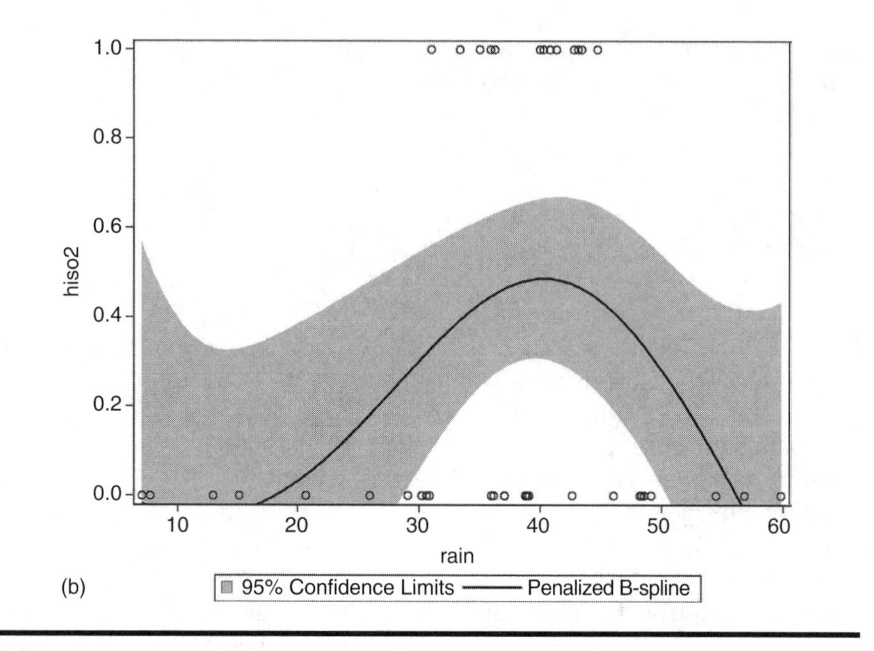

Figure 10.9 (continued)

a spline smooth, one of these degrees of freedom is allocated to the linear component. ODS graphics are used to generate the component plot that is shown in Figure 10.10. The output statement with the p option saves the predicted values in the gamout

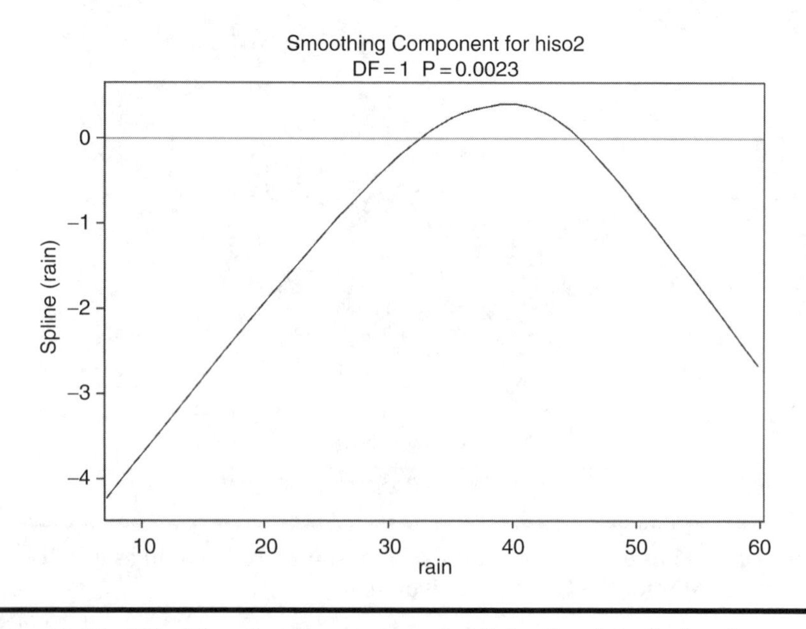

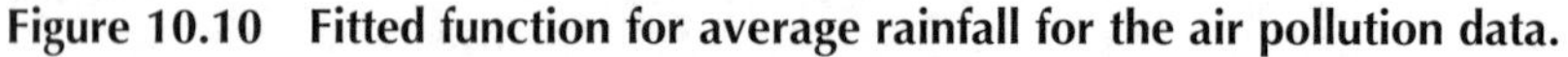
Figure 10.10 Fitted function for average rainfall for the air pollution data.

data set. The output is shown in Table 10.5, where we see that the linear component is not significant but that the non-linear smooth is. The predicted values (p_hiso2) in the output data set are the log odds of a high SO_2 value. The predicted probabilities can be calculated and plotted as follows:

```
data gamout;
  set gamout;
  odds = exp(P_hiso2);
  pred = odds/(1+odds);
run;
proc sgplot data = gamout;
  scatter y = pred x = rain;
run;
```

The resulting plot is shown in Figure 10.11.

Table 10.5 Results of Fitting a GAM Model to the Air Pollution Data

Dependent Variable: hiso2

Smoothing Model Component(s): spline(rain)

Summary of Input Data Set	
Number of Observations	40
Number of Missing Observations	0
Distribution	Binomial
Link Function	Logit

Response Profile		
Ordered Value	**hiso2**	**Total Frequency**
1	1	13
2	0	27

NOTE: PROC GAM is modeling the probability that hiso2=1. One way to change this to model the probability that hiso2=0 is to specify the response variable option EVENT='0'.

Iteration Summary and Fit Statistics	
Number of local scoring iterations	17
Local scoring convergence criterion	8.619345E-10
Final Number of Backfitting Iterations	1
Final Backfitting Criterion	6.0284602E-9
The Deviance of the Final Estimate	40.425572907

The local scoring algorithm converged.

Regression Model Analysis
Parameter Estimates

Parameter	Parameter Estimate	Standard Error	t Value	Pr > |t|
Intercept	-0.85734	2.05467	-0.42	0.6789
Linear(rain)	0.00886	0.05190	0.17	0.8654

Smoothing Model Analysis
Fit Summary for Smoothing Components

Component	Smoothing Parameter	DF	GCV	Num Unique Obs
Spline(rain)	0.999899	1.000000	29.622066	40

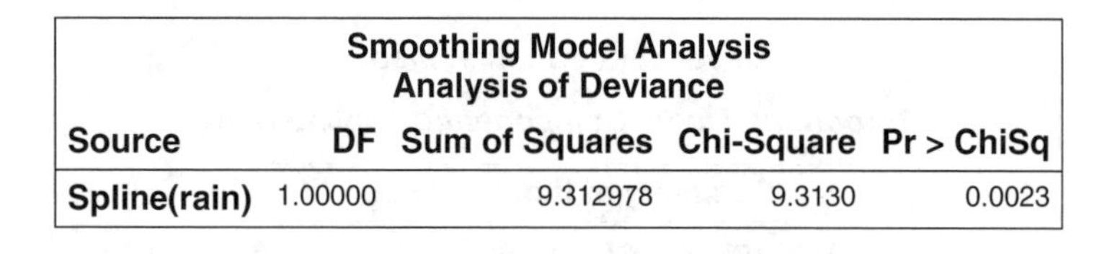

Smoothing Model Analysis
Analysis of Deviance

Source	DF	Sum of Squares	Chi-Square	Pr > ChiSq
Spline(rain)	1.00000	9.312978	9.3130	0.0023

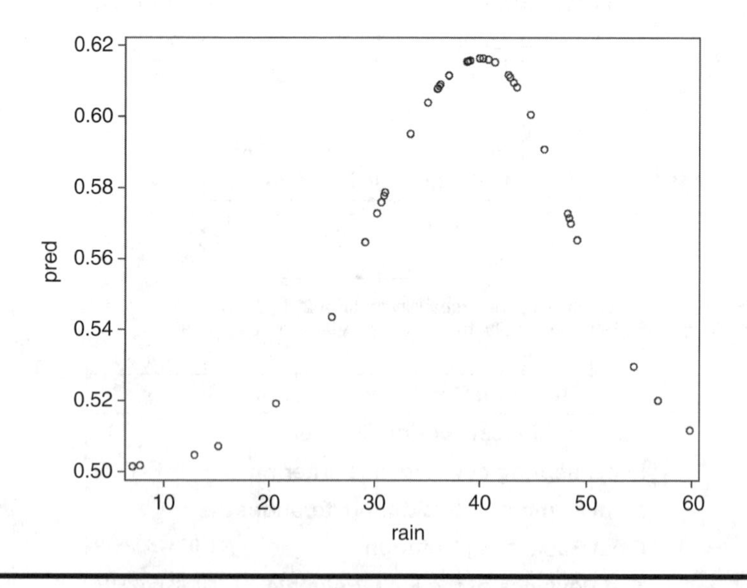

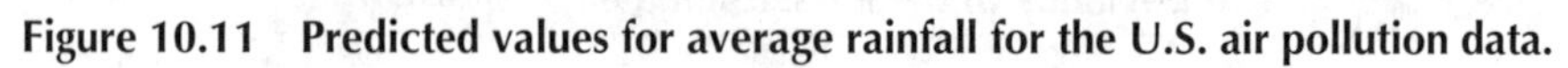

Figure 10.11 Predicted values for average rainfall for the U.S. air pollution data.

Exercises

1 Tyre Wear Data

1. Fit a further GAM to these data, one that uses cubic splines rather than loess fits.
2. Does the model fit better than the loess model considered in the text?

2 Air Pollution Data

1. Use the results from the fitting of GAM given in this chapter to select a possible logistic regression model for the data.
2. Fit your chosen model and interpret the results.

3 Insulin-Dependent Diabetes in Children

Data are given in Hastie and Tibshirani (1990) that arise from a study of the factors affecting patterns of insulin-dependent diabetes mellitus in children. The objective was to investigate the dependence of the level of serum C-peptide on various other factors in order to try to understand the patterns of residual insulin secretion.

The variables in insulin.dat are

- **ID:** Subject identifier
- **Age:** Age in years
- **Base:** Base deficit, a measure of acidity
- **Pep:** C-peptide concentration

1. Construct scatterplots of log(Pep) against age and base, showing on each the fitted linear regression and the loess fit.
2. Fit a GAM with log(Pep) as the response and age and base as explanatory variables, using loess fits for both explanatory variables.
3. Fit a model that includes a loess fit for age and a linear term for base. Compare the fit of this model with the one fitted above.
4. Comment on the results of the model fitting exercise.

Chapter 11

Analysis of Variance of Repeated Measures: Visual Acuity

11.1 Description of Data

The data to be used in this chapter are taken from Table 397 of *SDS*. They are reproduced in Table 11.1. Seven subjects had their response times measured when a light was flashed into each eye through lenses of powers 6/6, 6/18, 6/36 and 6/60. (A lens of power *a*/*b* means that the eye will perceive as being at *a* feet an object that is actually positioned at *b* feet.) Measurements are in milliseconds, and the question of interest was whether response time varied with lens strength.

11.2 Repeated Measures Data

In the data in Table 11.1 the response variable, reaction time, is recorded eight times for each subject in the study. Data involving several measures of the response variable on each subject often arise, particularly in the behavioural sciences and related disciplines, and are usually referred to as *repeated measures* data. Researchers typically adopt the repeated measures paradigm as a means of reducing error variability and as the natural way of measuring certain phenomena (e.g., developmental changes over time, learning and memory tasks). In this type of design, effects of experimental factors giving rise to the repeated measures are assessed relative to the average response made by a subject on all conditions or occasions. In essence,

Table 11.1 Visual Acuity and Lens Strength

	Left Eye				*Right Eye*			
	6/6	6/18	6/36	6/60	6/6	6/18	6/36	6/60
1	116	119	116	124	120	117	114	122
2	110	110	114	115	106	112	110	110
3	117	118	120	120	120	120	120	124
4	112	116	115	113	115	116	116	119
5	113	114	114	118	114	117	116	112
6	119	115	94	116	100	99	94	97
7	110	110	105	118	105	105	115	115

each subject serves as his or her own control and, accordingly, variability due to differences in average responsiveness of the subjects is eliminated from the extraneous error variance. A consequence of this is that the power to detect the effects of within-subject experimental factors is increased compared with testing in a between-subject design.

Unfortunately, the advantages of a repeated measures design come at a cost, and that is the probable lack of independence of the repeated measurements. Observations made under different conditions involving the same subjects will very likely be correlated rather than independent. This violates one of the assumptions of the analysis of variance procedures described in Chapters 4 and 5, and accounting for the dependence between observations in a repeated measures design requires some thought. In the visual acuity example, only within-subject factors occur, and it is possible, indeed likely, that the lens strengths under which a subject was observed were given in random order. But in examples, where time is the single within-subject factor, randomization is not, of course, an option. This makes the type of study where subjects are simply observed over time rather different from other repeated measures designs, and they are often given a different label, *longitudinal designs*. Due to their different nature we shall leave consideration of longitudinal data sets until Chapters 12–14.

11.3 Analysis of Variance for Repeated Measures Designs

Despite the lack of independence of the observations made within subjects in a repeated measures design, it remains possible to use relatively straightforward analysis of variance procedures to analyse the data if three particular assumptions about the observations are valid.

1. *Normality*: The data arise from populations with normal distributions.
2. *Homogeneity of variance*: The variances of the assumed normal distributions are equal.

3. *Sphericity*: The variances of the differences between all pairs of the repeated measurements are equal. This condition implies that the correlations between pairs of repeated measures are also equal, the so-called *compound symmetry* pattern.

It is the last assumption that is most critical for the validity of the analysis of variance F tests. When the sphericity assumption is not regarded as likely, there are two alternatives to a simple analysis of variance, the use of correction factors and multivariate analysis of variance. All three possibilities will be considered in this chapter.

11.3.1 Model for the Visual Acuity Data

We begin by considering a simple model for the visual acuity observations, y_{ijk}, where this represents the reaction time of the ith subject for eye j and lens strength k. The model assumed is

$$y_{ijk} = \mu + \alpha_j + \beta_k + (\alpha\beta)_{jk} + \gamma_i + (\gamma\alpha)_{ij} + (\gamma\beta)_{ik} + (\gamma\alpha\beta)_{ijk} + \varepsilon_{ijk}, \qquad (11.1)$$

where α_j represents the effect of eye j; β_k is the effect of the kth lens strength; $(\alpha\beta)_{jk}$ is the eye $\times$ lens strength interaction; γ_i is a constant associated with subject i; $(\gamma\alpha)_{ij}$, $(\gamma\beta)_{ik}$ and $(\gamma\alpha\beta)_{ijk}$ represent interaction effects of subject i with each factor and their interactions; and ε_{ijk} are error terms.

The terms α_j, β_k and $(\alpha\beta)_{ik}$ are assumed to be fixed effects, but the subject and error terms are assumed to be random variables from normal distributions with zero means and variances specific to each term, σ_1^2 for the subject effects and σ^2 for the error terms. This is an example of the mixed effects models considered in more detail in Chapter 13.

The model in Equation 11.1 implies that the variances of the eight repeated measurements are all equal to $\sigma_1^2 + \sigma^2$, and the covariances between each pair of repeated measurements are all equal to σ_1^2. The equal covariances (and also equal correlations) between the repeated measures arise as a consequence of the subject effects in this model, and if this structure is valid, a relatively straightforward analysis of variance of the data can be used. But when the investigator thinks the assumption of equal correlations is too strong there are two alternatives that can be used.

1. *Correction factors*: Greenhouse and Geisser (1959) and Huynh and Feldt (1976) considered the effects of departures from the sphericity assumption in a repeated measure analysis of variance. They demonstrated that the extent to which a set of repeated measures departs from the sphericity assumption can be summarized in terms of a parameter ε, which is a function of the variances

and covariances of the repeated measures. And an estimate of this parameter can be used to decrease the degrees of freedom of F tests for the within-subject effect to account for deviation from sphericity. In this way, larger F values will be needed to claim statistical significance than when the correction is not used, and so the increased risk of falsely rejecting the null hypothesis is removed. Formulae for the correction factors are given in Everitt (2001).

2. *Multivariate analysis of variance*: An alternative to the use of correction factors in the analysis of repeated measures data when the sphericity assumption is judged to be inappropriate is to use multivariate analysis of variance (MANOVA). The advantage is that no assumptions are now made about the pattern of correlations between the repeated measurements. A disadvantage of using MANOVA for repeated measures is often stated to be the technique's relatively low power when the assumption of compound symmetry is actually valid. But Davidson (1972) shows that this is really a problem only with small sample sizes.

11.4 Analysis Using SAS

Assuming the ASCII file visual.dat is in the current directory, the data may be read in as follows:

```
data vision;
  infile 'visual.dat' expandtabs;
  input idno x1-x8;
run;
```

The data are tab separated and the expandtabs option on the infile statement converts the tabs to spaces as the data are read allowing a simple list input statement to be used.

The glm procedure is used for the analysis.

```
proc glm data=vision;
  model x1-x8= / nouni;
  repeated eye 2, strength 4 /summary;
run;
```

The eight repeated measures per subject are all specified as response variables on the model statement and so appear on the left-hand side of the equation. There are no between-subject factors in the design, so the right-hand side of the equation is left blank. Separate univariate analyses of the eight measures are of no interest and so the nouni option is included to suppress them.

The repeated statement specifies the within-subject factor structure. Each factor is given a name followed by the number of levels that it has. Factor specifications are separated by commas. The order in which they occur implies a data structure in which the factors are nested from right to left; in this case one where lens strength is nested within eye. It is also possible to specify the type of contrasts to be used for each within-subject factor. The default is to contrast each level of the factor with the last. The summary option requests ANOVA tables for each contrast.

The output is shown in Table 11.2. Concentrating first on the univariate tests, we see that none of the effects, eye, strength or eye × strength, is significant, and this is so whichever p value is used, unadjusted, Greenhouse and Geisser (G–G) adjusted or Huynh–Feldt (H–F) adjusted. But the multivariate tests have a different story to tell; now the strength factor is seen to be highly significant. (For this example, the four test statistics available when applying multivariate analysis of variance, Wilks' lambda, Pillai's trace, Hotelling–Lawley trace and Roy's greatest root, are all equivalent; this will not always be so and readers are referred to Everitt, 2001, for more details.)

Because the strength factor is on an ordered scale, we might investigate it further by using orthogonal polynomial contrasts, here a linear, quadratic and cubic contrast.

Table 11.2 Results from Applying Proc glm to the Visual Acuity Data

The GLM Procedure

Number of Observations Read	7
Number of Observations Used	7

The GLM Procedure

Repeated Measures Analysis of Variance

Repeated Measures Level Information								
Dependent Variable	**x1**	**x2**	**x3**	**x4**	**x5**	**x6**	**x7**	**x8**
Level of eye	1	1	1	1	2	2	2	2
Level of strength	1	2	3	4	1	2	3	4

MANOVA Test Criteria and Exact F Statistics for the Hypothesis of no eye Effect
H = Type III SSCP Matrix for eye
E = Error SSCP Matrix

S=1 M=-0.5 N=2

Statistic	Value	F Value	Num DF	Den DF	Pr > F
Wilks' Lambda	0.88499801	0.78	1	6	0.4112
Pillai's Trace	0.11500199	0.78	1	6	0.4112
Hotelling-Lawley Trace	0.12994604	0.78	1	6	0.4112
Roy's Greatest Root	0.12994604	0.78	1	6	0.4112

MANOVA Test Criteria and Exact F Statistics for the Hypothesis of no strength Effect H = Type III SSCP Matrix for strength E = Error SSCP Matrix S=1 M=0.5 N=1					
Statistic	**Value**	**F Value**	**Num DF**	**Den DF**	**Pr > F**
Wilks' Lambda	0.05841945	21.49	3	4	0.0063
Pillai's Trace	0.94158055	21.49	3	4	0.0063
Hotelling-Lawley Trace	16.11758703	21.49	3	4	0.0063
Roy's Greatest Root	16.11758703	21.49	3	4	0.0063

MANOVA Test Criteria and Exact F Statistics for the Hypothesis of no eye*strength Effect H = Type III SSCP Matrix for eye*strength E = Error SSCP Matrix S=1 M=0.5 N=1					
Statistic	**Value**	**F Value**	**Num DF**	**Den DF**	**Pr > F**
Wilks' Lambda	0.70709691	0.55	3	4	0.6733
Pillai's Trace	0.29290309	0.55	3	4	0.6733
Hotelling-Lawley Trace	0.41423331	0.55	3	4	0.6733
Roy's Greatest Root	0.41423331	0.55	3	4	0.6733

The GLM Procedure

Repeated Measures Analysis of Variance

Univariate Tests of Hypotheses for Within Subject Effects

Source	DF	Type III SS	Mean Square	F Value	Pr > F
eye	1	46.4464286	46.4464286	0.78	0.4112
Error(eye)	6	357.4285714	59.5714286		

						Adj Pr > F	
Source	**DF**	**Type III SS**	**Mean Square**	**F Value**	**Pr > F**	**G - G**	**H - F**
strength	3	140.7678571	46.9226190	2.25	0.1177	0.1665	0.1528
Error(strength)	18	375.8571429	20.8809524				

Greenhouse-Geisser Epsilon	0.4966
Huynh-Feldt Epsilon	0.6229

Source	DF	Type III SS	Mean Square	F Value	Pr > F	Adj Pr > F G - G	Adj Pr > F H - F
eye*strength	3	40.6250000	13.5416667	1.06	0.3925	0.3700	0.3819
Error(eye*strength)	18	231.0000000	12.8333333				

Greenhouse-Geisser Epsilon	0.5493
Huynh-Feldt Epsilon	0.7303

The GLM Procedure

Repeated Measures Analysis of Variance

Analysis of Variance of Contrast Variables

eye_N represents the contrast between the nth level of eye and the last

Contrast Variable: eye_1

Source	DF	Type III SS	Mean Square	F Value	Pr > F
Mean	1	371.571429	371.571429	0.78	0.4112
Error	6	2859.428571	476.571429		

The GLM Procedure

Repeated Measures Analysis of Variance

Analysis of Variance of Contrast Variables

strength_N represents the contrast between the nth level of strength and the last

Contrast Variable: strength_1

Source	DF	Type III SS	Mean Square	F Value	Pr > F
Mean	1	302.2857143	302.2857143	5.64	0.0552
Error	6	321.7142857	53.6190476		

Contrast Variable: strength_2

Source	DF	Type III SS	Mean Square	F Value	Pr > F
Mean	1	175.0000000	175.0000000	3.55	0.1086
Error	6	296.0000000	49.3333333		

Contrast Variable: strength_3

Source	DF	Type III SS	Mean Square	F Value	Pr > F
Mean	1	514.2857143	514.2857143	5.57	0.0562
Error	6	553.7142857	92.2857143		

eye_N represents the contrast between the nth level of eye and the last

strength_N represents the contrast between the nth level of strength and the last

Contrast Variable: eye_1*strength_1

Source	DF	Type III SS	Mean Square	F Value	Pr > F
Mean	1	9.14285714	9.14285714	0.60	0.4667
Error	6	90.85714286	15.14285714		

Contrast Variable: eye_1*strength_2

Source	DF	Type III SS	Mean Square	F Value	Pr > F
Mean	1	11.5714286	11.5714286	0.40	0.5480
Error	6	171.4285714	28.5714286		

Contrast Variable: eye_1*strength_3

Source	DF	Type III SS	Mean Square	F Value	Pr > F
Mean	1	146.2857143	146.2857143	1.79	0.2291
Error	6	489.7142857	81.6190476		

Polynomial contrasts for lens strength can be obtained by resubmitting the previous glm step with the following repeated statement:

```
repeated eye 2, strength 4 (1 3 6 10) polynomial/summary;
```

The specification of the lens strength factor has been expanded: numeric values for the four levels of lens strength have been specified in parentheses and orthogonal polynomial contrasts requested. The values specified will be used as spacings in the calculation of the polynomials.

The edited results are shown in Table 11.3. None of the contrasts is significant although it has to be remembered here that the sample size is small so that the tests are not very powerful.

Table 11.3 Selected Results for Orthogonal Polynomial Contrasts of the Visual Acuity Data

Repeated Measures Analysis of Variance

Analysis of Variance of Contrast Variables

eye_N represents the contrast between the nth level of eye and the last

Contrast Variable: eye_1

Source	DF	Type III SS	Mean Square	F Value	Pr > F
Mean	1	371.571429	371.571429	0.78	0.4112
Error	6	2859.428571	476.571429		

strength_N represents the nth degree polynomial contrast for strength

Contrast Variable: strength_1

Source	DF	Type III SS	Mean Square	F Value	Pr > F
Mean	1	116.8819876	116.8819876	2.78	0.1468
Error	6	252.6832298	42.1138716		

Contrast Variable: strength_2

Source	DF	Type III SS	Mean Square	F Value	Pr > F
Mean	1	97.9520622	97.9520622	1.50	0.2672
Error	6	393.0310559	65.5051760		

Contrast Variable: strength_3

Source	DF	Type III SS	Mean Square	F Value	Pr > F
Mean	1	66.7016645	66.7016645	3.78	0.1000
Error	6	106.0000000	17.6666667		

eye_N represents the contrast between the nth level of eye and the last

strength_N represents the nth degree polynomial contrast for strength

Contrast Variable: eye_1*strength_1

Source	DF	Type III SS	Mean Square	F Value	Pr > F
Mean	1	1.00621118	1.00621118	0.08	0.7857
Error	6	74.64596273	12.44099379		

Contrast Variable: eye_1*strength_2

Source	DF	Type III SS	Mean Square	F Value	Pr > F
Mean	1	56.0809939	56.0809939	1.27	0.3029
Error	6	265.0789321	44.1798220		

Contrast Variable: eye_1*strength_3

Source	DF	Type III SS	Mean Square	F Value	Pr > F
Mean	1	24.1627950	24.1627950	1.19	0.3180
Error	6	122.2751052	20.3791842		

Exercises

1 Visual Acuity Data

1. Plot the left and right eye means for the different lens strengths. Include standard error bars on the plot.
2. Examine the raw data graphically in some way to assess whether there is any evidence of outliers. If there is, repeat the analyses described in the text.
3. Find the correlations between the repeated measures for the data used in this chapter.
4. Does the pattern of the observed correlations lead to an explanation in the different results produced by the univariate and multivariate treatment of these data?

2 Skin Resistance Data

In an experiment, 5 different types of electrodes were applied to the arms of 16 subjects and the resistance measured in kilo-ohms. The experiment was carried out to see whether all electrode types performed similarly (taken from Table 240 of *SDS*).

The variables in resist.dat are

- **ID:** Subject identifier
- **E1:** Resistance for electrode 1
- **E2:** Resistance for electrode 2
- **E3:** Resistance for electrode 3
- **E4:** Resistance for electrode 4
- **E5:** Resistance for electrode 5

1. Investigate whether there are any possible outliers in the data. If there are, suggest an explanation why they might have occurred.
2. Carry out a suitable analysis of variance to assess whether all electrode types perform similarly.

3 Quitting Smoking Experiment

In an investigation of cigarette smoking, three different procedures for quitting smoking – tapering off, immediate stopping, and aversion therapy – were compared. Subjects were randomly allocated to each treatment and were asked to rate on a 1–10 point scale their desire to smoke "right now" in two different environments, home versus work, both before and after quitting (Howell, 2002).

The variables in smoke.dat are

- **ID:** Subject identifier
- **Proc:** Quitting smoking procedure, 1 = tapering, 2 = immediate stopping, 3 = aversion therapy
- **Befh:** Rating at home before quitting
- **Befw:** Rating at work before quitting
- **Afth:** Rating at home after quitting
- **Aftw:** Rating at work after quitting:

1. Carry out a suitable univariate analysis of variance of these data, noting that the groups formed by the two between-subject factors, gender and treatment groups, have different numbers of subjects.
2. Now carry out a multivariate analysis of variance and compare the results with those from the univariate analysis. (This example illustrates that the four multivariate test criteria mentioned in the chapter do not always give equivalent results.)

Chapter 12

Longitudinal Data I: Treatment of Post-Natal Depression

12.1 Description of Data

The data set to be analysed in this chapter originates from a clinical trial of the use of oestrogen patches in the treatment of post-natal depression. Full details of the study are given in Gregoire et al. (1996). In total, 61 women with major depression, which began within 3 months of childbirth and persisted for up to 18 months post-natally, were allocated randomly to the active treatment or a placebo (a dummy patch); 34 received the former and the remaining 27 received the latter. The women were assessed twice pre-treatment and then monthly for 6 months after treatment on the Edinburgh post-natal depression scale (EPDS), higher values of which indicate increasingly severe depression. The first five observations in each treatment group are shown in Table 12.1. In this table a value of −9 indicates that the corresponding observation was not made for some reason.

12.2 Analyses of Longitudinal Data

The data in Table 12.1 consist of repeated observations over time on each of the 61 patients; they are a particular form of repeated measures data (see Chapter 11), with time as the single within-subjects factor. The analysis of variance methods

Table 12.1 Data from a Clinical Trial of Oestrogen Patches for Treating Post-natal Depression

Group 0 = *Placebo* 1 = *Treatment*	*EPDS Score*							
	Month							
	Pre1	*Pre2*	*1*	*2*	*3*	*4*	*5*	*6*
0	18	18	17	18	15	17	14	15
0	25	27	26	23	18	17	12	10
0	19	16	17	14	−9	−9	−9	−9
0	24	17	14	23	17	13	12	12
0	19	15	12	10	8	4	5	5
1	21	21	13	12	9	9	13	6
1	27	27	8	17	15	7	5	7
1	24	15	8	12	10	10	6	5
1	28	24	14	14	13	12	18	15
1	19	15	15	16	11	14	12	8

described in Chapter 11 could be, and frequently are, applied to such data, but in the case of longitudinal data, the sphericity assumption is very unlikely to be plausible since observations closer together in time are very likely more highly correlated than those taken further apart. Consequently, other methods are generally more useful than the analysis of variance for these types of data. In this chapter, we will consider a number of relatively simple approaches to the analysis of longitudinal data including:

- Graphical displays
- Summary measure or response feature analysis

In the next two chapters, we shall discuss more formal modelling techniques that can be used to analyse longitudinal data.

12.3 Analysis Using SAS

Data sets for longitudinal and repeated measures data may be structured in two ways. In the first form there is one observation per subject and the repeated measurements are held in separate variables. Alternatively, there may be separate observations for each measurement, with variables indicating to which subject and occasion it belongs. When analysing longitudinal data both formats may be needed. This is typically achieved by reading the raw data into a data set in one format and then using a second data step to reformat it. In the example below, both types of data sets are created in the one data step.

```
data pndep(keep=idno group x1-x8) pndep2(keep=idno
  group time dep);
  infile 'c:\handbook3\datasets\channi.dat';
  input group x1-x8;
  idno=_n_;
  array xarr {8} x1-x8;
  do i=1 to 8;
  if xarr{i}=-9 then xarr{i}=.;
    time=i;
    dep=xarr{i};
    output pndep2;
  end;
  output pndep;
run;
```

The data statement contains the names of two data sets, pndep and pndep2, indicating that two data sets are to be created. The keep= option in parentheses specifies which variables are to be retained in each. The input statement reads the group information and the eight depression scores. The raw data comprise 61 such lines, so the automatic SAS variable _n_ will increment from 1 to 61 accordingly. The variable idno is assigned its value to use as a case identifier because _n_ is not stored in the data set itself.

The eight depression scores are declared an array, and a do loop processes them individually. The value −9 in the data indicates a missing value and these are reassigned to the SAS missing value by the if-then statement. The variable time records the measurement occasion as 1–8, and dep contains the depression score for that occasion. The output statement writes an observation to the data set pndep2. From the data statement we can see that this data set will contain the subject identifier, idno, plus group, time and dep. Because this output statement is within the do loop, it will be executed for each iteration of the do loop, that is, eight times.

The second output statement writes an observation to the pndep data set. This data set contains idno, group and the eight depression scores in the variables x1 to x8.

Having run this data step, the SAS log confirms that pndep has 61 observations and pndep2, 488 (i.e., 61×8).

Table 12.2 Summary Statistics for the Post-natal Depression Data

	group					
	0			1		
	dep			dep		
	Mean	Var	N	Mean	Var	N
time						
1	21.92	10.15	26	21.94	10.54	34
2	20.78	15.64	27	21.24	12.61	34
3	16.48	27.87	27	13.35	30.84	34
4	15.86	37.74	22	11.71	43.01	31
5	14.12	24.99	17	9.10	30.02	29
6	12.18	34.78	17	8.80	21.71	28
7	11.35	20.24	17	7.29	33.10	28
8	10.82	22.15	17	6.46	22.48	28

To begin, let us look at some means and variances of the observations. Proc means, proc summary or proc univariate could all be used for this, but proc tabulate gives particularly neat and concise output. The second format, one case per measurement, allows a simpler specification of the table, which is shown in Table 12.2.

```
proc tabulate data=pndep2 f=6.2;
  class group time;
  var dep;
  table time,
    group*dep*(mean var n);
run;
```

There is a general decline in the EPDS over time in both groups, with the values in the active treatment group (group = 1) being consistently lower.

12.3.1 Graphical Displays

Often, a useful preliminary step in the analysis of longitudinal data is to graph the observations in some way. The aim is to highlight two particular aspects of the data, how they evolve over time and how the measurements made at different times are related. A number of graphical displays might be helpful here including:

- Separate plots of each subject's response against time, differentiating in some way between subjects in different groups
- Box plots of the observations at each time point for each group
- Plot of means and standard errors by treatment group for every time point
- Scatterplot matrix of the repeated measurements

12.3.1.1 Separate Plots of Each Subject's Response over Time

These plots can be produced in SAS as follows:

```
proc sgpanel data=pndep2 noautolegend;
  panelby group/rows=2;
  series y=dep x=time/group=idno break lineattrs=(pattern=solid);
run;
```

Line plots with sgplot were introduced in Chapter 1. Here proc sgpanel is used. The group=idno option is used to produce a plot with a separate line for each subject. The line pattern is specified explicitly. In this instance, the data are already in the correct order for the plot. Otherwise, they would need to be sorted appropriately. The result is shown in Figure 12.1.

The graphs in Figure 12.1, although somewhat "messy," demonstrate the variability in the data and also indicate the general decline in the depression scores in both groups, with those in the active group remaining generally lower.

12.3.1.2 Box Plots of the Observations at Each Time Point

The required box plots can be produced as follows:

```
proc sort data=pndep2;
  by group;
run;
proc sgplot data=pndep2;
  vbox dep /category=time;
  by group;
run;
```

To use the by statement to produce separate box plots for each group the data must be sorted by group. The results are shown in Figures 12.2 and 12.3. Again the decline in depression scores in both groups is clearly seen in the two graphs.

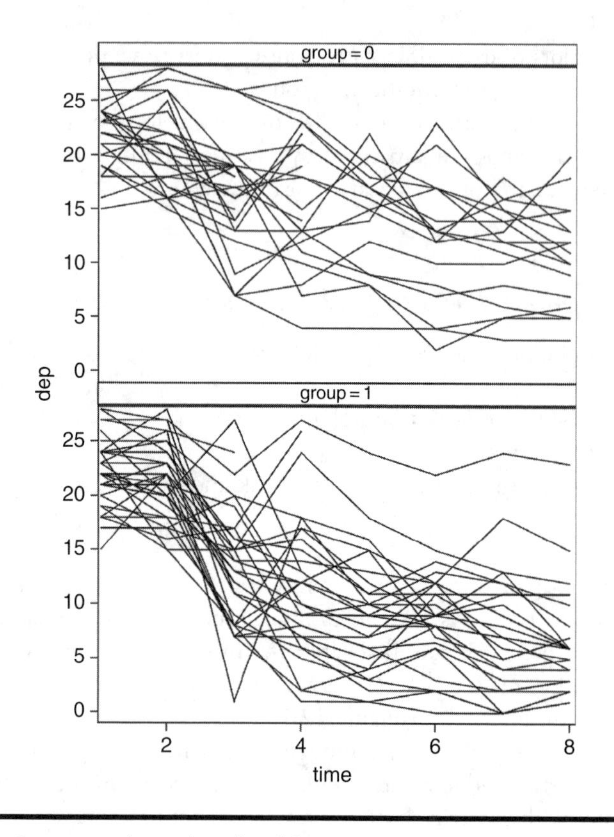

Figure 12.1 Separate plots of each subject's response over time in the oestrogen patch data.

12.3.1.3 Plot of Means and Standard Errors

The code to produce this plot is as follows:

```
proc sgplot data=pndep2;
  vline time/response=dep stat=mean limitstat=stderr group=group;
run;
```

The use of proc sgplot to create plots of summary statistics was introduced in Chapter 1. The default would be frequency counts for each time point. To specify means and standard errors the response, stat and limitstat options are all needed. The result is shown in Figure 12.4. This diagram shows that from the first visit after randomization, the depression score in the active group is lower than that in the control group, a situation that continues for the remainder of the trial.

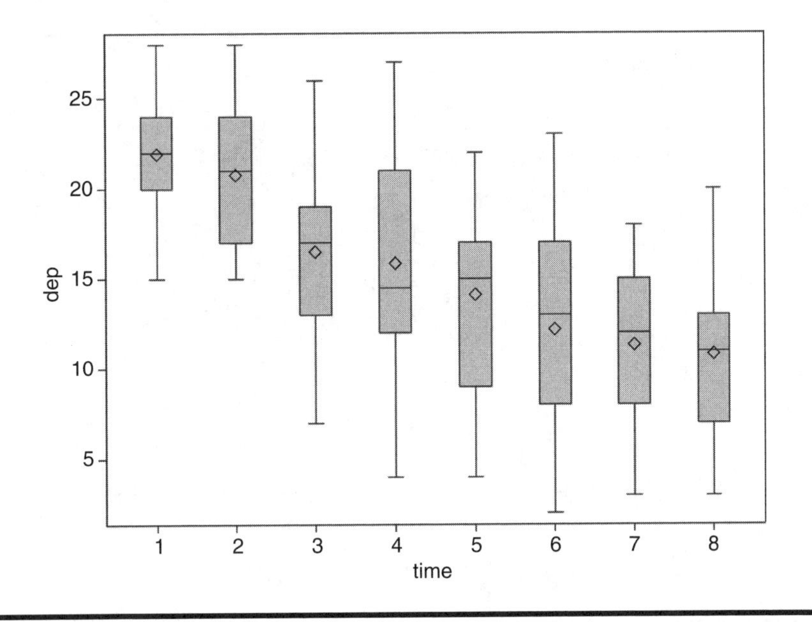

Figure 12.2 Box plots of the observations at each time point in the placebo group.

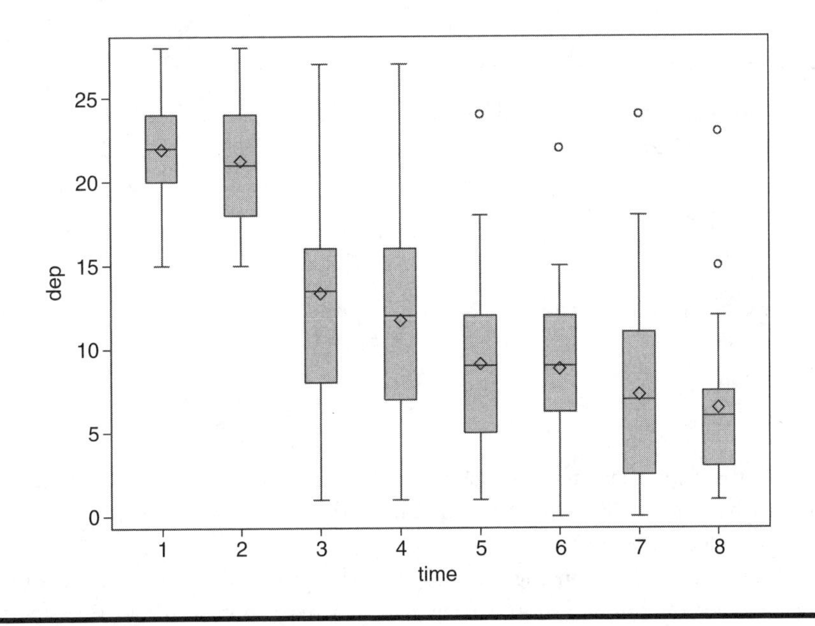

Figure 12.3 Box plots of the observations at each time point in the oestrogen patch group.

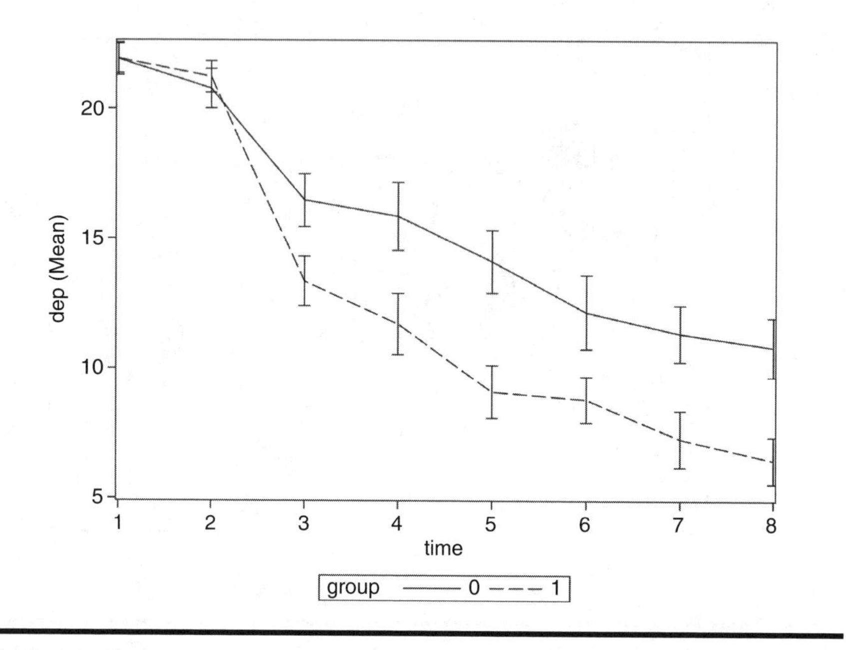

Figure 12.4 Plot of mean profiles and standard errors for placebo and active groups.

12.3.1.4 Scatterplot Matrix

Scatterplot matrices can be produced using the statistical graphics procedure proc sgscatter as follows:

```
proc sort data=pndep;
  by group;
run;
proc sgscatter data=pndep;
  matrix x1-x8;
  by group;
run;
```

Because we want separate scatterplot matrices for each group, we first sort the data set by group. Within proc sgscatter the matrix statement specifies the variables to form the scatterplot matrix.

The result is shown in Figure 12.5. Clearly, observations made on occasions close together in time are more strongly related than those made further apart, a phenomenon that has implications for more formal modelling of the data (see Chapter 11).

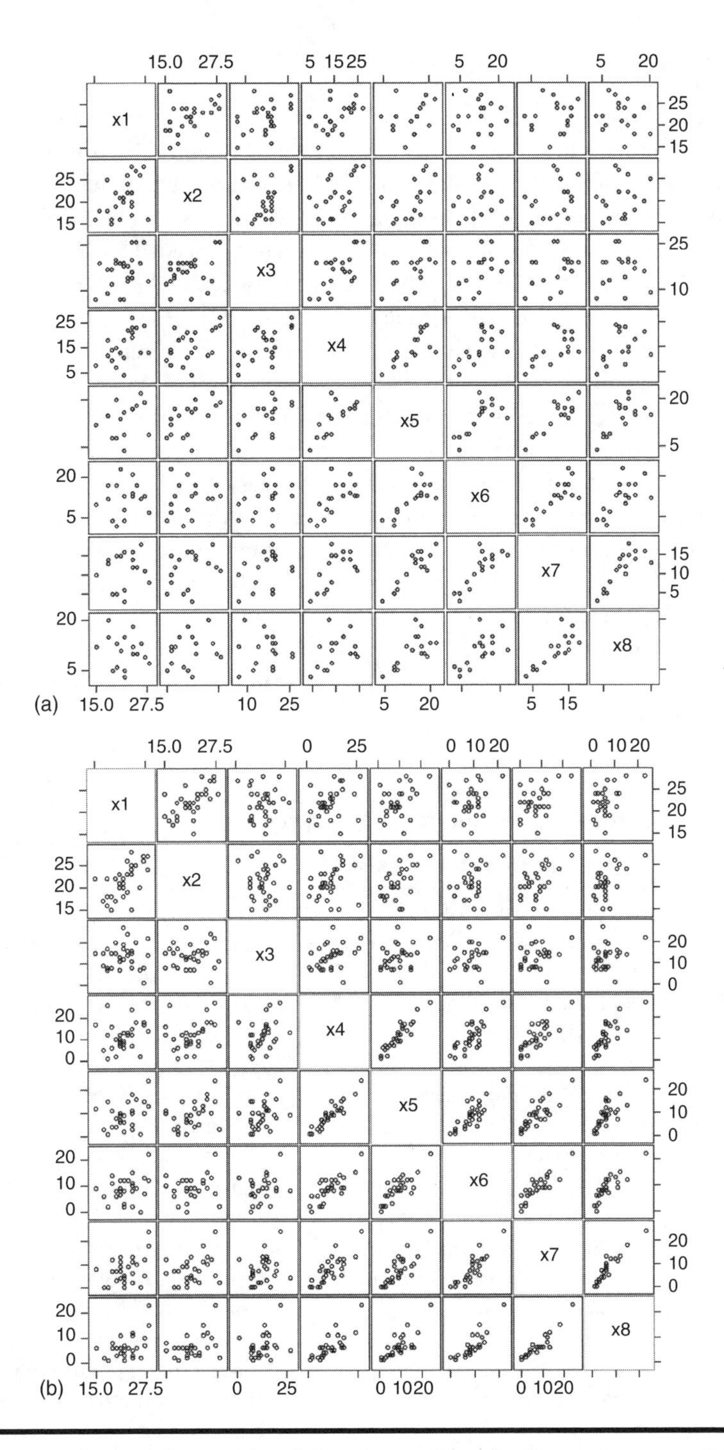

Figure 12.5 Scatterplot matrix of the observations in the oestrogen patch data. (a) TAU, (b) BtB.

12.3.2 Response Feature Analysis

A relatively straightforward approach to the analysis of longitudinal data involves the use of summary measures, sometimes known as response feature analysis. The repeated observations on a subject are used to construct a single number that characterizes some relevant aspect of the subject's response profile. (In some situations, more than a single summary measure may be needed to characterize the profile adequately.) The summary measure to be used does, of course, need to be decided upon before the analysis of the data.

The most commonly used summary measure is the mean of the responses over time since many investigations, for example, clinical trials, are most concerned with differences in overall level rather than more subtle effects. But other summary measures might be considered more relevant in particular circumstances, and Table 12.3 lists a number of alternative possibilities.

Having identified a suitable summary measure, the analysis of the repeated measures data reduces to a simple univariate test of group differences on the chosen measure. In the case of two groups this will involve the application of a two-sample *t*-test or perhaps its non-parametric equivalent.

Returning to the oestrogen patch data we shall use the mean as the chosen summary measure, but before undertaking the analysis there are two further problems to consider:

Table 12.3 Some Possible Summary Measures

Type of Data	*Questions of Interest*	*Summary Measure*
Peaked	Is overall value of outcome variable the same in different groups?	Overall mean (equal time intervals) or area under curve (unequal intervals)
Peaked	Is maximum (minimum) response different between groups?	Maximum (minimum) value
Peaked	Is time to maximum (minimum) response different between groups?	Time to maximum (minimum) response
Growth	Is rate of change of outcome different between groups?	Regression coefficient
Growth	Is eventual value of outcome different between groups?	Final value of outcome or difference between last and first values or percentage change between first and last values
Growth	Is response in one group delayed relative to the other?	Time to reach a particular value (e.g., a fixed percentage of baseline)

1. How to deal with the missing values?
2. How to incorporate the pre-treatment measurements into an analysis?

The missing values can be dealt with in at least three ways:

1. Take the mean over the available observations for a subject, that is, if a subject has only four post-treatment values recorded use the mean of these.
2. Include in the analysis only those subjects with all six post-treatment observations.
3. Impute the missing values in some way, for example, use the last observation carried forward (LOCF) approach popular in the pharmaceutical industry.

The pre-treatment value might be incorporated into an analysis by calculating change scores, that is, post-treatment mean – pre-treatment mean value, or as a covariate in an analysis of covariance of the post-treatment means. Let us begin however, by simply ignoring the pre-treatment values and dealing only with the post-treatment means.

The three possibilities of calculating the mean summary measure can be implemented as follows:

```
data pndep;
  set pndep;
  array xarr {8} x1-x8;
  array locf {8} locf1-locf8;
  do i=3 to 8;
    locf{i}=xarr{i};
    if xarr{i}=. then locf{i}=locf{i-1};
end;
  mnbase=mean(x1,x2);
  mnresp=mean(of x3-x8);
  mncomp=(x3+x4+x5+x6+x7+x8)/6;
  mnlocf=mean(of locf3–locf8);
  chscore=mnbase-mnresp;
run;
```

The summary measures are to be included in the pndep data set, so this is named in the data statement. The set statement indicates that the data are to be read from the current version of pndep. The eight depression scores x1–x8 are declared as an array, and another array is declared for the LOCF values. Eight variables are declared although only six will be used. The do loop assigns LOCF values for those occasions when the depression score is missing. The mean of the two baseline measures is then computed using the SAS mean function. The next statement computes the mean of the recorded follow-up scores. When a variable list is used with the mean function it must be preceded with of. The mean function will result in a missing value only if all the variables are missing. Otherwise it computes the mean of the non-missing values. So the mnresp variable will contain the mean of the available follow-up scores for a subject. Because an arithmetic operation involving a missing value results in a missing value, mncomp will be assigned a missing value if any of the variables is missing.

A *t*-test can now be applied to assess the difference between treatments for each of the three procedures. The tabular results are shown in Table 12.4.

Table 12.4 Summary Measure Results for the Oestrogen Patch Data

Variable: mnresp

group	N	Mean	Std Dev	Std Err	Minimum	Maximum
0	27	14.7284	4.5889	0.8831	4.1667	26.5000
1	34	10.5172	5.3664	0.9203	2.0000	24.0000
Diff (1-2)		4.2112	5.0386	1.2988		

group	Method	Mean	95% CL Mean		Std Dev	95% CL Std Dev	
0		14.7284	12.9131	16.5437	4.5889	3.6138	6.2888
1		10.5172	8.6447	12.3896	5.3664	4.3284	7.0637
Diff (1-2)	Pooled	4.2112	1.6123	6.8102	5.0386	4.2709	6.1454
Diff (1-2)	Satterthwaite	4.2112	1.6586	6.7639			

Method	Variances	DF	t Value	Pr > |t|
Pooled	Equal	59	3.24	0.0020
Satterthwaite	Unequal	58.644	3.30	0.0016

Equality of Variances				
Method	Num DF	Den DF	F Value	Pr > F
Folded F	33	26	1.37	0.4148

Variable: mnlocf

group	N	Mean	Std Dev	Std Err	Minimum	Maximum
0	27	14.9259	4.6868	0.9020	4.1667	26.8333
1	34	10.6299	5.5711	0.9554	2.0000	24.5000
Diff (1-2)		4.2960	5.2000	1.3404		

group	Method	Mean	95% CL Mean		Std Dev	95% CL Std Dev	
0		14.9259	13.0719	16.7800	4.6868	3.6910	6.4230
1		10.6299	8.6861	12.5737	5.5711	4.4935	7.3331
Diff (1-2)	Pooled	4.2960	1.6138	6.9782	5.2000	4.4077	6.3422
Diff (1-2)	Satterthwaite	4.2960	1.6666	6.9254			

Method	Variances	DF	t Value	Pr > |t|
Pooled	Equal	59	3.20	0.0022
Satterthwaite	Unequal	58.777	3.27	0.0018

Equality of Variances				
Method	Num DF	Den DF	F Value	Pr > F
Folded F	33	26	1.41	0.3676

Variable: mncomp

group	N	Mean	Std Dev	Std Err	Minimum	Maximum
0	17	13.3333	4.3104	1.0454	4.1667	18.1667
1	28	9.2589	4.5737	0.8643	2.0000	23.6667
Diff (1-2)		4.0744	4.4775	1.3767		

group	Method	Mean	95% CL Mean		Std Dev	95% CL Std Dev	
0		13.3333	11.1171	15.5496	4.3104	3.2103	6.5602
1		9.2589	7.4854	11.0324	4.5737	3.6160	6.2254
Diff (1-2)	Pooled	4.0744	1.2980	6.8508	4.4775	3.6994	5.6731
Diff (1-2)	Satterthwaite	4.0744	1.3220	6.8268			

Method	Variances	DF	t Value	Pr > |t|
Pooled	Equal	43	2.96	0.0050
Satterthwaite	Unequal	35.516	3.00	0.0049

Equality of Variances				
Method	Num DF	Den DF	F Value	Pr > F
Folded F	27	16	1.13	0.8237

```
proc ttest data=pndep;
class group;
var mnresp mnlocf mncomp;
run;
```

Here the results are similar and the conclusion in each case the same; namely that there is a substantial difference in overall level in the two treatment groups. The confidence intervals for the treatment effect given by each of the three procedures are as follows:

1. Using mean of available observations (1.612,6.810)
2. Using LOCF (1.614,6.978)
3. Using only complete cases (1.298,6.851)

All three approaches lead, in this example, to the conclusion that the active treatment considerably lowers depression. But, in general, using only subjects with a complete set of measurements or using LOCF is not to be recommended. Using only complete observations can produce bias in the results unless the missing observations are missing completely at random (see Chapter 13 and Everitt and Pickles, 2004). And the LOCF procedure has little in its favour since it makes highly unlikely assumptions, for example, that the expected values of the (unobserved) remaining observations remains at their last recorded values. Even using the mean of the values actually recorded is not without its problems (see Matthews, 1993), but it does appear, in general, to be the least objectionable of the three alternatives.

Now let us consider analyses that make use of the pre-treatment values available for each woman in the study. The change score analysis and the analysis of variance using the mean of available post-treatment values as the summary can be applied as follows:

```
proc glm data=pndep;
  class group;
  model chscore=group /solution;
proc glm data=pndep;
  class group;
  model mnresp=mnbase group /solution;
run;
```

We use proc glm for both analyses for comparability although we could also have used a *t*-test for the change scores. The results are shown in Tables 12.5 and 12.6. In both cases for this example, the group effect is highly significant, confirming the difference in depression scores of the active and control groups found in the previous analysis. The box plot associated with Table 12.5 illustrates graphically the greater change in the active group, and the plot associated with Table 12.6 demonstrates it for a given pre-treatment score; members of the active group have, on average, lower post-treatment scores than those in the placebo group.

Table 12.5 Using Change Scores in the Summary Measure Analysis of the Oestrogen Patch Data

Class Level Information		
Class	**Levels**	**Values**
group	2	0 1

Number of Observations Read	61
Number of Observations Used	61

Source	DF	Sum of Squares	Mean Square	F Value	Pr > F
Model	1	310.216960	310.216960	12.17	0.0009
Error	59	1503.337229	25.480292		
Corrected Total	60	1813.554189			

R-Square	Coeff Var	Root MSE	chscore Mean
0.171055	55.70617	5.047801	9.061475

Source	DF	Type I SS	Mean Square	F Value	Pr > F
group	1	310.2169603	310.2169603	12.17	0.0009

Source	DF	Type III SS	Mean Square	F Value	Pr > F
group	1	310.2169603	310.2169603	12.17	0.0009

Parameter		Estimate		Standard Error	t Value	Pr > |t|
Intercept		11.07107843	B	0.86569068	12.79	<.0001
group	**0**	-4.54021423	B	1.30120516	-3.49	0.0009
group	**1**	0.00000000	B	.	.	.

NOTE: The X'X matrix has been found to be singular, and a generalized inverse was used to solve the normal equations. Terms whose estimates are followed by the letter "B" are not uniquely estimable.

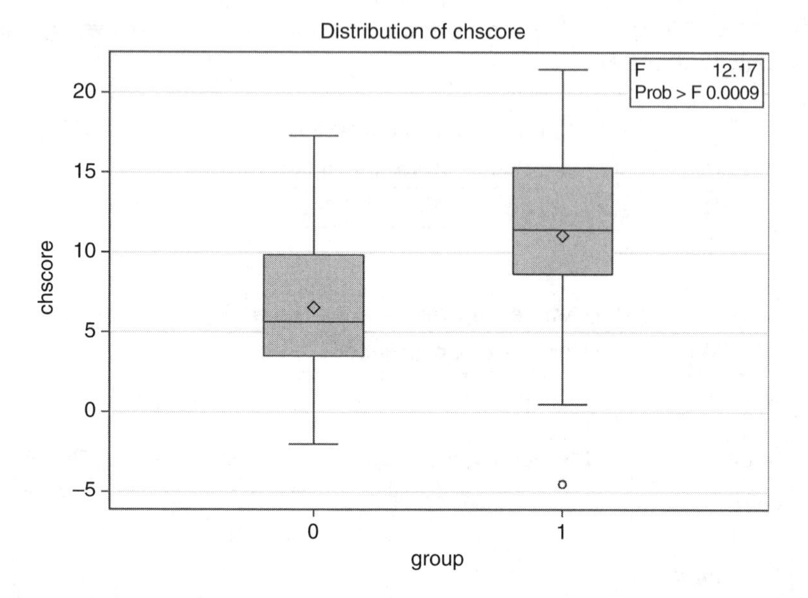

Table 12.6 Using Pre-treatment Score as a Covariate in the Analysis of the Oestrogen Patch Data

Class Level Information		
Class	Levels	Values
group	2	0 1

Number of Observations Read	61
Number of Observations Used	61

Source	DF	Sum of Squares	Mean Square	F Value	Pr > F
Model	2	404.398082	202.199041	8.62	0.0005
Error	58	1360.358293	23.454453		
Corrected Total	60	1764.756375			

R-Square	Coeff Var	Root MSE	mnresp Mean
0.229152	39.11576	4.842980	12.38115

Source	DF	Type I SS	Mean Square	F Value	Pr > F
mnbase	1	117.2966184	117.2966184	5.00	0.0292
group	1	287.1014634	287.1014634	12.24	0.0009

Source	DF	Type III SS	Mean Square	F Value	Pr > F
mnbase	1	137.5079844	137.5079844	5.86	0.0186
group	1	287.1014634	287.1014634	12.24	0.0009

Although in the oestrogen patch example the change score approach and the analysis of covariance approach lead to similar conclusions, in general, the analysis of covariance approach is to be preferred for reasons outlined in Senn (1997) and Everitt and Pickles (2004).

Exercises

1 Post-natal Depression

1. The graph in Figure 12.1 illustrates the phenomenon known as "tracking", the tendency of women with higher depression scores at the beginning of the trial to be those with the higher scores at the end. This phenomenon becomes more visible if standardized scores are plotted, that is, (depression scores – visit mean)/visit S.D. Calculate and plot these scores, differentiating on the plot the two treatment groups.
2. Apply the response feature approach described in the text, but now using the slope of each woman's depression score on time as the summary measure.

2 Phosphate Data

Data were collected on the plasma inorganic phosphate levels at various times after a glucose challenge for 33 subjects, 20 of whom were controls and 13 of whom had been classified as obese (Davis, 2002). The aim of the study was to see if phosphate levels differed in any respects in the two groups.

The variables in phos.dat are as follows:

- **ID:** Subject identifier
- **Group:** 1 = control, 2 = obese
- **P0:** Plasma inorganic phosphate level 0 h after glucose challenge

- **P05:** Plasma inorganic phosphate level 0.5 h after glucose challenge
- **P1:** Plasma inorganic phosphate level 1 h after glucose challenge
- **P15:** Plasma inorganic phosphate level 1.5 h after glucose challenge
- **P2:** Plasma inorganic phosphate level 2 h after glucose challenge
- **P3:** Plasma inorganic phosphate level 3 h after glucose challenge
- **P4:** Plasma inorganic phosphate level 4 h after glucose challenge
- **P5:** Plasma inorganic phosphate level 5 h after glucose challenge

1. Produce plots of the individual profiles of the 33 subjects differentiating the controls and the obese subjects.
2. Construct a plot of the mean profile of the control subjects and of the obese subjects along with standard error bars.
3. Using the plots produced in 1 and 2 as a guide to a suitable summary statistic, investigate whether the phosphate levels of the controls and the obese subjects differ in any way.

Chapter 13

Longitudinal Data II: Linear Mixed Models. Computerized Delivery of Cognitive Behavioural Therapy – Beat the Blues

13.1 Introduction

Depression is a major public health problem around the world. Antidepressants are the frontline treatment, but many patients either do not respond to them or do not like taking them. The main alternative treatment is psychotherapy, and the modern "talking treatments" such as *cognitive behavioural therapy* (CBT) have been shown to be as effective as drugs, and probably more so when it comes to relapse. But there is a problem, namely availability – there are simply not enough skilled therapists to meet the demand, and little prospect at all of correcting this situation.

A number of alternative modes of delivery of CBT have been explored, including interactive systems making use of the new computer technologies. The principles of CBT lend themselves reasonably well to computerisation, and, perhaps surprisingly, patients adapt well to this procedure and do not seem to miss the physical presence of the therapist as much as one might expect. The data to be used in this chapter arise from a clinical trial of an interactive multimedia program known as "Beat the Blues" (BtB) designed to deliver CBT to depressed patients via a computer. Full details are given in Proudfoot et al. (2002), but in essence BtB is an interactive program using multimedia techniques, in particular video vignettes. The computer-based intervention consists of nine sessions, followed by eight therapy sessions, each lasting about 50 min. Nurses are used to explain how the program works, but they are instructed to spend no more than 5 min with each patient at the start of each session and are there simply to assist with the technology.

In a randomized controlled trial of the program, patients with depression recruited in primary care were randomized to either the Beating the Blues program or "Treatment as Usual (TAU)." Patients randomized to BtB also received pharmacology and general GP support and practical/social help, offered as part of TAU, with the exception of any face-to-face counselling or psychological intervention. Patients allocated to TAU received whatever treatment their GP prescribed. The latter included, besides any medication, discussion of problems with GP, provision of practical/social help, referral to a counsellor, referral to a practice nurse, referral to mental health professionals (psychologist, psychiatrist, community psychiatric nurse, counsellor) or further physical examination.

A number of outcome measures were used in the trial, but here we concentrate on the Beck Depression Inventory II (Beck et al., 1996). Measurements on this variable were made on the following five occasions:

- Prior to treatment
- Two months after treatment began
- At 1, 3 and 6 months follow-up, that is, at 3, 5 and 8 months after treatment was begun

The data to be used in this chapter are a subset of size 100 from the original data collected, and are used with the kind permission of the organizers of the study, in particular Dr. Judy Proudfoot. The observations made on the first five subjects are shown in Table 13.1. In addition to treatment and the values of the Beck Depression Inventory, these observations also include a binary variable indicating whether a patient was taking anti-depressant drugs and a further binary variable

Table 13.1 Data for the First Five Participants in the BtB Clinical Trial

				Beck Depression Inventory				
Subject	*Drug*	*Duration (Months)*	*Treatment*	*Pre*	*2*	*3*	*5*	*8*
1	N	>6	TAU	29	2	2	.	.
2	Y	>6	BtB	32	16	24	17	20
3	Y	<6	TAU	25	20	.	.	.
4	N	>6	BtB	21	17	16	10	9
5	Y	>6	BtB	26	23	.	.	.

coding whether the length of the current episode of depression was 6 months or more or less than 6 months. The main question of interest about these data is whether there is any evidence of a treatment effect?

13.2 Linear Mixed Models for Longitudinal Data

The distinguishing feature of a longitudinal study such as the BtB trial is that the response variable of interest and a set of explanatory variables are measured several times on each individual in the study. The main objective in such a study is to characterize change in the repeated values of the response variable and to determine which explanatory variables are the most associated with any change. Due to the repeated measurement of the response variable on each individual in the study, the observations are likely to be correlated rather than independent even after conditioning on the explanatory variables. So suitable models for the data need to include parameters linking the explanatory variables to the repeated response values, that is, parameters analogous to those in the usual multiple regression model (see Chapter 7) and, in addition, parameters that aim to account for the correlational structure of the repeated measurements. It is the former parameters that are generally of most interest to the investigator (these parameters are generally known as *fixed effects*), with the latter often being regarded as *nuisance parameters*. But providing an adequate description for the correlations between the repeated measures is usually necessary if misleading inferences about the more interesting fixed effects parameters are to be avoided.

Linear mixed models introduce the needed correlations by formalizing the idea that an individual's pattern of responses is likely to depend on many characteristics of that individual, including some that are unobserved. These unobserved variables are then included in the model as random variables, that is, *random effects*. The essential feature of such models is that correlation among the repeated measurements on the same individual arises from shared unobserved variables, but conditional on the values of the random effects, the repeated measurements are assumed to be independent, the so-called *local independence* assumption. For more details see Diggle et al. (2002) and Davis (2002).

So in linear mixed effects models the mean response is modelled as a combination of population characteristics that are assumed to be shared by all individuals (the fixed effects) and subject-specific effects that are unique to a particular individual (the random effects).

Two simple linear mixed effects models will be used in the chapter, the *random intercepts model*, and the *random intercepts and random slopes model*. In the description that follows, we simplify matters by assuming that the only explanatory variable is time and that the observation of the response variable made at time, t_j, on individual i is represented as y_{ij}.

13.2.1 Random Intercept Model

The random intercept model can be written as

$$y_{ij} = \beta_0 + \beta_1 t_j + u_i + \varepsilon_{ij}. \tag{13.1}$$

Here, the total residual that would be present in the usual linear regression model (see Chapter 7) has been partitioned into a subject-specific random component u_i that is constant over time plus a residual ε_{ij} that varies randomly over time. The former is assumed to be normally distributed with mean zero and variance σ_u^2 and the latter normally distributed with zero mean and variance σ^2. The subject random effects and the error random effects are assumed to be independent of each other and of time.

(As described, the random intercept model also implies that the response variable has, conditional on the explanatory variables, a normal distribution, so such models are suitable only for responses where this is at least approximately true; models for non-normal responses will be the subject of the next chapter.)

The model in Equation 13.1 allows the repeated measurements for an individual to vary about that individual's *own* regression line, which may differ in intercept but not in slope from the regression lines of other individuals. The random effects, u_i, model possible heterogeneity in the intercepts of the individuals whereas time has a fixed effect, β_1, hence the name "mixed effects models". Because the mean of the random effects is assumed to be zero, u_i represents the deviation of the ith individual's intercept ($\beta_0 + u_i$) from the population intercept, β_0.

The random intercept model implies that the total variance of each repeated measurement is $\sigma_u^2 + \sigma^2$ and also implies that the covariance at two time points j and k in the same individual is σ_u^2. So a random intercept model constrains the variance of each repeated measurement to be the same and the covariance between any pair of measurements to be equal. This is the compound symmetry structure met previously in Chapter 11.

13.2.2 Random Intercept and Slope Model

For most longitudinal data sets, measures taken closer to each other in time are likely to be more highly correlated than those taken further apart, and it is also often the case that the variances of the later repeated measurements are greater than those of the earlier observations. Where either or both of these features occur (as they do in many of longitudinal data sets), the random intercept model with its associated constraint of compound symmetry for the correlation structure of the repeated measurements will not provide an adequate model for the data, and an alternative that allows heterogeneity in both slopes and intercepts may be preferable. This model, the random intercept and random slope model, can be written as follows:

$$y_{ij} = \beta_0 + \beta_1 t_j + u_i + v_i t_j + \varepsilon_{ij}. \tag{13.2}$$

The model now includes two types of subject random effects, u_i, again modelling the heterogeneity in the intercepts and the new random effects, v_i, modelling heterogeneity in the slopes. Here, v_i represents the deviation of the *i*th individual's slope $(\beta_1 + v_i)$ from the population average, β_1.

The two random effects u_i and v_i are assumed to have a bivariate normal distribution with zero means for both variables, variances σ_u^2 and σ_v^2, and covariance σ_{uv}. With this model, the variance of a measurement at time, t_j, is $\sigma_u^2 + 2\sigma_{uv}t_j + \sigma_v^2 t_j^2 + \sigma^2$, and the covariance between two time points, t_j and t_k, is $\sigma_u^2 + \sigma_{uv}(t_j - t_k) + \sigma_v^2 t_j t_k$. So we see that this model does not require equal variances at each time point or equal covariances for each pair of time points. (The model does, of course, imply conditional normality for the response as with the simpler random intercept model.)

The parameters in both the random intercept model and the random intercept and random slope model can be estimated by maximum likelihood. However, this method tends to underestimate the variances of the random effects. As a consequence, a modified version of maximum likelihood, known as restricted maximum likelihood, which provides consistent estimates of the variance components, is often recommended as an alternative approach. Details are given in Diggle et al. (2002) and Longford (1993). Competing linear mixed effects models can be compared using a likelihood ratio test. If, however, the models have been estimated by restricted maximum likelihood, this test can only be used if both models have the same set of fixed effects (see Longford).

(It should be noted that re-estimating either of the models described above after adding or subtracting a constant from t_j, for example, the mean time, will lead to different variance and covariance estimates for the random effects, but it will not affect the fixed effects. Adding or subtracting a constant in this way is sometimes known as *centering* the times of measurement and can have implications in the interpretation of the random effects and their variances, see Fitzmaurice et al., 2004, for details.)

13.3 Analysis Using SAS

To begin, we shall construct some useful graphics to obtain some initial insights into the structure of the BtB data. The first graphic will be a series of box plots for each treatment group, one for each of the five observations of the BDI in the data. We begin by reading the data in and restructuring it so that we have both wide and long formats available.

```
data btb;
    infile 'c:\handbook3\datasets\BtB.dat' expandtabs missover;
    input idno drug$ Duration$ Treatment$ BDIpre BDI2m
      BDI3m BDI5m BDI8m;
run;
data btbl;
  set btb;
  array bdis {*} BDIpre -- BDI8m;
  array t {*} t1-t5 (0 2 3 5 8);
  do i=1 to 5;
    bdi=bdis{i};
    time=t{i};
    if i<5 and bdis{i+1}~=. then next=1;
    else next=0;
    if bdi~=. then output;
  end;
  drop BDI2m -- BDI8m t1-t5 i;
run;
```

We saw in the previous chapter how to use the array, iterative do and output statements to re-structure a data set. In that case, the observations were made at equal, monthly, intervals, whereas here they are not. To construct a time variable with the right values we set up a second array that contains those values. The array statement declares an array named t and creates five new variables, t1-t5, with their initial values specified in parentheses. These are then used within the do loop to assign the appropriate value to the variable time. For each time point, an additional variable, next, is set to 1 if the person attended on the following visit and 0 otherwise. This will be used later when we assess the impact of dropout. It is also worth noting that the pre-treatment assessment is represented in the resulting data set both as an observation at time 0 and by the original variable, BDIpre.

```
proc sort data=btbl;
  by treatment time;
run;
proc boxplot data=btbl;
  plot bdi*time/boxstyle=schematic continuous;
  by treatment;
run;
```

Elsewhere, we have used proc sgplot for box plots. Here, we use proc boxplot because the *x*-axis values are not equally spaced and boxplot has the continuous option on the plot statement to cater for this. The data are first sorted by treatment and time to produce separate plots for each treatment group and because the boxplot procedure also needs the data sorted by the *x*-axis values.

The resulting plot is shown in Figure 13.1. In both treatment groups, the Beck depression score decreases over time with some indication that the decrease is more in the BtB group. The depression scores of the patients in the BtB group are less

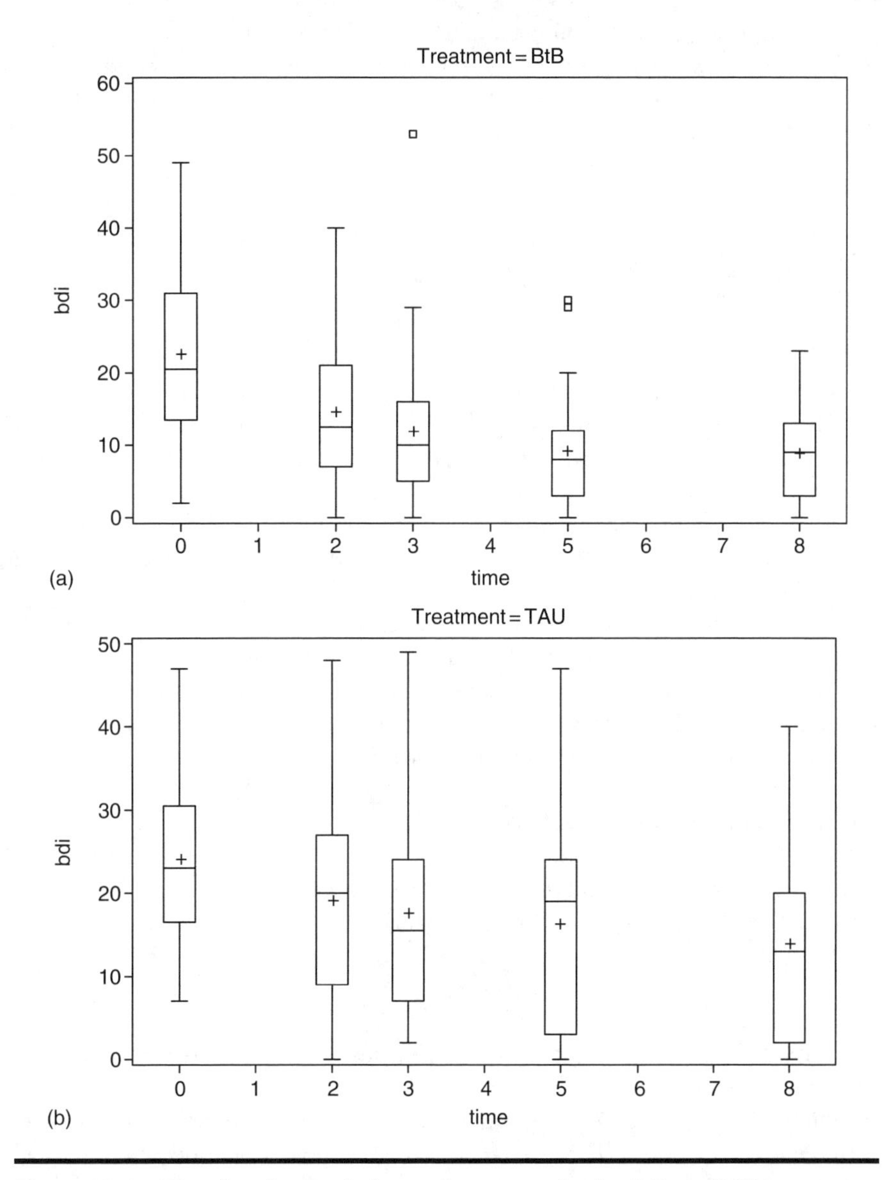

Figure 13.1 Box plots for Beck depression scores in the BtB and TAU treatment groups. (a) BtB, (b) TAU.

variable than those in the TAU group particularly later in the trial. At 3 and 5 months there is some evidence of outlier observations in the BtB group.

Next we will look at a scatterplot matrix of the five observations of the BDI because this may be helpful in suggesting what form the correlations between the repeated measures might take. To produce a scatterplot matrix for each treatment group, we use the "wide" format of the data set and sort it by treatment. The statistical graphics procedure, proc sgscatter, produces grids of comparative plots. For a scatterplot matrix, the matrix statement is used with a list of the variables to be plotted.

```
proc sort data=btb;
  by treatment;
run;
proc sgscatter data=btb;
  matrix BDIpre -- BDI8m;
  by treatment;
run;
```

Scatterplot matrices can also be produced using proc corr with ODS graphics.

The resulting plot is shown in Figure 13.2. In both groups, there appears to be a greater degree of correlation for the post-randomization measurements closer together in time than for those taken further apart.

We shall fit both random intercept and random intercept and slope models to the data, including the pre-BDI values, time, treatment group, drugs and duration as fixed effect covariates.

The data contain a number of missing values, and in applying proc mixed to the long form of the data set, these will be dropped from the analysis. But notice it is only the missing values that are removed, not participants who have at least one missing value. All the available data are used in the model fitting process. We begin by fitting the random intercept and slope model:

```
proc mixed data=btbl covtest noclprint=3;
  class drug duration treatment idno;
  model bdi=bdipre drug duration treatment time/s ddfm=bw;
  random int time /subject=idno type=un;
  where time>0;
run;
```

The covtest option in the proc statement requests significance tests for the covariance parameters (which is how SAS refers to random effects). The noclprint=3 option suppresses printing of class level information for class variables with three or more levels. This is useful because the subject identifier, idno, is listed as one of the class variables. The model statement specifies the regression equation in terms of the fixed effects. The specification of effects is the same as that for proc glm.

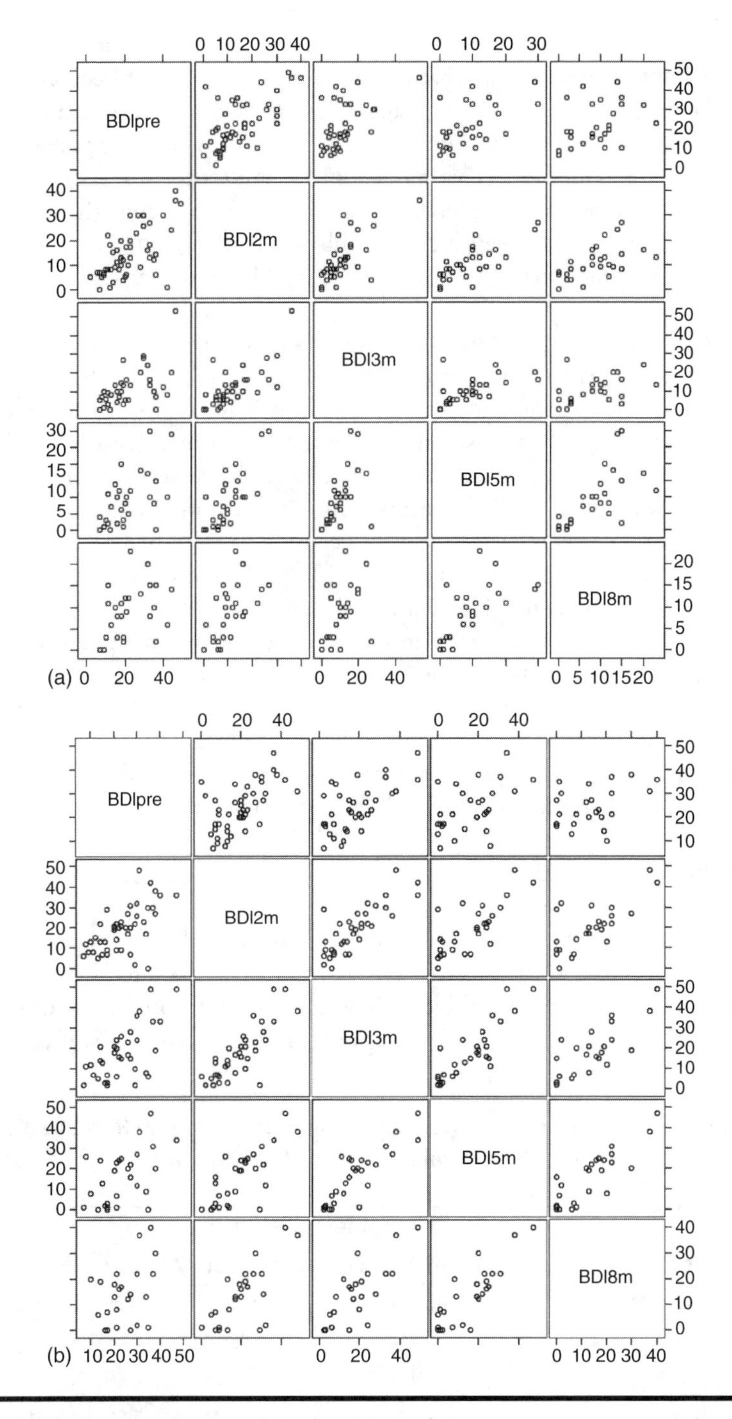

Figure 13.2 Scatterplot matrices for BtB and TAU treatment groups in the BtB trial. (a) BtB, (b) TAU.

The s option requests the solution, that is, the parameter estimates for the fixed effects, and ddfm = bw specifies that the denominator degrees of freedom for the *F* tests of the fixed effects should be calculated using the between–within method.

The random statement specifies which random effects are to be included in the model. For the random intercepts model, int (or intercept) is specified. Random slopes are specified by including time. The subject = option names the variable that identifies the subjects in the data set. If the subject identifier, idno in this case, is not declared in the class statement, the data set should be sorted into subject identifier order.

The type option in the random statement specifies the structure of the covariance matrix of the parameter estimates for the random effects. The default structure is type = vc (variance components), which models a different variance component for each random effect but constrains the covariances to zero. Unstructured covariances, type = un, allow separate estimation of each element of the covariance matrix. In this example, it allows an intercept–slope covariance to be estimated as a random effect, whereas the default would constrain this to be zero.

Because the pre-treatment value of the BDI is included as a covariate in the model, the observations for time 0 are excluded from the analysis by the where statement. The estimates of the variances and the covariance of the random effects and associated Wald tests for this model are shown in Table 13.2.

The results suggest that the variance of the random slope and the covariance of the two random effects can be assumed to be zero indicating that a simpler model with only a random intercept is probably adequate for these data. This random intercept model can be fitted by amending the random statement to

```
random int /subject = idno;
```

The results from fitting this model are given in Table 13.3. From a practical point of view, it is the estimates of the fixed effects parameters that are of most interest. Examining the relevant part of Table 13.3, we find that only pre-BDI and time have effects significant at the 5% level. The negative time effect reflects the decline in the

Table 13.2 Estimates of Variances and Covariances of Random Effects in the Random Intercept and Slope Model Fitted to the BtB Data

Covariance Parameter Estimates					
Cov Parm	**Subject**	**Estimate**	**Standard Error**	**Z Value**	**Pr Z**
UN(1,1)	idno	54.2469	13.7242	3.95	<.0001
UN(2,1)	idno	-0.6455	1.7013	-0.38	0.7044
UN(2,2)	idno	0.2101	0.3028	0.69	0.2439
Residual		23.9980	3.0027	7.99	<.0001

Table 13.3 Output from Proc Mixed when Fitting a Random Intercept Model for the BtB Data

Model Information	
Data Set	WORK.BTBL
Dependent Variable	bdi
Covariance Structure	Variance Components
Subject Effect	idno
Estimation Method	REML
Residual Variance Method	Profile
Fixed Effects SE Method	Model-Based
Degrees of Freedom Method	Between-Within

Class Level Information		
Class	**Levels**	**Values**
drug	2	n y
Duration	2	<6m >6m
Treatment	2	BtheB TAU
idno	97	not printed

Dimensions	
Covariance Parameters	2
Columns in X	9
Columns in Z Per Subject	1
Subjects	97
Max Obs Per Subject	4

Number of Observations	
Number of Observations Read	280
Number of Observations Used	280
Number of Observations Not Used	0

Iteration History			
Iteration	**Evaluations**	**-2 Res Log Like**	**Criterion**
0	1	1996.97865893	
1	2	1866.90340100	0.00000121
2	1	1866.90257247	0.00000000

Convergence criteria met.

Covariance Parameter Estimates					
Cov Parm	**Subject**	**Estimate**	**Standard Error**	**Z Value**	**Pr > Z**
Intercept	idno	51.9296	9.2405	5.62	<.0001
Residual		25.2867	2.6318	9.61	<.0001

Fit Statistics	
-2 Res Log Likelihood	1866.9
AIC (smaller is better)	1870.9
AICC (smaller is better)	1870.9
BIC (smaller is better)	1876.1

Solution for Fixed Effects								
Effect	**drug**	**Duration**	**Treatment**	**Estimate**	**Standard Error**	**DF**	**t Value**	**Pr > |t|**
Intercept				2.9368	3.1158	92	0.94	0.3484
BDIpre				0.6404	0.07992	92	8.01	<.0001
Drug	n			2.8160	1.7729	92	1.59	0.1156
Drug	y			0	.	.	.	.
Duration		<6m		-0.1790	1.6816	92	-0.11	0.9154
Duration		>6m		0	.	.	.	.
Treatment			BtheB	-2.3151	1.7151	92	-1.35	0.1804
Treatment			TAU	0	.	.	.	.
time				-0.7016	0.1469	182	-4.77	<.0001

Type 3 Tests of Fixed Effects				
Effect	**Num DF**	**Den DF**	**F Value**	**Pr > F**
BDIpre	1	92	64.21	<.0001
drug	1	92	2.52	0.1156
Duration	1	92	0.01	0.9154
Treatment	1	92	1.82	0.1804
time	1	182	22.80	<.0001

BDI scores in both TAU and BtB groups seen in the box plots in Figure 13.1. The positive pre-BDI effect simply demonstrates that pre-BDI score is significantly related to the post-treatment BDI scores. The effects for treatment, drugs and duration are not significant. So, in particular, there is no clear evidence for a treatment effect although the estimated treatment effect of -2.32 associated with the BtB group is perhaps suggestive that the treatment may be contributing to some small decrease in average depression for the patients in this group. The two random effects in the models have variances that differ significantly from zero.

We now need to consider briefly how the dropouts may affect the analyses reported above. To understand the problems that patients' dropping out can cause for the analysis of data from a longitudinal trial, we need to consider a classification of dropout mechanisms first introduced by Rubin (1976). The type of mechanism involved has implications for which approaches to analysis are suitable and which are not. Rubin's suggested classification involves three types of dropout mechanisms:

1. *Dropout completely at random* (DCAR): The probability that a patient drops out does not depend on either the observed or missing values of the response. Consequently, the observed (non-missing) values effectively constitute a simple random sample of the values for all subjects. Possible examples include missing laboratory measurements because of a dropped test tube (if it was not dropped because of the knowledge of any measurement), the accidental death of a participant in a study or a participant's moving to another area. Intermittent missing values in a longitudinal data set, whereby a patient misses a clinic visit for transitory reasons ("went shopping instead" or the like), can reasonably be assumed to be DCAR. Completely random dropout causes the fewest problem for data analysis, but it is a strong assumption.
2. *Dropout at random* (DAR): The DAR mechanism occurs when the probability of dropping out depends on the outcome measures that have been observed in the past, but given this information is conditionally independent of all the future (unrecorded) values of the outcome variable following dropout. Here, "missingness" depends only on the observed data, with the distribution of future values for a subject who drops out at a particular time being the same as the distribution of the future values of a subject who remains in at that time, if they have the same covariates and the same past history of outcome up to and including the specific time point. Murray and Findlay (1988) provide an example of this type of missing value from a study of hypertensive drugs in which the outcome measure was diastolic blood pressure. The protocol of the study specified that the participant was to be removed from the study when his or her blood pressure got too high. Here blood pressure at the time of dropout was observed before the participant dropped out, so the dropout mechanism is not DCAR because it depends on the values of blood pressure, but it is DAR

because dropout depends only on the observed part of the data. A further example of a DAR mechanism is provided by Heitjan (1997) and involves a study in which the response measure is body mass index (BMI). Suppose that the measure is missing because subjects who had high BMI values at earlier visits avoided being measured at later visits out of embarrassment, regardless of whether they had gained or lost weight in the intervening period. The missing values here are DAR but not DCAR; consequently, methods applied to the data that assumed the latter might give misleading results.

3. *Non-ignorable* (sometimes referred to as informative): The final type of dropout mechanism is one where the probability of dropping out depends on the unrecorded missing values – observations are likely to be missing when the outcome values that would have been observed had the patient not dropped out are systematically higher or lower than usual (corresponding perhaps to their condition's becoming worse or improving). A non-medical example is when individuals with lower income levels or very high incomes are less likely to provide their personal income in an interview. In a medical setting, possible examples are a participant dropping out of a longitudinal study when his or her blood pressure became too high and this value was not observed, or when his or her pain become intolerable and we did not record the associated pain value. For the BDI example introduced earlier, if subjects were more likely to avoid being measured when they had put on extra weight since the last visit, then the data are non-ignorably missing. Dealing with data containing missing values that result from this type of dropout mechanism is difficult. The correct analyses for such data must estimate the dependence of the missingness probability on the missing values. Models and software that attempt this are available (see, e.g., Diggle and Kenward, 1994), but their use is not routine and, in addition, it must be remembered that the associated parameter estimates can be unreliable.

Under what type of dropout mechanism are the mixed effects models considered in this chapter valid? The good news is that such models can be shown to give valid results under the relatively weak assumption that the dropout mechanism is DAR (see Carpenter et al., 2002). When the missing values are thought to be informative, any analysis is potentially problematical. But Diggle and Kenward (1994) have developed a modelling framework for longitudinal data with informative dropouts, in which random or completely random dropout mechanisms are also included as explicit models. The essential feature of the procedure is a logistic regression model for the probability of dropping out, in which the explanatory variables can include previous values of the response variable and, in addition, the unobserved value at dropout as a latent variable (i.e., an unobserved variable). In other words, the dropout probability is allowed to depend on both the observed measurement history and the unobserved value at dropout. This allows both a formal assessment of the type of dropout mechanism in the data and the estimation of effects of interest, for example, treatment effects under different assumptions about the dropout mechanism. A full technical account of the model is given in Diggle and Kenward (1994), and a detailed example that uses the approach is described in Carpenter et al.

One of the problems for an investigator struggling to identify the dropout mechanism in a data set is that there are no routine methods to help, although a number of largely ad hoc graphical procedures can be used, as described in Diggle (1998), and Carpenter et al. (2002). One very simple procedure for assessing the dropout mechanism suggested in Carpenter et al. involves plotting the values of the response variable for the subjects in each for each treatment group, at each time point (including pre-randomization), differentiating on each plot between two categories of patient, those who do and those who do not attend their next scheduled visit. Any clear difference between the distributions of values for the two categories indicates that dropout is not completely at random.

```
data btbl;
  set btbl;
  nexttime = time+next*.2;
run;
proc sgpanel data = btbl;
  panelby treatment/rows = 2;
  scatter y = bdi x = nexttime/group = next;
  where time < 8;
run;
```

First we create a new variable, nexttime, which combines the time of the visit and whether the patient attended the subsequent visit. This will serve to separate the plot values slightly to avoid their overlapping. Then sgpanel is used for the two scatterplots using the nexttime as the *x*-axis variable and separate plotting by attendance or not at the next visit.

Figure 13.3 suggests little difference in the distribution of BDI scores between patients who drop out at the next visit and those who do not, a pattern that is consistent with a DCAR mechanism giving some reason for encouragement that the results from fitting the random effects models are valid.

It should be mentioned here that one approach to the dropout problem in longitudinal data is imputation, that is, to substitute or fill-in the values that were not recorded with imputed values. The attraction of this approach is that once a filled-in data set has been constructed, standard methods for complete data can be applied. But this seeming advantage can be illusionary, unless the method of imputation is appropriate (see Fitzmaurice et al., 2004, for a full discussion).

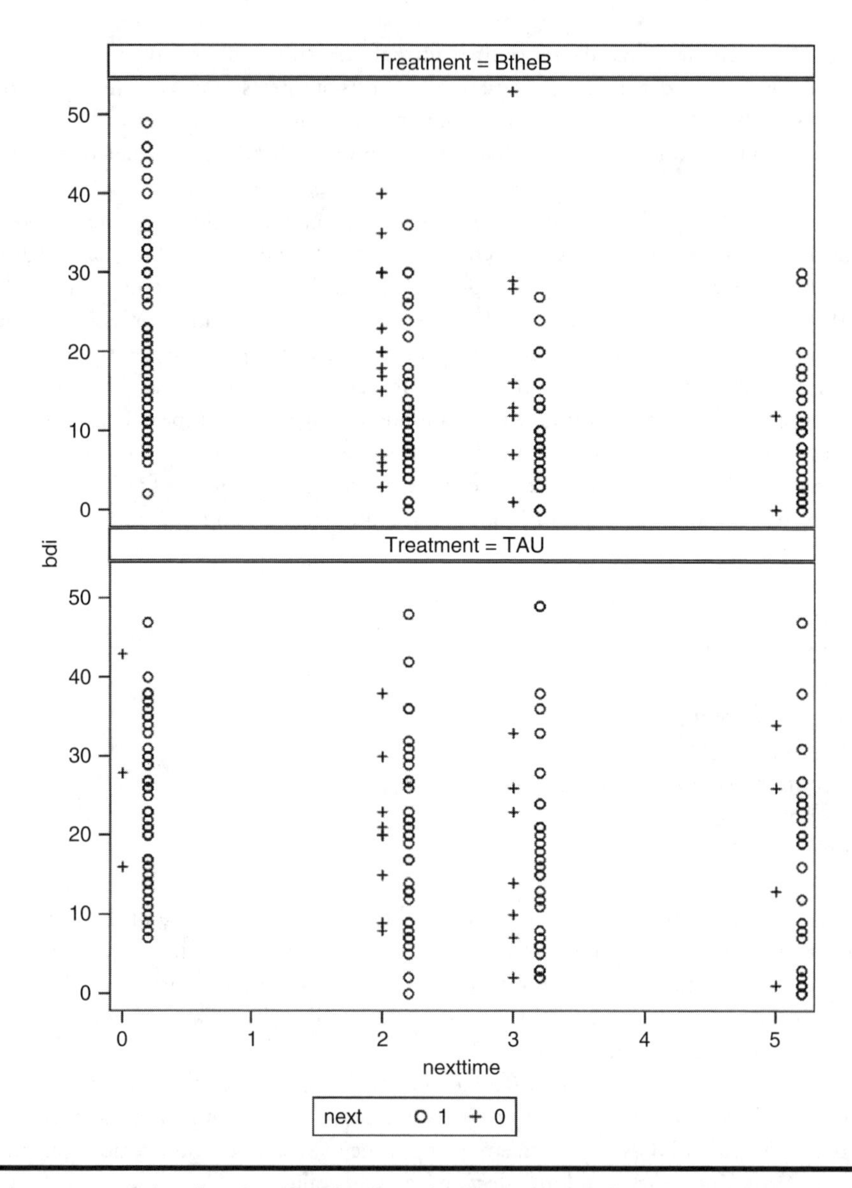

Figure 13.3 Plot showing distribution of BDI scores for those patients who drop out of the study at the next visit compared with those who do not drop out.

The analysis of longitudinal data is not really complete without an examination of suitable residuals. But residuals for longitudinal data require more careful attention than those in use when fitting independent observations, say, by multiple regression (see Chapters 6 and 7), and so we will simply refer readers to Fitzmaurice et al. (2004) for a description of the possibilities.

Exercises

1 Beat the Blues

1. Use `proc glm` to fit a model to the data that assume the repeated measurements are independent. Compare the results to those from fitting the random intercept model as described in the chapter.
2. Investigate whether there is any evidence of an interaction between treatment and time for the BtB data.

2 Phosphate Data

1. Using the plots on the phosphate data asked for in Section 12.4.2, fit what you think may be sensible linear mixed effects models for the data.
2. Comment on your results.

3 Post-natal Depression

1. Fit a random intercept and slope model to the post-natal depression data given in Chapter 12.
2. Summarize your conclusions.

Chapter 14

Longitudinal Data III: Generalized Estimating Equations and Generalized Mixed Models: Treating Toenail Infection

14.1 Introduction

De Backer et al. (1998) describe a clinical trial to compare two competing oral anti-fungal treatments for toenail infection (*dermatophyte onychomycosis*). A total of 378 patients were randomly allocated into two treatment groups, one group receiving 250 mg/day of terbinafine and the other group 200 mg/day of itraconazole. Patients were evaluated at seven visits, intended to be at weeks 0, 4, 8, 12, 24, 36 and 48, for the degree of separation of the nail plate from the nail bed (*onycholysis*) dichotomized into "moderate or severe" (coded as 1) and "none or mild" (coded as 0). But patients did not always arrive exactly at the scheduled time, and the exact time in months that they did attend was recorded. The data are not balanced since not all patients attended for all seven planned visits; the second subject in Table 14.1, for

Table 14.1 Data for the First Two Patients in the Clinical Trial of Treatments for Toenail Infection

Patient	*Outcome*	*Treatment*	*Month*	*Visit*
1	1	1	0	1
1	1	1	0.8571429	2
1	1	1	3.535714	3
1	0	1	4.535714	4
1	0	1	7.535714	5
1	0	1	10.03571	6
1	0	1	13.07143	7
2	0	0	0	1
2	0	0	0.9642857	2
2	1	0	2	3
2	1	0	3.035714	4
2	0	0	6.5	5
2	0	0	9	6

Note: Treatment: 0 = itraconazole, 1 = terbinafine; Month = exact timing of visit in months.

example, only attended for the first six visits. "Long form" data for the first two patients are shown in Table 14.1. The main question of interest about the data is whether there is any difference in the responses of patients in the two treatment groups.

14.2 Methods for Analysing Longitudinal Data Where the Response Variable Cannot Be Assumed to Have a Normal Distribution

The longitudinal data generated by the clinical trial of treatments for toenail infection differ in an important respect from the longitudinal data considered in the previous chapter, namely that the response variable is dichotomous making models assuming normality for this variable of little use. In the linear mixed models for Gaussian responses described in Chapter 13, estimation of the regression parameters linking explanatory variables to the response variable and their standard errors needed to take account of the correlational structure of the data, but their interpretation could be undertaken independent of this structure. When modelling non-normal responses, this independence of estimation and interpretation no longer holds. Different assumptions about how the correlations are generated can lead to regression coefficients with different interpretations. The essential difference is between *marginal models* and *conditional models*.

14.2.1 Marginal Models

Longitudinal data can be considered as a series of cross-sections, and marginal models for such data use the generalized linear model (GLM, see Chapter 9) to fit each cross-section. In this approach, the relationship of the marginal mean and the explanatory variables is modelled separately from the within-subject correlation. The marginal regression coefficients have the same interpretation as coefficients from a cross-sectional analysis, and marginal models are natural analogues for correlated data of generalized linear models for independent data. Fitting marginal models to non-normal longitudinal data involves the use of a procedure known as *generalized estimating equations* (GEE), introduced by Liang and Zeger (1986). This approach may be viewed as a multivariate extension of the generalized linear model and the quasi-likelihood method (see Chapter 9). But the problem with applying a direct analogue of the generalized linear model to longitudinal data with non-normal responses is that there is usually no suitable likelihood function with the required combination of the appropriate link function, error distribution and correlation structure. To overcome this problem, Liang and Zeger introduced a general method for incorporating within-subject correlation in GLMs, which is essentially an extension of the quasi-likelihood approach mentioned briefly in Chapter 9. As in conventional generalized linear models, the variances of the responses, given the covariates, are assumed to be of the form $V(y) = \phi V(\mu)$, where the variance function is determined by the choice of distribution family (see Chapter 9). Because overdispersion is common in longitudinal data, the dispersion parameter ϕ is typically estimated even if the distribution requires $\phi = 1$. The feature of GEE that differs from the usual generalized linear model is that different responses on the same individual are allowed to be correlated, given the covariates. These correlations are assumed to have a relatively simple structure parameterized by a small number of parameters. The following correlation structures are commonly used (Y_{ij} represents the value of the jth repeated measurement of the response variable on subject i):

1. Identity matrix leading to the independence working model in which the generalized estimating equation reduces to the univariate estimating equation given in Chapter 9, obtained by assuming that the repeated measurements are independent.
2. Exchangeable correlation matrix with a single parameter similar to that described in Chapter 12. Here the correlation between each pair of repeated measurements is assumed to be the same, that is, $\text{corr}(Y_{ij}, Y_{ik}) = \alpha$.
3. AR-1 autoregressive correlation matrix, also with a single parameter, but in which $\text{corr}(Y_{ij}, Y_{ik}) = \alpha^{|k-j|}$, $j \neq k$. This can allow the correlations of measurements taken further apart to be less than those taken closer to one another.
4. Unstructured correlation matrix with $T(T-1)/2$ parameters, where T is the number of repeated measurements and $\text{corr}\ (Y_{ij}, Y_{ik}) = \alpha_{jk}$.

For given values of the regression parameters $\beta_1, \ldots, \beta_p$, the α-parameters of the working correlation matrix can be estimated along with the dispersion parameter ϕ (see Zeger and Liang, 1986, for details). These estimates can then be used in the so-called generalized estimating equations to obtain estimates of the regression parameters. The GEE algorithm proceeds by iterating between (1) estimation of the regression parameters using the correlation and dispersion parameters from the previous iteration and (2) estimation of the correlation and dispersion parameters using the regression parameters from the previous iteration.

The estimated regression coefficients are "robust," in the sense that they are consistent from misspecified correlation structures assuming that the mean structure is correctly specified. Note, however, that the GEE estimates of marginal effects are not robust against misspecified regression structures, such as omitted covariates.

14.2.2 Conditional Models

The random effects approach described in the previous chapter can be extended to non-normal responses, although the resulting models can be difficult to estimate because the likelihood involves integrals over the random effects distribution that generally do not have closed forms. A consequence is that it is often only possible to fit relatively simple models. In these models, estimated regression coefficients have to be interpreted, *conditional* on the random effects. The regression parameters in the model are said to be *subject specific*, and such effects will differ from the marginal or population-averaged effects estimated using GEE, except when using an identity link function and a normal error distribution.

For a set of longitudinal data in which y_{ij} is the value of a binary response for individual i at, say, time t_j, the logistic regression model (see Chapter 8) for the response is now written as

$$\text{logit}[\Pr(y_{ij} = 1|u_i)] = \beta_0 + \beta_1 t_j + u_i, \tag{14.1}$$

where u_i is a random effect assumed to be normally distributed with zero mean and variance σ_u^2. This is a simple example of a generalized linear mixed model because it is a generalized linear model with both a fixed effect, β_1, and a random effect, u_i.

Here the regression parameter β_1 again represents the change in the log odds per unit change in time, but this is now conditional on the random effect, u_i. We can illustrate this difference graphically by simulating the model in Equation 14.1; the result is shown in Figure 14.1.

Here the thin curves represent subject-specific relationships between the probability that the response equals one and a covariate x for model in Equation 14.1. The horizontal shifts are due to different values of the random intercept. The thick curve represents the population-averaged relationship, formed by averaging the thin

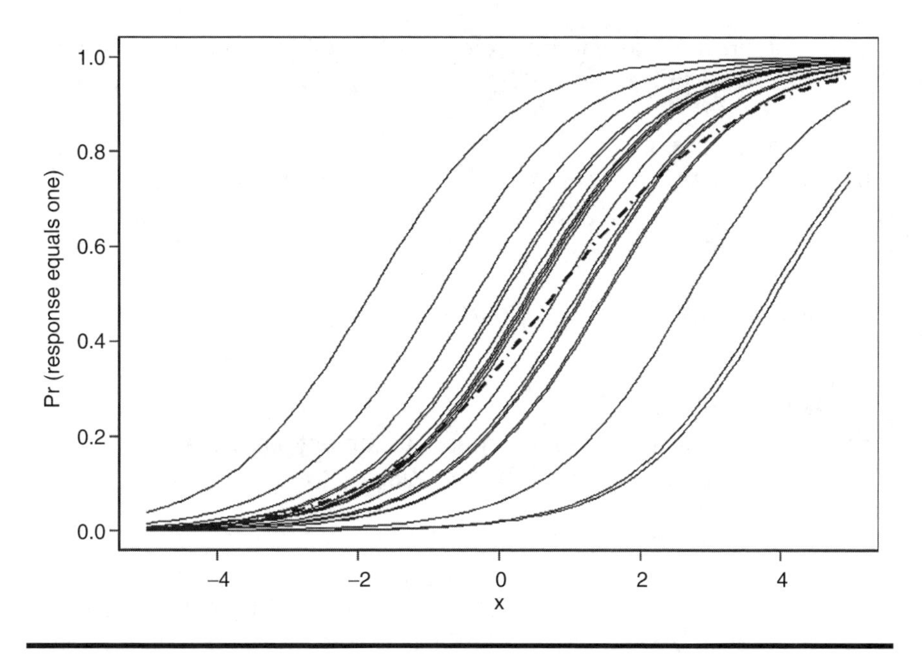

Figure 14.1 Simulation of a positive response in a random intercept logistic regression model for 20 subjects. The dotted line is the average over all 20 subjects.

curves for each value of x. It is, in effect, the dotted line that would be estimated in a marginal model (see Section 14.2.1). The population-averaged regression parameters tend to be attenuated (closest to zero) relative to the subject-specific regression parameters. A marginal regression model does not address questions concerning heterogeneity between individuals. Estimating the parameters in a random effects model is again generally undertaken by maximum likelihood.

14.3 Analysis Using SAS

As always, it will be useful to begin our investigation of the toenail data by plotting the data as a line graph, plotting the observed proportions at each visit against time, the average time associated with each visit.

```
data toenail;
  infile 'c:\handbook3\datasets\toenail.dat' expandtabs;
  input patient outcome treatment month visit;
run;
```

The raw data are in the long format, so they can be read straightforwardly without the need for restructuring.

```
proc sort data=toenail;
  by treatment visit;
run;
proc means data=toenail;
  var month outcome;
  output out=means mean=;
  by treatment visit;
run;
proc sgplot data=means;
  series y=outcome x=month/group=treatment markers;
  yaxis label='proportion with onycholysis';
run;
```

To obtain the proportion with onycholysis and the average time of each visit for the two treatment groups, we first sort the data by treatment and visit. Together, these two variables define 14 groups in the data set: 7 visits for each of the two treatment groups. Using proc means with the same by statement calculates the means separately for each of these groups. The var statement specifies the variables for which means are to be calculated, the default being all numeric variables. The output statement creates a data set, means, and the mean= option specifies that it is to contain variables with the means in (other summary statistics could be calculated). The = may be followed by new names for the output variables – by default they are given the same names as the original variables. Note that the = is required even if the default names are being used. The by variables are automatically included in the output data set, which can then be used with proc sgplot to produce the plot. The yaxis statement is used to label the vertical axis.

The resulting plot is shown in Figure 14.2. The diagram appears to indicate that the proportion of patients with onycholysis decreases with time in both treatment groups and that the proportion is less in the group treated with terbinafine, particularly after the fourth visit.

14.3.1 Fitting a Logistic Regression Model Assuming Independence of the Responses

We shall begin modelling the toenail data with an unrealistic logistic model, namely one that assumes the responses for a given subject are mutually independent after controlling for the included covariates. Such a model, including both explanatory variables, treatment and time of visit, can be fitted in the same way as described in Chapter 8 using proc logistic applied to the long form of the toenail data.

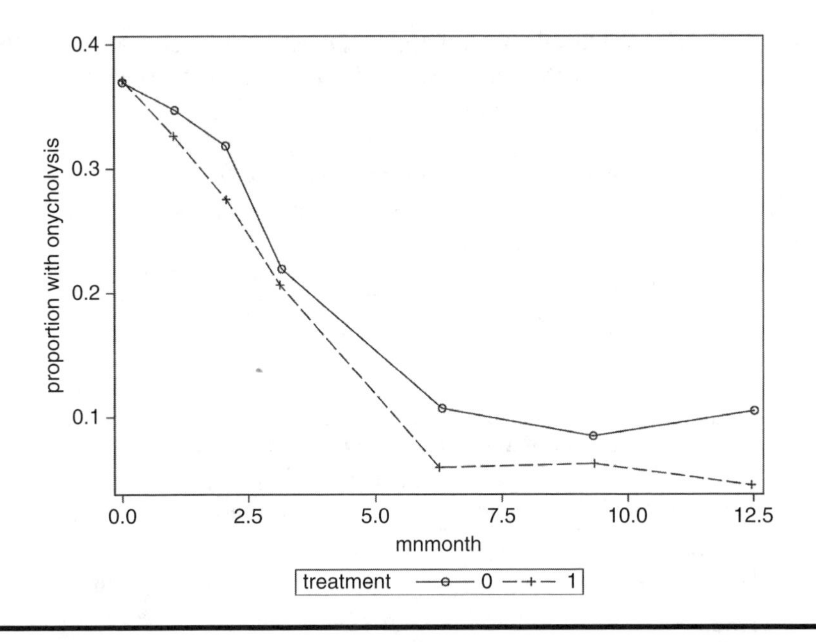

Figure 14.2 Line plot of patients with toenail infection by visit and treatment group.

```
proc logistic data=toenail desc;
  model outcome=treatment month/clodds=wald;
run;
```

The results are shown in Table 14.2. The three tests that assess the hypothesis that both regression coefficients in the model are zero each have a very small associated p value indicating that the hypothesis should be rejected. Examination of the significance tests for each regression coefficient suggests that treatment group is not significant but time since randomization is. This is confirmed by looking at the confidence intervals for the odds ratios extracted from the estimated regression coefficients (see Chapter 8). The confidence interval for treatment group includes the value 1, so there is no evidence of a treatment effect. But the corresponding interval for month does not include the value 1 and it appears that the odds of onycholysis during any month is between 0.790 and 0.849 times the odds for the previous month.

We might also be interested in a model that includes a treatment × time interaction; this model can be written explicitly as

$$\text{logit}\{\Pr(y_{ij} = 1|x_{2j}, x_{3ij})\} = \beta_0 + \beta_1 x_{2j} + \beta_2 x_{3ij} + \beta_3 x_{2j} x_{3ij}, \qquad (14.1)$$

where y_{ij} is the outcome at visit i for patient j, x_{2j} is the dummy variable for treatment group, and x_{3ij} is the variable month.

Table 14.2 Selected Results from Fitting a Logistic Regression Model with Covariates Time and Treatment Group to the Toenail Data Assuming Independence of a Patient's Responses

Testing Global Null Hypothesis: BETA=0			
Test	**Chi-Square**	**DF**	**Pr > ChiSq**
Likelihood Ratio	161.1995	2	<.0001
Score	138.3987	2	<.0001
Wald	120.6816	2	<.0001

Analysis of Maximum Likelihood Estimates					
Parameter	**DF**	**Estimate**	**Standard Error**	**Wald Chi-Square**	**Pr > ChiSq**
Intercept	1	-0.4690	0.0969	23.4167	<.0001
treatment	1	-0.1891	0.1162	2.6470	0.1037
month	1	-0.1995	0.0183	118.6294	<.0001

Odds Ratio Estimates			
		95% Wald	
Effect	**Point Estimate**	**Confidence Limits**	
treatment	0.828	0.659	1.039
Month	0.819	0.790	0.849

This model allows for a difference between groups at baseline and linear changes in the log odds of onycholysis over time with slope β_2 in the itraconazole group and slope $\beta_2 + \beta_3$ in the terbinafine-treated group. So β_3, the difference in the rate of improvement (on the log odds scale) between treatment groups can be viewed as the treatment effect, terbinafine versus itraconazole. (Remember that here the model considered is making the unrealistic assumption that the responses for a given patient are independent, given the covariates.)

The interaction model can be fitted as follows:

```
proc logistic data = toenail desc;
  model outcome = month|treatment/clodds = wald;
run;
```

Table 14.3 Selected Results from Fitting a Logistic Regression Model with Covariates Time, Treatment Group and Their Interaction to the Toenail Data Assuming Independence of a Patient's Responses

Testing Global Null Hypothesis: BETA=0			
Test	**Chi-Square**	**DF**	**Pr > ChiSq**
Likelihood Ratio	164.4721	3	<.0001
Score	138.7971	3	<.0001
Wald	118.7319	3	<.0001

Analysis of Maximum Likelihood Estimates					
Parameter	**DF**	**Estimate**	**Standard Error**	**Wald Chi-Square**	**Pr > ChiSq**
Intercept	1	-0.5566	0.1090	26.0959	<.0001
month	1	-0.1703	0.0236	51.9890	<.0001
treatment	1	-0.00059	0.1561	0.0000	0.9970
month*treatment	1	-0.0672	0.0375	3.2089	0.0732

The results of fitting this model are shown in Table 14.3. The results for the individual regression coefficients in the model show that there is no evidence that β_1 differs from zero, strong evidence that β_2 does differ from zero and perhaps some weak evidence that β_3 also differs from zero. If we exponentiate the estimate of β_2, we get a value of 0.844 for the corresponding odds ratio, and doing the same for β_3 we obtain an odds ratio of 0.935. Consequently, the odds ratio is estimated as 0.844 per month in the itraconazole group and $0.844 \times 0.935 = 0.836$ in the terbinafine group. These estimates correspond to what we have seen in the line plot given in Figure 14.2, namely an increasing difference in the proportions of patients with onycholysis in the two treatment groups as time goes on. We have to remember, however, that the 95% confidence interval for β_3 is (0.869, 1.006) (see Chapter 8 for details of the calculations needed), so it might well be that the odds ratios per month are the same for both treatment groups (i.e., if β_3 is taken as 1) and that there is really no convincing evidence of a treatment $\times$ month interaction.

14.3.2 Using the GEE Approach on the Toenail Data

Now we can move on to using GEE methodology on the toenail data. We will fit the logistic regression model that includes both treatment and time and their interaction, and allow for departure from independence of the repeated measurements by specifying an exchangeable correlation structure. The necessary SAS code is

```
proc sort data=toenail;
  by patient visit;
run;
proc genmod data=toenail desc;
  class patient treatment visit;
  model outcome=month|treatment/d=b;
  repeated subject=patient/type=exch within=visit;
run;
```

The use of proc genmod to fit generalized linear models was described in Chapter 9. Extending the procedure to cover GEE models is done by including the repeated statement and specifying the variable that identifies the subjects, patient in this case. This variable must be named in a class statement. The structure of the working correlation matrix is specified with type= option. Other structures commonly used for longitudinal data are autoregressive (ar) and unstructured (un). The order in which the repeated measurements were made should be specified with the withinsubject (within=) option and the variable must also be named in a class statement. When the data are in the correct order, as they are here, this option might still be important if subjects had measurements missing in the middle of a sequence.

The results are shown in Table 14.4. The parameter estimates given in Table 14.4 are very similar to those for the independence model (see Table 14.3), but their

Table 14.4 Results from Using GEE to Fit a Logistic Regression Model Including Treatment, Time and Their Interaction and an Exchangeable Correlation Structure to the Toenail Data

Model Information	
Data Set	WORK.TOENAIL
Distribution	Binomial
Link Function	Logit
Dependent Variable	outcome

Number of Observations Read	1908
Number of Observations Used	1908
Number of Events	408
Number of Trials	1908

Class Level Information		
Class	**Levels**	**Values**
patient	294	1 2 3 4 6 7 9 10 11 12 13 15 16 17 18 19 20 21 22 23 24 25 28 29 30 31 33 35 37 38 39 40 41 45 48 49 50 51 52 53 54 55 56 58 59 60 61 63 64 65 66 68 69 70 72 73 75 76 78 79 80 81 82 83 84 85 86 87 88 89 90 93 94 95 96 97 99 101 102 104 105 106 107 108 ...
treatment	2	0 1
visit	7	1 2 3 4 5 6 7

Response Profile		
Ordered Value	**outcome**	**Total Frequency**
1	1	408
2	0	1500

PROC GENMOD is modeling the probability that outcome='1'.

Parameter Information		
Parameter	**Effect**	**treatment**
Prm1	Intercept	
Prm2	month	
Prm3	treatment	0
Prm4	treatment	1
Prm5	month*treatment	0
Prm6	month*treatment	1

Algorithm converged.

GEE Model Information	
Correlation Structure	Exchangeable
Within-Subject Effect	visit (7 levels)
Subject Effect	Patient (294 levels)
Number of Clusters	294
Correlation Matrix Dimension	7
Maximum Cluster Size	7
Minimum Cluster Size	1

Algorithm converged.

Exchangeable Working Correlation	
Correlation	0.421202978

GEE Fit Criteria	
QIC	1838.3787
QICu	1824.5878

Analysis Of GEE Parameter Estimates
Empirical Standard Error Estimates

Parameter		Estimate	Standard Error	95% Confidence Limits		Z	Pr > |Z|
Intercept		-0.5747	0.1942	-0.9553	-0.1940	-2.96	0.0031
month		-0.2490	0.0450	-0.3373	-0.1607	-5.53	<.0001
treatment	0	-0.0072	0.2595	-0.5157	0.5013	-0.03	0.9779
treatment	1	0.0000	0.0000	0.0000	0.0000	.	.
month*treatment	0	0.0777	0.0541	-0.0283	0.1838	1.44	0.1509
month*treatment	1	0.0000	0.0000	0.0000	0.0000	.	.

estimated standard errors are somewhat larger. Assuming a patient's responses are independent, given the covariates in the model, leads to unrealistic precision in the estimated regression parameters. But, this said, the conclusions from the GEE model are similar to those from the independence model as readers will be able to see if they go through the calculation of odds ratios, etc., as detailed earlier for the independence model.

14.3.3 Random Effects Model for the Toenail Data

In this section we shall fit the random effects logistic regression model described earlier in the chapter to the toenail data. Again we shall include treatment, time and their interaction in the model. Here the random intercept represents the combined effect of all omitted subject-specific (time-constant) covariates that cause some subjects to be more prone to onycholysis than others. The model is applied as follows:

```
proc glimmix data=toenail noclprint;
  class patient;
  model outcome(desc)=month|treatment/d=binary s ddfm=bw;
  random int/subject=patient;
run;
```

Proc glimmix has a similar syntax to proc mixed with additional options to cover distributions other than the normal distribution. The distribution is specified with the d = option in the model statement. The solution (s) option gives the parameter estimates for the fixed effects, and ddfm = bw specifies the between–within method of calculating the denominator degrees of freedom, which is suitable for longitudinal data. For binary responses, the desc option can be specified in parentheses after the response variable to reverse the default ordering. The noclprint option in the proc statement is used to suppress the listing of the patient ids.

The results are shown in Table 14.5. Concentrating on the estimates of the regression coefficients and transforming these to odds ratios in the usual way we find the following:

- Month: OR = 0.757, 95% CI [0.711,0.806]
- Treatment: OR = 0.974, 95% CI [0.504,1.88]
- Treatment × month: OR = 0.909, 95% CI [0.822,1.004]

Table 14.5 Results of Fitting a Random Intercept Logistic Regression Model Including Time, Treatment, and Their Interaction to the Toenail Data

Model Information	
Data Set	WORK.TOENAIL
Response Variable	outcome
Response Distribution	Binary
Link Function	Logit
Variance Function	Default
Variance Matrix Blocked By	patient
Estimation Technique	Residual PL
Degrees of Freedom Method	Between-Within

Number of Observations Read	1908
Number of Observations Used	1908

Response Profile		
Ordered Value	**outcome**	**Total Frequency**
1	1	408
2	0	1500
The GLIMMIX procedure is modelling the probability that outcome='1'.		

Dimensions	
G-side Cov. Parameters	1
Columns in X	4
Columns in Z per Subject	1
Subjects (Blocks in V)	294
Max Obs per Subject	7

Optimization Information	
Optimization Technique	Newton-Raphson with Ridging
Parameters in Optimization	1
Lower Boundaries	1
Upper Boundaries	0
Fixed Effects	Profiled
Starting From	Data

Iteration History

Iteration	Restarts	Subiterations	Objective Function	Change	Max Gradient
0	0	5	8519.193785	0.95204444	8.203E-7
1	0	4	9475.4897865	0.47981085	0.000014
2	0	4	10396.052137	0.20593141	1.261E-9
3	0	3	10932.849196	0.07453779	1.14E-7
4	0	2	11106.554516	0.02771321	0.000012
5	0	2	11147.349584	0.00667530	4.32E-8
6	0	1	11156.554229	0.00146148	0.000048
7	0	1	11158.602561	0.00032446	2.402E-6
8	0	1	11159.059349	0.00007181	1.181E-7
9	0	1	11159.160547	0.00001588	5.779E-9
10	0	1	11159.182933	0.00000351	2.83E-10
11	0	1	11159.187884	0.00000078	1.39E-11
12	0	0	11159.188978	0.00000000	3.673E-6

Convergence criterion (PCONV=1.11022E-8) satisfied.

Fit Statistics	
-2 Res Log Pseudo-Likelihood	11159.19
Generalized Chi-Square	1489.85
Gener. Chi-Square / DF	0.78

Covariance Parameter Estimates			
Cov Parm	**Subject**	**Estimate**	**Standard Error**
Intercept	Patient	4.7095	0.6024

Solutions for Fixed Effects					
Effect	**Estimate**	**Standard Error**	**DF**	**t Value**	**Pr > ltl**
Intercept	-0.7204	0.2370	292	-3.04	0.0026
month	-0.2782	0.03222	1612	-8.64	<.0001
treatment	-0.02594	0.3360	292	-0.08	0.9385
month*treatment	-0.09583	0.05105	1612	-1.88	0.0607

Type III Tests of Fixed Effects				
Effect	**Num DF**	**Den DF**	**F Value**	**Pr > F**
month	1	1612	74.57	<.0001
treatment	1	292	0.01	0.9385
month*treatment	1	1612	3.52	0.0607

The significance of the effects as estimated by this random effects model and by the GEE model used previously is generally similar although there are substantial differences in the parameter estimates themselves and their associated confidence intervals. This is to be expected since in the GEE model it is marginal effects that are being estimated and in the random effects model, effects conditional on each patient's random effect (see Figure 14.1). The regression coefficients estimated in the random effects logistic regression model measure the change in the log odds of response per unit increase in the corresponding covariate for any given individual having some unobservable underlying propensity to respond positively, u_i. Generalized linear mixed effect models are most useful when the main scientific objective is to make inferences about individuals rather than the population averages; the population averages are the targets of inference in marginal models.

Exercises

1 *Toenail Data*

1. Fit GEE models to the data with alternative correlational structures to the one used in the text and compare results.
2. Fit a generalized linear mixed model to the data including both a random intercept and random slope.

2 *Respiratory Data*

Davis (1991) reports the results from a clinical trial comparing two treatments for a respiratory illness. In each of two centres, eligible patients were randomly assigned to active treatment or placebo. During treatment, the respiratory status (categorized as 0 = poor, 1 = good) was determined at each of the four visits. A total of 111 patients were involved in the trial, 54 in the active group and 57 in the placebo group. The sex and age of each participant was also recorded along with a baseline respiratory status.

The variables in resp.dat are

- **ID:** Subject identifier
- **Cent:** Centre where patient was treated
- **Treat:** Treatment patient received, 1 = placebo, 2 = active
- **Sex:** Sex of patients, 1 = male, 2 = female
- **Age:** Age of patient in years
- **BL:** Baseline respiratory status
- **V1:** Respiratory status at first visit after treatment has begun
- **V2:** Respiratory status at second visit after treatment has begun
- **V3:** Respiratory status at third visit after treatment has begun
- **V4:** Respiratory status at fourth visit after treatment has begun

1. Fit a GEE model to these data including all explanatory variables but no interactions assuming independence.
2. Next fit a GEE model again with no interactions but now assuming an exchangeable correlation structure.
3. Interpret the results from whichever model you think more sensible.
4. Fit a random intercept logistic model to the data again including all explanatory variables but none of their interactions.
5. Comment on the results from the GEE and the random intercept logistic model.

3 *Epilepsy Data*

In a clinical trial reported in Thail and Vail (1990), 59 patients with epilepsy were randomized to receive either the anti-epileptic drug progabide or a placebo in addition to standard chemotherapy. The number of seizures was counted over 4 two-week periods. In addition, a baseline seizure rate was recorded for each patient, based on the 8-week pre-randomization seizure count.

The variables in epil.dat are

- **ID:** Subject identifier
- **P1:** Number of seizures in first period
- **P2:** Number of seizures in second period
- **P3:** Number of seizures in third period
- **P4:** Number of seizures in fourth period
- **Treat:** Treatment, 0 = placebo, 1 = progabide
- **BL:** Baseline seizure count
- **Age:** Age in years

1. Construct box plots of seizure counts at each time point for each treatment group. Identify any observations that you consider may be outliers and decide how to deal with them before beginning formal modelling of the data.
2. Fit a Poisson regression model to the data assuming that the repeated measures are independent.
3. Fit a Poisson model allowing for an exchangeable correlation structure between the repeated measures.
4. Compare the results from (2) and (3) and draw whatever conclusions you think are appropriate.

Chapter 15

Survival Analysis: Gastric Cancer, the Treatment of Heroin Addicts and Heart Transplants

15.1 Introduction

In this chapter, we shall analyse three data sets. The first shown in Table 15.1 involves the survival times of two groups of 45 patients suffering from gastric cancer. Group 1 received chemotherapy and radiation, and group 2 only chemotherapy. An asterisk denotes censoring, that is, the patient was still alive at the time the study ended. Interest lies in comparing the survival times of the two groups. (These data are given in Table 467 of *SDS*.)

But "survival times" do not always involve the endpoint death. This is so for the second data set that we will consider in this chapter; part of the data is shown in Table 15.2. Given in this display are the times that heroin addicts remained in a clinic for methadone maintenance treatment. Here the endpoint of interest is not death, but termination of treatment. Some subjects were still in the clinic at the time these data were recorded, and this is indicated by the variable status, which is equal to 1 if the person had departed from the clinic on completion of treatment and 0 otherwise. Possible explanatory variables for time to complete treatment are

Table 15.1 Gastric Cancer Data

Group 1			*Group 2*		
17	185	542	1	383	778
42	193	567	63	383	786
44	195	577	105	388	797
48	197	580	125	394	955
60	208	795	182	408	968
72	234	855	216	460	977
74	235	1174*	250	489	1245
95	254	1214	262	523	1271
103	307	1232*	301	524	1420
108	315	1366	301	535	1460*
122	401	1455*	342	562	1516*
144	445	1585*	354	569	1551
167	464	1622*	356	675	1690*
170	484	1626*	358	676	1694
183	528	1736*	380	748	

maximum methadone dose, whether the addict had a criminal record and the clinic in which the addict was being treated. (These data are given in Table 354 of *SDS*.)

Part of the third data set to be considered in this chapter is shown in Table 15.3. Here survival time of potential heart transplant recipients is from their date of acceptance into the Stanford heart transplant programme. With these data, a patient is part of the control group until a suitable donor is located and transplantation is carried out, at which time the patient joins the treatment group. In this example, treatment is a time-varying covariate. Apart from treatment, the other covariates of interest are age (in years minus 48), whether the patient had previous

Table 15.2 Heroin Addiction Data

ID	*Clinic*	*Status*	*Time*	*Prison*	*Dose*	*ID*	*Clinic*	*Status*	*Time*	*Prison*	*Dose*
1	1	1	428	0	50	132	2	0	633	0	70
2	1	1	275	1	55	133	2	1	661	0	40
3	1	1	262	0	55	134	2	1	232	1	70
4	1	1	183	0	30	135	2	1	13	1	60
5	1	1	259	1	65	137	2	0	563	0	70
⋮						⋮					
127	2	1	26	0	40	262	2	1	540	0	80
128	2	0	72	1	40	263	2	0	551	0	65
129	2	0	641	0	70	264	1	1	90	0	40
131	2	0	367	0	70	266	1	1	47	0	45

Table 15.3 Stanford Heart Transplant Data

ID	*Start*	*Stop*	*Event*	*Age*	*Year*	*Surgery*	*Transplant*
1	0.0	50.0	1	−17.155	0.123	0	0
2	0.0	6.0	1	3.836	0.255	0	0
3	0.0	1.0	0	6.297	0.266	0	0
3	1.0	16.0	1	6.297	0.266	0	1
4	0.0	36.0	0	−7.737	0.490	0	0
4	36.0	39.0	1	−7.737	0.490	0	1
5	0.0	18.0	1	−27.214	0.608	0	0
⋮							
100	0.0	38.0	0	−12.939	6.396	1	0
100	38.0	39.0	0	−12.939	6.396	1	1
101	0.0	31.0	0	1.517	6.418	0	0
102	0.0	11.0	0	−7.608	6.472	0	0
103	0.0	6.0	1	−8.684	−0.049	0	0

Note: Surgery: 0 = no previous surgery, 1 = previous surgery; transplant: 0 = no transplant, 1 = transplant; event: 0 = censored, 1 = died.

heart surgery and the waiting time for acceptance into the programme (years since October 1, 1967).

For the gastric cancer data, the main question of interest is whether the survival time differs in the two treatment groups, and for the methadone data, the possible effects of the explanatory variables on time to completion of treatment are of concern. For the heart transplant data the effect of the covariates on the survival times of patients is of most concern. It might be thought that such concerns could be addressed by some of the techniques covered in earlier chapters (e.g., *t*-tests or multiple regression). Survival times however, require, special methods of analysis for two main reasons:

1. They are restricted to being positive so that familiar parametric assumptions, for example, normality, may not be justifiable.
2. The data often contain censored observations, that is, observations for which, at the end of the study, the event of interest (death in the first data set, completion of treatment in the second) has not occurred; all that can be said about a censored survival time is that the unobserved, uncensored value would have been greater than the value recorded.

15.2 Describing Survival Data

Of central importance in the analysis of survival time data are two functions used to describe their distribution, namely the survival function and the hazard function.

15.2.1 Survival Function

Using T to denote survival time, the survival function, $S(t)$, is defined as the probability that an individual survives longer than t.

$$S(t) = \Pr(T > t). \tag{15.1}$$

The graph of $S(t)$ against t is known as the survival curve and is useful in assessing the general characteristics of a set of survival times.

Estimating $S(t)$ from sample data is straightforward when there are no censored observations, when $\hat{S}(t)$ is simply the proportion of survival times in the sample greater that t. When, as is generally the case, the data do contain censored observations, estimation of $S(t)$ is more complex. The most usual estimator is now the Kaplan–Meier or product-limit estimator. This involves first ordering the survival times from the smallest to the largest, $t_{(1)} \leq t_{(2)} \leq \cdots \leq t_{(n)}$ and then applying the following formula to obtain the required estimate:

$$\hat{S}(t) = \prod_{j|t_{(j)} \leq t} \left[1 - \frac{d_j}{r_j}\right], \tag{15.2}$$

where r_j is the number of individuals at risk just before $t_{(j)}$, and d_j is the number who experience the event of interest at $t_{(j)}$ (individuals censored at $t_{(j)}$ are included in r_j).

The variance of the Kaplan–Meir estimator can be estimated as

$$Var[\hat{S}(t)] = [\hat{S}(t)]^2 \sum_{j||t_{(j)} \leq t} \frac{d_j}{r_j(r_j - d_j)}. \tag{15.3}$$

Plotting estimated survival curves for different groups of observations (e.g., males and females, treatment A and treatment B) is a useful initial procedure for comparing the survival experience of the groups. More formally, the difference in survival experience can be tested by either a log-rank test or Mantel–Haenszel test. These tests essentially compare the observed number of "deaths" occurring at each particular time point with the number to be expected if the survival experience of the groups is the same. (Details of both the log-rank test and the Mantel–Haenszel test are given in Hosmer and Lemeshow, 1999.)

15.2.2 Hazard Function

The hazard function, $h(t)$, is defined as the probability that an individual experiences the event of interest in a small time interval s, given that the individual has survived up to the beginning of this interval. In mathematical terms,

$$h(t) = \lim_{s \to 0} \Pr \frac{(\text{event in}(t, t + s), \text{given survival up to } t)}{s}. \tag{15.4}$$

The hazard function is also known as the instantaneous failure rate or age-specific failure rate. It is a measure of how likely an individual is to experience an event as a function of the age of the individual, and is used to assess which periods have the highest and which the lowest chance of "death," among those people alive at the time. In the very old, for example, there is a high risk of dying each year, among those entering that stage of their life. The probability of any individual dying in their 100th year is, however, small because very few individuals live to be 100 years old.

The hazard function can also be defined in terms of the cumulative distribution and probability density function of the survival times as follows:

$$h(t) = \frac{f(t)}{1 - F(t)} = \frac{f(t)}{S(t)}. \tag{15.5}$$

It then follows that

$$h(t) = -\frac{d}{dt}\{\ln S(t)\} \tag{15.6}$$

and so

$$S(t) = \exp\{-H(t)\}, \tag{15.7}$$

where $H(t)$ is the integrated or cumulative hazard given by

$$H(t) = \int_0^t h(u)\, du. \tag{15.8}$$

The hazard function can be estimated as the proportion of individuals experiencing the event of interest in an interval per unit time, given that they have survived to the beginning of the interval, that is,

$$\hat{h}(t) = \frac{\text{number of individuals experiencing an event in the interval beginning at time } t}{(\text{number of patients surviving at } t) \times (\text{interval width})}. \tag{15.9}$$

In practice, the hazard function may increase, decrease, remain constant or indicate a more complicated process. The hazard function for deaths in humans has, for example, the "bath-tub" shape shown in Figure 15.1. It is relatively high immediately after birth, declines rapidly in the early years and the n remains approximately constant before beginning to rise again during late middle age.

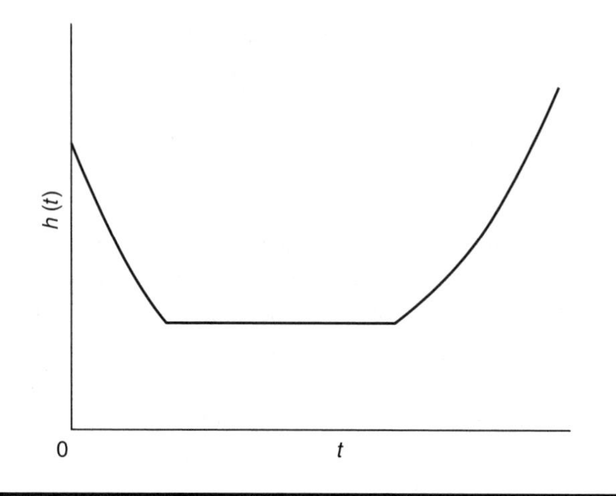

Figure 15.1 Bath-tub hazard function for death in humans.

15.3 Cox's Regression

When the response variable of interest is a possibly censored survival time, we need special regression techniques for modelling the relationship of the response to explanatory variables on interest. A number of procedures are available but the most widely used by some margin is known as *Cox's proportional hazards model* or *Cox's regression*, for short. Introduced by Sir David Cox in 1972, the method has become one of the most widely used, particularly in medical statistics, and the original paper, one of the most heavily cited.

Cox's regression is a *semi-parametric* approach to survival analysis, semi-parametric because the method does not require the probability distribution of the survival times to be specified; however, unlike most non-parametric methods, Cox's regression does use regression parameters in the same way as generalized linear models. With this approach, the hazard function is used as the vehicle for the modelling process because it does not involve the cumulative history of events. But modelling the hazard function directly as a linear function of explanatory variables is not appropriate as $h(t)$ is restricted to being positive. A more suitable model might be

$$\log[h(t)] = \beta_0 + \beta_1 x_1 + \cdots + \beta_p x_p, \qquad (15.10)$$

where $x_1, x_2, \ldots, x_p$ are the explanatory variables. But the model in Equation 15.10 would only be suitable for a hazard function that is constant over time; such a model would be very restrictive because hazards that increase or decrease with time, or have some more complex form, are far more likely to occur in practice (see Figure 15.1, for example). The problem is overcome in the model suggested by Cox by allowing the form of dependence of $h(t)$ on t to remain unspecified, so that the model for the hazard function becomes

$$\log[h(t)] = \log[h_0(t)] + \boldsymbol{\beta}'\mathbf{x}, \tag{15.11}$$

where $\boldsymbol{\beta}$ is a vector of regression parameters, and $\mathbf{x}$ a vector of covariate values.

The set of parameters, $h_0(t)$, is called the *baseline hazard function* and can be thought of as nuisance parameters whose purpose is merely to control the parameters of interest, namely the elements of $\boldsymbol{\beta}$, for any changes in the hazard over time.

The model for Cox's regression can also be written as

$$h(t) = h_0(t)\exp(\boldsymbol{\beta}'\mathbf{x}). \tag{15.12}$$

Written in this way, we can see that the model forces the hazard ratio between two individuals with vectors of covariates x_1 and x_2 to be constant over time since

$$\frac{h(t|\mathbf{x}_1)}{h(t|\mathbf{x}_2)} = \frac{\exp(\boldsymbol{\beta}'\mathbf{x}_1)}{\exp(\boldsymbol{\beta}'\mathbf{x}_2)}. \tag{15.13}$$

In other words, if an individual has a risk of death at some initial time point that is twice as high as another individual, then at all later times the risk of death remains twice as high. Hence, the term proportional hazards.

The regression parameters in $\boldsymbol{\beta}$ are estimated by maximizing the partial log likelihood given by

$$\sum_f \log\left(\frac{\exp(\boldsymbol{\beta}'\mathbf{x}_f)}{\sum_{i\in r(f)}\exp(\boldsymbol{\beta}'\mathbf{x}_i)}\right), \tag{15.14}$$

where the first summation is overall failures f and the second summation is overall subjects $r(f)$ still alive (and therefore "at risk") at the time of failure. It can be shown that this log likelihood is a log profile likelihood (i.e., the log of the likelihood in which the nuisance parameters have been replaced by functions of $\boldsymbol{\beta}$ which maximize the likelihood for fixed $\boldsymbol{\beta}$).

The parameters in a Cox model are interpreted in a similar fashion to those in other regression models met in earlier chapters; that is, the estimated coefficient for an explanatory variable gives the change in the logarithm of the hazard function when the variable changes by one conditional on the other explanatory variables not changing. A more appealing interpretation is achieved by exponentiating the coefficient, giving the effect in terms of the hazard function. An additional aid to interpretation is to calculate

$$100[\exp(\text{coefficient}) - 1]. \tag{15.15}$$

The resulting value gives the percentage change in the hazard function with each unit change in the explanatory variable.

The baseline hazards may be estimated by maximizing the full log likelihood with the regression parameters evaluated at their estimated values. These hazards are non-zero only when a failure occurs. Integrating the hazard function gives the cumulative hazard function

$$H(t) = H_0(t)\exp(\boldsymbol{\beta}'\mathbf{x}), \tag{15.16}$$

where $H_0(t)$ is the integral of $h_0(t)$. The survival curve may be obtained from $H(t)$ using Equation 15.7.

It follows from Equation 15.7 that the survival curve for a Cox model is given by

$$S(t) = S_0(t)^{\exp(\boldsymbol{\beta}'\mathbf{x})}. \tag{15.17}$$

The log of the cumulative hazard function predicted by the Cox model is given by

$$\log[H(t)] = \log H_0(t) + \boldsymbol{\beta}'\mathbf{x} \tag{15.18}$$

so that the log cumulative hazard functions of any two subjects i and j are parallel with constant difference given by $\boldsymbol{\beta}'(\mathbf{x}_i - \mathbf{x}_j)$.

If the subjects fall into different groups and we are not sure whether we can make the assumption that the group's hazard functions are proportional to each other, we can estimate separate log cumulative hazard functions for the groups using a stratified Cox model. These curves may then be plotted to assess whether they are sufficiently parallel. For a stratified Cox model, the partial likelihood has the same form as in Equation 15.14, except that the risk set for a failure is not confined to subjects in the same stratum.

15.3.1 Time-Varying Covariates

Studies in which survival data are collected often include covariates with values that do not remain fixed over time. Individuals might, for example, have laboratory measurements made repeatedly during the time they were observed. A small hypothetical data set of this kind is shown in Table 15.4.

Table 15.4 Hypothetical Survival Data with a Time-Varying Covariate

	Laboratory Measurement (day)			*Survival Time*	*Status*
Individual	0	60	120		
1	0.5	0.7	0.8	130	1
2	0.2	0.6	0.3	190	1
3	0.2	0.4	–	70	0

Note: Status: 0 = censored, 1 = dead.

Table 15.5 Re-arranged Data from Table 15.4

Individual	*Interval* (T_1, T_2)	*Lab Measurement*	*Status*
1	0, 60	0.5	0
1	61, 120	0.7	0
1	121, 130	0.8	1
2	0, 60	0.2	0
2	61, 120	0.6	0
2	121, 130	0.3	1
3	0, 60	0.2	0
3	61, 170	0.4	1

Note: The survival time for each interval is calculated as $T_2 - T_1$.

How can we now fit a Cox's regression model that makes allowance for the changes in such variables? In essence, it is very simple; the survival period of each individual is divided into a sequence of shorter survival spells, each characterized by an entry and exit time, and within which covariate values remain fixed. Thus, the data for each individual are represented by a number of shorter censored intervals and possibly one interval ending with the occurrence of the event of interest (death, for example). Table 15.5 shows the data in Table 15.4 rearranged in this way.

It may be thought that the observations in Table 15.4 that arise from the same individual are "correlated" and so not suitable for Cox's regression, as described in Section 15.3. Fortunately, this is not an issue, since the partial likelihood on which estimation is based has a term for each unique death or event time, and involves sums over those observations that are available or at risk at the actual event date. Since the intervals for a particular individual do not overlap, the likelihood will involve at most only one of the observations for the individual, and so will still be based on independent observations. The values of the covariates between-event times do not enter the partial likelihood. So applying Cox's model to survival data with time-varying covariates is little more complex than that for time-fixed covariates.

15.3.2 Diagnostics for Cox's Regression

As with the other regression models discussed in earlier chapters, an application of Cox's regression needs to be followed by the examination of suitable residual and diagnostic plots for evidence of outliers, influential observations or departures from assumptions such as those of proportional hazards. Model checking for proportional hazard models is complicated by the fact that easy-to-use residuals such as those discussed in Chapter 6 for linear regression models are not available. But several possible diagnostics do exist and a number of these are now described.

- Cox–Snell residual is defined as

$$r_{Ci} = \exp(\hat{\beta}' \hat{H}_0(t_i)), \tag{15.19}$$

where $\hat{H}_0(t_i)$ is the estimated integrated hazard function at time t_i, the observed survival time of subject i. If the model is correct, r_{Ci} follows an exponential distribution with mean 1.
- Martingale residual is the difference between the event indicator δ_i (equal to 1 if the person died and 0 otherwise) and the Cox–Snell residual

$$r_{Mi} = \delta_i - r_{Ci}. \tag{15.20}$$

- Deviance residual is defined as

$$r_{Di} = \text{sign}\left(r_{Mi} \sqrt{-2[r_{Mi} + \delta_i \log (\delta_i - r_{Mi})]} \right). \tag{15.21}$$

Subjects with large positive or negative deviance residuals are poorly predicted by the model.
- Partial score residual, also known as Schoenfeld residual or efficient score residual, is defined as the first derivative of the partial log-likelihood function with respect to an explanatory variable. For the jth explanatory variable,

$$r_{Sij} = \frac{x_{ij} - \Sigma_{r(i)} x_{rj} \exp(\beta' \mathbf{x}_r)}{\Sigma_{r(i)} x_{rj} \exp(\beta' \mathbf{x}_r)}.$$

Score residual is large in absolute value if a case's explanatory variable differs substantially from the explanatory variables of subjects whose estimated risk of failure is large at the case's time of failure or censoring. This residual can be used to detect potentially influential points.
- For each explanatory variable, there is one score residual for each event. A re-scaled score residual can be defined whose expected value is equal to the corresponding regression coefficient (Grambsch and Therneau, 1994). If the hazards are proportional, the regression coefficient is constant over time and the plot of the rescaled Schoenfeld residuals against time should form a horizontal line. Grambsch and Therneau proposed a test of proportional hazards based on these rescaled residuals.

Influence statistics are defined as for linear regression. For example, we can compute the change in scaled coefficient associated with the removal of each observation from the data set. All aspects of survival analysis are described in more detail in Collett (2003b) and Hosmer and Lemeshow (1999).

15.4 Analysis Using SAS

15.4.1 Gastric Cancer

The data shown in Table 15.1 consist of 89 survival times. There are six values per line, except the last line, which has five. The first three values on each line belong to patients in the first treatment group and the remainder to patients in the second group. The following data step constructs a suitable SAS data set:

```
data cancer;
  infile 'c:\handbook3\datasets\time.dat' expandtabs missover;
  do i = 1 to 6;
  input temp $ @;
  censor = (index(temp,'*') > 0);
    temp = substr(temp,1,4);
  days = input(temp,4.);
  group = i > 3;
  if days > 0 then output;
  end;
  drop temp i;
run;
```

The infile statement gives the full path name of the file containing the ASCII data. The values are tab separated so the expandtabs option is used. The missover option prevents SAS from going to a new line if the input statement contains more variables than data values, as is the case for the last line. In this case, the variable for which there is no corresponding data is set to missing.

Reading and processing the data takes place within an iterative do loop. The input statement reads one value into a character variable, temp. A character variable is used to allow for processing of the asterisks, which indicate censored values, as there is no space between the number and the asterisk. The trailing @ holds the line for further data to be read from it.

If temp contains an asterisk, the index function gives its position, if not, the result is zero. The censoring variable is set accordingly. The substr (sub-string) function takes the first four characters of temp, and the input function reads this into a numeric variable, days.

If the value of days is greater than zero, an observation is output to the data set. This has the effect of excluding the missing value generated because the last line contains only five values.

Finally, the character variable temp and the loop index variable, i, are dropped from the data set, as they are no longer needed.

With a complex data step such as this, it would be wise to check the resulting data set, for example, with proc print.

Proc lifetest can be used to estimate and compare the survival functions of the two groups of patients as follows:

```
proc lifetest data=cancer plots=(s);
  time days*censor(1);
  strata group;
run;
```

The plots=(s) option in the proc statement specifies that survival curves be plotted. Log survival (ls), log–log survival (lls), hazard (h) and PDF (p) are other functions that may be plotted as well as a plot of censored values by strata (c). A list of plots may be specified, for example, plots=(s,ls,lls).

The time statement specifies the survival time variable followed by an asterisk and the censoring variable, with the value(s) indicating a censored observation in parentheses. The censoring variable must be numeric with non-missing values for both censored and uncensored observations.

The strata statement indicates the variable, or variables, that determines the strata levels.

The output is shown in Table 15.6, and the associated plot in Figure 15.2. From Table 15.6, we find that the median survival time in group 1 (coded 0) is 254 with 95% confidence interval of (193, 484). In group 2 (coded 1), the corresponding values are 506 and (383, 676). The log-rank test for a difference in the survival curves of the two groups has an associated p value of 0.4521. There is no evidence of a difference. (The Wilcoxon test mentioned in Table 15.6 is described in Kalbfleisch and Prentice, 2002, and the likelihood ratio test in Lawless, 2002.)

15.4.2 Methadone Treatment of Heroin Addicts

The data on treatment of heroin addiction shown in Table 15.2 can be read in with the following data step:

```
data heroin;
  infile 'c:\handbook3\datasets\heroin.dat' expandtabs;
  input id clinic status time prison dose @@;
run;
```

Each line contains the data values for two observations, but there is no relevant difference between those that occur first and second. This being the case, the data can be read using list input and a double trailing @. This holds the current line for further data to be read from it. The difference between the double trailing @ and the single trailing @, illustrated above, is that the double @ will hold the line across iterations of the data step. SAS will only go on to a new line when it runs out of data on the current line.

The SAS log will contain the message 'NOTE: SAS went to a new line when INPUT statement reached past the end of a line', which is not a cause for concern in

Table 15.6 Abbreviated Results from Proc Lifetest for the Gastric Cancer Data

Stratum 1: group = 0

Product-Limit Survival Estimates							
days		Survival	Failure	Survival Standard Error	Number Failed	Number Left
0.00		1.0000	0	0	0	45
17.00		0.9778	0.0222	0.0220	1	44
42.00		0.9556	0.0444	0.0307	2	43
44.00		0.9333	0.0667	0.0372	3	42
48.00		0.9111	0.0889	0.0424	4	41
.						
.						
.						
795.00		0.2222	0.7778	0.0620	35	10
855.00		0.2000	0.8000	0.0596	36	9
1174.00	*	.	.	.	36	8
1214.00		0.1750	0.8250	0.0572	37	7
1232.00	*	.	.	.	37	6
1366.00		0.1458	0.8542	0.0546	38	5
1455.00	*	.	.	.	38	4
1585.00	*	.	.	.	38	3
1622.00	*	.	.	.	38	2
1626.00	*	.	.	.	38	1
1736.00	*	0.1458	0.8542	.	38	0

NOTE: The marked survival times are censored observations.

Stratum 1: group = 0

Summary Statistics for Time Variable days

Quartile Estimates				
			95% Confidence Interval	
Percent	Point Estimate	Transform	[Lower	Upper)
75	580.00	LOGLOG	464.00	.
50	254.00	LOGLOG	185.00	484.00
25	144.00	LOGLOG	72.00	193.00

Mean	Standard Error
491.84	71.01

NOTE: The mean survival time and its standard error were underestimated because the largest observation was censored and the estimation was restricted to the largest event time.

Stratum 2: group = 1

Product-Limit Survival Estimates

days		Survival	Failure	Survival Standard Error	Number Failed	Number Left
0.00		1.0000	0	0	0	44
1.00		0.9773	0.0227	0.0225	1	43
63.00		0.9545	0.0455	0.0314	2	42
105.00		0.9318	0.0682	0.0380	3	41
125.00		0.9091	0.0909	0.0433	4	40
182.00		0.8864	0.1136	0.0478	5	39
.						
.						
.						
1420.00		0.1136	0.8864	0.0478	39	5
1460.00	*	.	.	.	39	4
1516.00	*	.	.	.	39	3
1551.00		0.0758	0.9242	0.0444	40	2
1690.00	*	.	.	.	40	1
1694.00		0	1.0000	0	41	0

NOTE: The marked survival times are censored observations.

Stratum 2: group = 1

Summary Statistics for Time Variable days

Quartile Estimates

Percent	Point Estimate	95% Confidence Interval Transform	[Lower	Upper)
75	876.00	LOGLOG	569.00	1420.00
50	506.00	LOGLOG	380.00	676.00
25	348.00	LOGLOG	182.00	383.00

Mean	Standard Error
653.22	72.35

Summary of the Number of Censored and Uncensored Values					
Stratum	group	Total	Failed	Censored	Percent Censored
1	0	45	38	7	15.56
2	1	44	41	3	6.82
Total		89	79	10	11.24

Testing Homogeneity of Survival Curves for days over Strata

Rank Statistics		
group	Log-Rank	Wilcoxon
0	3.3043	502.00
1	-3.3043	-502.00

Covariance Matrix for the Log-Rank Statistics		
group	0	1
0	19.3099	-19.3099
1	-19.3099	19.3099

Covariance Matrix for the Wilcoxon Statistics		
group	0	1
0	58385.0	-58385.0
1	-58385.0	58385.0

Test of Equality over Strata			
Test	Chi-Square	DF	Pr > Chi-Square
Log-Rank	0.5654	1	0.4521
Wilcoxon	4.3162	1	0.0378
-2Log(LR)	0.3574	1	0.5500

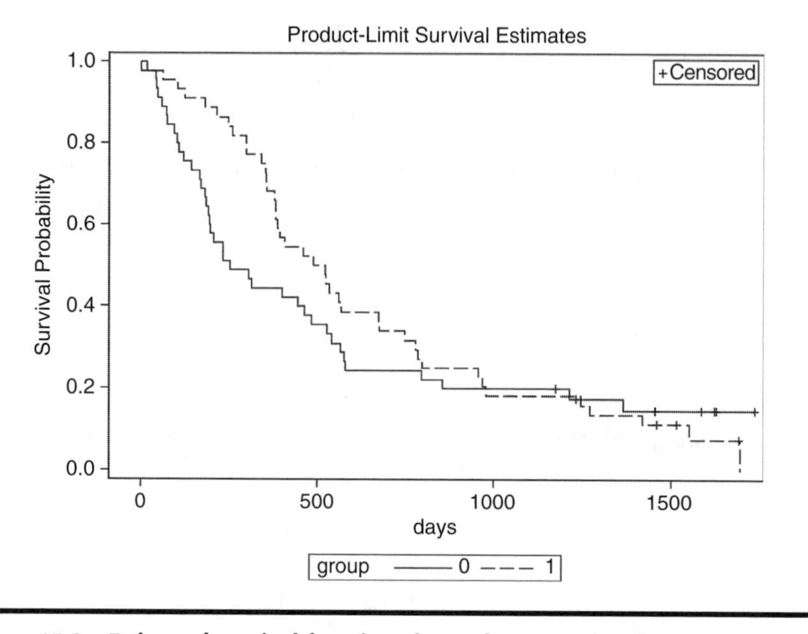

Figure 15.2 Estimated survival functions for each group of patients in the gastric cancer data.

this case. It is also worth noting that although the ID variable ranges from 1 to 266, there are actually 238 observations in the data set.

Cox regression is implemented within SAS in the phreg procedure.

The data come from two different clinics and it is possible, indeed likely, that these clinics have different hazard functions that may well not be parallel. A Cox regression model with clinics as strata and the other two variables, dose and prison, as explanatory variables can be fitted in SAS using the phreg procedure.

```
proc phreg data=heroin;
  model time*status(0)=prison dose/rl;
  strata clinic;
run;
```

On the model statement the response variable, that is, the failure time, is followed by an asterisk, the name of the censoring variable, and a list of censoring values in parentheses. The rl (risklimits) option requests confidence limits for the hazard ratio. By default these are the 95% limits. The strata statement specifies a stratified analysis with clinics forming the strata.

The output is shown in Table 15.7. Examining the maximum likelihood estimates we find that the parameter estimate for prison is 0.38877 and that for

Table 15.7 Results from Applying Proc Phreg to the Heroin Addicts Data

Model Information	
Data Set	WORK.HEROIN
Dependent Variable	time
Censoring Variable	status
Censoring Value(s)	0
Ties Handling	BRESLOW

Number of Observations Read	238
Number of Observations Used	238

Summary of the Number of Event and Censored Values					
Stratum	**clinic**	**Total**	**Event**	**Censored**	**Percent Censored**
1	1	163	122	41	25.15
2	2	75	28	47	62.67
Total		238	150	88	36.97

Convergence Status
Convergence criterion (GCONV=1E-8) satisfied.

Model Fit Statistics		
Criterion	**Without Covariates**	**With Covariates**
-2 LOG L	1229.367	1195.428
AIC	1229.367	1199.428
SBC	1229.367	1205.449

Testing Global Null Hypothesis: BETA=0			
Test	**Chi-Square**	**DF**	**Pr > ChiSq**
Likelihood Ratio	33.9393	2	<.0001
Score	33.3628	2	<.0001
Wald	32.6858	2	<.0001

Analysis of Maximum Likelihood Estimates

Parameter	DF	Parameter Estimate	Standard Error	Chi-Square	Pr > ChiSq	Hazard Ratio	95% Hazard Ratio Confidence Limits	
Prison	1	0.38877	0.16892	5.2974	0.0214	1.475	1.059	2.054
Dose	1	-0.03514	0.00647	29.5471	<.0001	0.965	0.953	0.978

dose −0.03514. Interpretation becomes simpler if we concentrate on the exponentiated versions of those given under Hazard Ratio. Using the approach given in Equation 15.15 we see first that subjects with a prison history are 47.5% more likely to complete treatment than those without a prison history. And for every increase in methadone dose by one unit (1 mg), the hazard is multiplied by 0.965. This coefficient is very close to 1, but this may be because 1 mg of methadone is not a large quantity. In fact, subjects in this study differ from each other by 10–15 units, and so it may be more informative to find the hazard ratio of two subjects differing by a standard deviation unit. This can be done simply by rerunning the analysis with dose standardized to zero mean and unit variance.

The analysis can be repeated with dose standardized to zero mean and unit variance as follows:

```
proc stdize data=heroin out=heroin2;
  var dose;
proc phreg data=heroin2;
  model time*status(0)=prison dose/rl;
  strata clinic;
  baseline out=phout loglogs=lls/method=ch;
proc sgplot data=phout;
  step y=lls x=time/group=clinic;
run;
```

The stdize procedure is used to standardize dose (proc standard could also have been used). Zero mean and unit variance is the default method of standardization. The resulting data set is given a different name with the out= option and the variable to be standardized is specified with the var statement.

The phreg step uses this new data set to repeat the analysis. The baseline statement is added to save the log cumulative hazards in the data set phout. Loglogs=lls specifies that the log of the negative log of survival is to be computed

Table 15.8 Selected Results from Applying Proc Phreg to the Data on Heroin Addicts after Standardizing Dose

Analysis of Maximum Likelihood Estimates								
							95% Hazard Ratio Confidence Limits	
Parameter	**DF**	**Parameter Estimate**	**Standard Error**	**Chi-Square**	**Pr > ChiSq**	**Hazard Ratio**		
Prison	1	0.38877	0.16892	5.2974	0.0214	1.475	1.059	2.054
Dose	1	-0.50781	0.09342	29.5471	<.0001	0.602	0.501	0.723

and stored in the variable lls. The product-limit estimator is the default, and method = ch requests the alternative empirical cumulative hazard estimate.

Proc sgplot is then used to plot the log cumulative hazard with separate lines for each clinic.

The output from the phreg step is shown in Table 15.8 and the plot in Figure 15.3. The coefficient of dose is now −0.50781 and the hazard ratio is 0.602. This may be interpreted as indicating a decrease in the hazard by 40% when

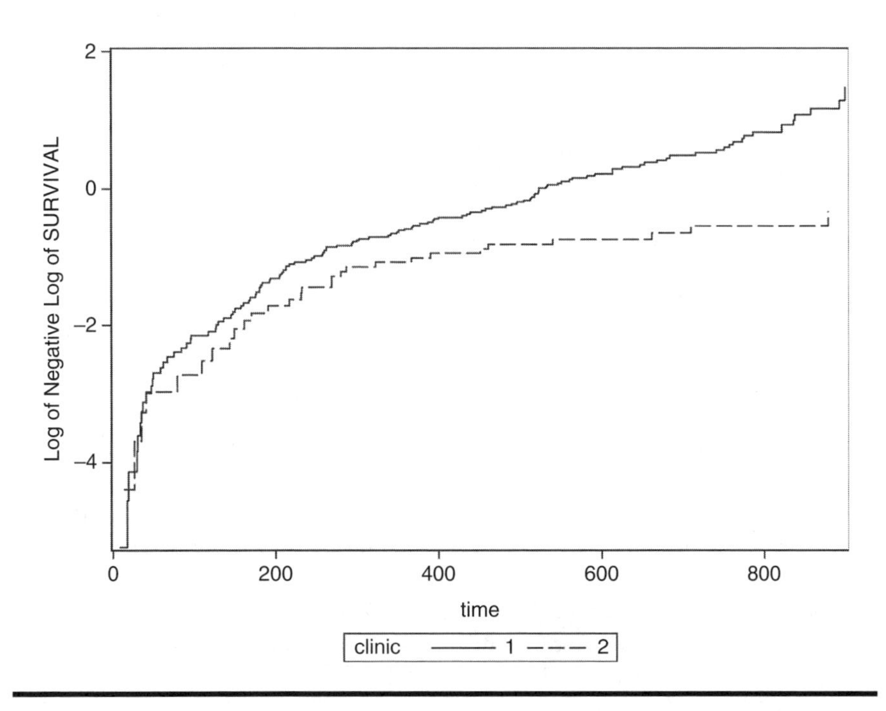

Figure 15.3 Log(−log survival) curves for the two clinics in the heroin addiction data.

the methadone dose increases by one standard deviation unit. Clearly, an increase in methadone dose decreases the likelihood of the addict's completing treatment.

In Figure 15.3, the increment at each event represents the estimated logs of the hazards at that time. Clearly the curves are not parallel, underlying that treating the clinics as strata was sensible.

15.4.3 Heart Transplant Data

We will now illustrate the use of Cox's regression with time-varying covariates using the set of survival times of potential heart transplant recipients from their date of acceptance into the Stanford heart transplant program given in the introduction of this chapter (see Table 15.3). These data are in the form described previously in Section 15.3.1. So, for example, patient 3 waited a single day for a transplant and then died after 15 days. In these data, patients change treatment status during the course of the study. Specifically, a patient is part of the control group until a suitable donor is located and transplantation takes place, at which time he joins the treatment group. So treatment is a time-dependent covariate. The other covariates to be considered are age (in years −48), whether the patient had had previous heart surgery and waiting time for acceptance into the program (years since October 1, 1967).

The necessary SAS code to read in the data and apply a Cox's regression with all four covariates and an interaction between Year and Transplant is as follows:

```
data SHTD;
  infile 'c:\handbook3\datasets\SHTD.dat';
  input ID Start Stop Event Age Year Surgery Transplant;
  duration = stop-start;
run;
proc phreg data = SHTD;
  model (start,stop)*event(0) = Age Year Surgery Trans-
    plant transplant*year/rl;
run;
```

Proc phreg has an alternative version of the model statement designed for data in this format, which SAS refers to as the "counting process style of input". Instead of a single variable for the survival time, two variables are named (in parentheses), which define the beginning and end of a period during which the subject is at risk. The example also illustrates the use of the rl (risklimits) option to include the confidence limits in the output. From version 9.2, proc phreg supports categorical predictor variables via the class statement and the specification of interactions with the asterisk. Prior to version 9.2 proc tphreg had these capabilities. The output is shown in Table 15.9.

Table 15.9 Results from Applying Proc Phreg to the Stanford Heart Transplant Data

Model Information	
Data Set	WORK.SHTD
Dependent Variable	Start
Dependent Variable	Stop
Censoring Variable	Event
Censoring Value(s)	0
Ties Handling	BRESLOW

Number of Observations Read	172
Number of Observations Used	172

Summary of the Number of Event and Censored Values			
Total	**Event**	**Censored**	**Percent Censored**
172	75	97	56.40

Convergence Status
Convergence criterion (GCONV=1E-8) satisfied.

Model Fit Statistics		
Criterion	**Without Covariates**	**With Covariates**
-2 LOG L	596.651	579.569
AIC	596.651	589.569
SBC	596.651	601.156

Testing Global Null Hypothesis: BETA=0			
Test	**Chi-Square**	**DF**	**Pr > ChiSq**
Likelihood Ratio	17.0823	5	0.0043
Score	16.6844	5	0.0051
Wald	15.9359	5	0.0070

Analysis of Maximum Likelihood Estimates					
Parameter	DF	Parameter Estimate	Standard Error	Chi-Square	Pr > ChiSq
Age	1	0.02988	0.01374	4.7307	0.0296
Year	1	-0.25211	0.10482	5.7848	0.0162
Surgery	1	-0.66270	0.36811	3.2410	0.0718
Transplant	1	-0.62153	0.53092	1.3704	0.2417
Year*Transplant	1	0.19697	0.13944	1.9953	0.1578

Analysis of Maximum Likelihood Estimates				
Parameter	Hazard Ratio	95% Hazard Ratio Confidence Limits		Variable Label
Age	1.030	1.003	1.058	
Year	.	.	.	
Surgery	0.515	0.251	1.061	
Transplant	.	.	.	
Year*Transplant	.	.	.	Year * Transplant

The results appear to imply that survival time depends on the time of acceptance into the study; as this increases, the hazard function for the death of a patient decreases. But this claim becomes less clear-cut if we examine the transplant × time of acceptance interaction which approaches significance, with an effect that is in the opposite direction. According to Kalbfleisch and Prentice (2002), taken together these results imply that the overall quality of patient being admitted to the study may be improving with time (possibly due to the relaxation of admission requirements or to improving patient management), but the survival time of the transplanted patients is not improving at the same rate.

A further model that might be considered is one which allows separate baseline hazards for the after transplanted and before or not transplanted patients, but common coefficients in each group. To fit this model we can use the following code:

```
proc phreg data=SHTD;
  model (start,stop)*event(0)=Age Year Surgery/rl;
  strata transplant;
run;
```

The results are shown in Table 15.10.

Table 15.10 Results from Stratified Analysis of Stanford Heart Transplant Data

Model Information	
Data Set	WORK.SHTD
Dependent Variable	Start
Dependent Variable	Stop
Censoring Variable	Event
Censoring Value(s)	0
Ties Handling	BRESLOW

Number of Observations Read	172
Number of Observations Used	172

Summary of the Number of Event and Censored Values					
Stratum	**Transplant**	**Total**	**Event**	**Censored**	**Percent Censored**
1	0	103	30	73	70.87
2	1	69	45	24	34.78
Total		172	75	97	56.40

Convergence Status
Convergence criterion (GCONV=1E-8) satisfied.

Model Fit Statistics		
Criterion	**Without Covariates**	**With Covariates**
-2 LOG L	525.608	510.000
AIC	525.608	516.000
SBC	525.608	522.952

Testing Global Null Hypothesis: BETA=0			
Test	**Chi-Square**	**DF**	**Pr > ChiSq**
Likelihood Ratio	15.6082	3	0.0014
Score	15.4010	3	0.0015
Wald	14.8433	3	0.0020

Analysis of Maximum Likelihood Estimates					
Parameter	DF	Parameter Estimate	Standard Error	Chi-Square	Pr > ChiSq
Age	1	0.02931	0.01391	4.4370	0.0352
Year	1	-0.15201	0.07099	4.5849	0.0323
Surgery	1	-0.61681	0.37067	2.7690	0.0961

Analysis of Maximum Likelihood Estimates			
Parameter	Hazard Ratio	95% Hazard Ratio Confidence Limits	
Age	1.030	1.002	1.058
Year	0.859	0.747	0.987
Surgery	0.540	0.261	1.116

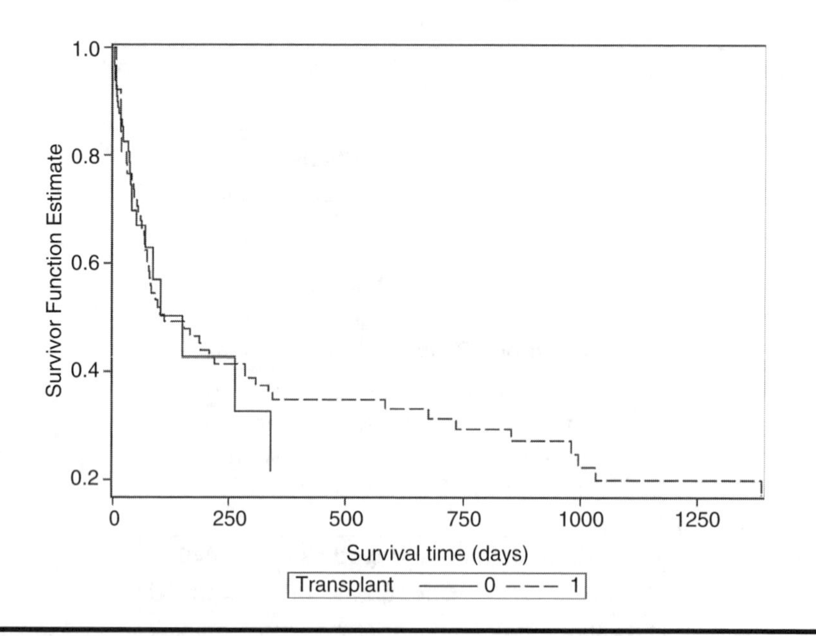

Figure 15.4 Predicted survival curves for heart transplant data.

Plotting the graph of the predicted survival curves in each stratum, with all covariates equal to their mean values using the following code, results in Figure 15.4.

```
proc phreg data = SHTD;
  model (start,stop)*event(0) = Age Year Surgery/rl;
  strata transplant;
baseline out = phout survival = sfest/method = ch;
run;
proc sgplot data = phout;
  step y = sfest x = stop/group = transplant;
  xaxis label = 'Survival time (days)';
run;
```

The baseline statement saves the survival function estimates as the variable sfest in the phout data set. The product-limit estimator is the default and method = ch requests the alternative empirical cumulative hazard estimate. The estimates are then plotted separately for the transplant and control group. The xaxis statement labels the horizontal axis. Figure 15.4 demonstrates the longer survival experience of patients who have had a transplant.

Exercises

1 Heroin Addiction Data

1. In the original analyses of the data in this chapter (see Caplehorn and Bell, 1991), it was judged that the hazards were approximately proportional for the first 450 days. Consequently, the data for this time period were analysed using clinic as a covariate rather than by stratifying on clinic. Repeat this analysis using clinic, prison and standardized dose as covariates.
2. Following Caplehorn and Bell (1991), repeat the analyses in the chapter but now treating dose as a categorical variable with three levels (<60, 60–79, ≥ 80) and plot the predicted survival curves for the three dose categories when prison takes the value 0 and clinic the value 1.
3. Test for an interaction between clinic and methadone dose using both continuous and categorical scales for dose.

2 Malignant Glioma

In a non-randomized clinical trial investigating a novel radioimmunotherapy in malignant glioma patients (see Grana et al., 2002), a control group underwent the standard therapy and another group of patients was treated with radioimmunotherapy in addition.

The variables in glioma.dat are

- **ID:** Subject identifier
- **Age:** Age in years
- **Sex:** Male or female
- **Group:** RIT = radioimmunotherapy, control = standard therapy
- **Event:** Whether the patient had died (true) or not (false) by the end of the study
- **Time:** Survival time in weeks

1. Plot the survival curves for the two treatment groups and find a confidence interval for the difference in the median survival times.
2. Test formally whether the patients treated with the novel radioimmunotherapy survived for a longer time on average than the patients in the control group.

3 Breast Cancer Survival

The effects of hormonal treatment with Tamoxifen in women suffering from node-positive breast cancer were investigated in a randomized clinical (Schumacher et al., 1994). Data drawn from randomized patients from this trial and additional non-randomized patients (from the German Breast Cancer Study Group 2, GBSG2) are used by Sauerbrei and Royston (1999). The data are in the file GBSG2.csv, comma separated with the variable names in the first row.

The variables are

- **Horm:** Dichotomous variable indicating whether hormonal therapy was applied
- **Age:** Age in years
- **Menstat:** Menopausal status, post- or pre-
- **Tsize:** Tumour size
- **Tgrade:** Tumour grade
- **Pnodes:** Number of positive lymph nodes
- **Progrec:** Progesterone receptor
- **Estrec:** Oestrogen receptor
- **Time:** Survival time in days
- **Status:** Whether alive (0) or dead (1) at end of study

1. Plot the survival curves for those women who received and those who did not receive hormonal therapy.
2. Fit a Cox regression to the data and interpret the results.
3. Investigate the assumptions of the Cox regression by plotting suitable diagnostics.

Chapter 16

Principal Components Analysis and Factor Analysis: Olympic Decathlon and Statements about Pain

16.1 Introduction

In this chapter we shall be concerned with two data sets; the first, given in Table 16.1 (*SDS* 357), involves the results for the men's decathlon in the 1988 Olympics, and the second, shown in Table 16.2, arises from a study concerned with the development of a standardized scale to measure beliefs about controlling pain (Skevington, 1990). Here a sample of 123 people suffering from extreme pain were asked to rate nine statements about pain on a scale of 1–6 ranging from disagreement to agreement. It is the correlations between these statements that appear in Table 16.3 (*SDS* 492). The nine statements used were as follows:

1. Whether or not I am in pain in the future depends on the skills of the doctors.
2. Whenever I am in pain, it is usually because of something I have done or not done.

Table 16.1 Results of Olympic Decathlon (1988)

Athlete	*100 m*	*Long Jump*	*Shot*	*High Jump*	*400 m*	*110 m Hurd.*	*Disc.*	*Pole Vlt.*	*Jav.*	*1500 m*	*Score*
Schenk	11.25	7.43	15.48	2.27	48.90	15.13	49.28	4.7	61.32	268.95	8488
Voss	10.87	7.45	14.97	1.97	47.71	14.46	44.36	5.1	61.76	273.02	8399
Steen	11.18	7.44	14.20	1.97	48.29	14.81	43.66	5.2	64.16	263.20	8328
Thompson	10.62	7.38	15.02	2.03	49.06	14.72	44.80	4.9	64.04	285.11	8306
Blondel	11.02	7.43	12.92	1.97	47.44	14.40	41.20	5.2	57.46	256.64	8286
Plaziat	10.83	7.72	13.58	2.12	48.34	14.18	43.06	4.9	52.18	274.07	8272
Bright	11.18	7.05	14.12	2.06	49.34	14.39	41.68	5.7	61.60	291.20	8216
De Wit	11.05	6.95	15.34	2.00	48.21	14.36	41.32	4.8	63.00	265.86	8189
Johnson	11.15	7.12	14.52	2.03	49.15	14.66	42.36	4.9	66.46	269.62	8180
Tarnovetsky	11.23	7.28	15.25	1.97	48.60	14.76	48.02	5.2	59.48	292.24	8167
Keskitalo	10.94	7.45	15.34	1.97	49.94	14.25	41.86	4.8	66.64	295.89	8143
Gaehwiler	11.18	7.34	14.48	1.94	49.02	15.11	42.76	4.7	65.84	256.74	8114
Szabo	11.02	7.29	12.92	2.06	48.23	14.94	39.54	5.0	56.80	257.85	8093
Smith	10.99	7.37	13.61	1.97	47.83	14.70	43.88	4.3	66.54	268.97	8083
Shirley	11.03	7.45	14.20	1.97	48.94	15.44	41.66	4.7	64.00	267.48	8036

Poelman	11.09	7.08	14.51	2.03	49.89	14.78	43.20	4.9	57.18	268.54	8021
Olander	11.46	6.75	16.07	2.00	51.28	16.06	50.66	4.8	72.60	302.42	7869
Freimuth	11.57	7.00	16.60	1.94	49.84	15.00	46.66	4.9	60.20	286.04	7860
Warming	11.07	7.04	13.41	1.94	47.97	14.96	40.38	4.5	51.50	262.41	7859
Hraban	10.89	7.07	15.84	1.79	49.68	15.38	45.32	4.9	60.48	277.84	7781
Werthner	11.52	7.36	13.93	1.94	49.99	15.64	38.82	4.6	67.04	266.42	7753
Gugler	11.49	7.02	13.80	2.03	50.60	15.22	39.08	4.7	60.92	262.93	7745
Penalver	11.38	7.08	14.31	2.00	50.24	14.97	46.34	4.4	55.68	272.68	7743
Kruger	11.30	6.97	13.23	2.15	49.98	15.38	38.72	4.6	54.34	277.84	7623
Lee Fu-An	11.00	7.23	13.15	2.03	49.73	14.96	38.06	4.5	52.82	285.57	7579
Mellado	11.33	6.83	11.63	2.06	48.37	15.39	37.52	4.6	55.42	270.07	7517
Moser	11.10	6.98	12.69	1.82	48.63	15.13	38.04	4.7	49.52	261.90	7505
Valenta	11.51	7.01	14.17	1.94	51.16	15.18	45.84	4.6	56.28	303.17	7422
O'Connell	11.26	6.90	12.41	1.88	48.24	15.61	38.02	4.4	52.68	272.06	7310
Richards	11.50	7.09	12.94	1.82	49.27	15.56	42.32	4.5	53.50	293.85	7237
Gong	11.43	6.22	13.98	1.91	51.25	15.88	46.18	4.6	57.84	294.99	7231
Miller	11.47	6.43	12.33	1.94	50.30	15.00	38.72	4.0	57.26	293.72	7016
Kwang-Ik	11.57	7.19	10.27	1.91	50.71	16.20	34.36	4.1	54.94	269.98	6907
Kunwar	12.12	5.83	9.71	1.70	52.32	17.05	27.10	2.6	39.10	281.24	5339

Table 16.2 Correlations between Statements about Pain

1.0								
−.0385	1.0							
.6066	−.0693	1.0						
.4507	−.1167	.5916	1.0					
.0320	.4881	.0317	−.0802	1.0				
−.2877	.4271	−.1336	−.2073	.4731	1.0			
−.2974	.3045	−.2404	−.1850	.4138	.6346	1.0		
.4526	−.3090	.5886	.6286	−.1397	−.1329	−.2599	1.0	
.2952	−.1704	.3165	.3680	−.2367	−.1541	−.2893	.4047	1.0

Table 16.3 Output from Proc Univariate for Olympic Decathlon Data

Variable: score

Moments			
N	34	Sum Weights	34
Mean	7782.85294	Sum Observations	264617
Std Deviation	594.582723	Variance	353528.614
Skewness	-2.2488675	Kurtosis	7.67309194
Uncorrected SS	2071141641	Corrected SS	11666444.3
Coeff Variation	7.63964997	Std Error Mean	101.970096

Basic Statistical Measures			
Location		Variability	
Mean	7782.853	Std Deviation	594.58272
Median	7864.500	Variance	353529
Mode	.	Range	3149
		Interquartile Range	663.00000

Tests for Location: Mu0=0				
Test	Statistic		p Value	
Student's t	t	76.32486	Pr > |t|	<.0001
Sign	M	17	Pr >= |M|	<.0001
Signed Rank	S	297.5	Pr >= |S|	<.0001

Quantiles (Definition 5)	
Quantile	**Estimate**
100% Max	8488.0
99%	8488.0
95%	8399.0
90%	8306.0
75% Q3	8180.0
50% Median	7864.5
25% Q1	7517.0
10%	7231.0
5%	6907.0
1%	5339.0
0% Min	5339.0

Extreme Observations					
Lowest			**Highest**		
Value	**name**	**Obs**	**Value**	**name**	**Obs**
5339	Kunwar	34	8286	Blondel	5
6907	Kwang-Ik	33	8306	Thompson	4
7016	Miller	32	8328	Steen	3
7231	Gong	31	8399	Voss	2
7237	Richards	30	8488	Schenk	1

3. Whether or not I am in pain depends on what the doctors do for me.
4. I cannot get any help for my pain unless I go to seek medical advice.
5. When I am in pain I know that it is because I have not been taking proper exercise or eating the right food.
6. People's pain results from their own carelessness.
7. I am directly responsible for my pain.
8. Relief from pain is chiefly controlled by the doctors.
9. People who are never in pain are just plain lucky.

For the decathlon data we will investigate ways of displaying the data graphically, and, in addition, see how a statistical approach to assigning an overall score agrees with the score shown in Table 16.1, which is calculated using a series of standard conversion tables for each event. For the pain data, the main question of interest is regarding what is the underlying structure of the pain statements.

16.2 Principal Components Analysis and Factor Analysis

Two methods of analysis will be the subjects of this chapter, principal components analysis and factor analysis. In very general terms, both can be seen as approaches to summarizing and uncovering any patterns in a set of multivariate data. The details behind each method are, however, quite different.

16.2.1 Principal Components Analysis

Principal components analysis is among the oldest and most widely used of multivariate techniques. Originally introduced by Pearson (1901) and independently by Hotelling (1933), the basic idea of the method is to describe the variation in a set of multivariate data in terms of a set of new, uncorrelated variables, each of which is defined to be a particular linear combination of the original variables. In other words, principal components analysis is a transformation from the observed variables, $x_1, \ldots x_p$, to variables, $y_1, \ldots, y_p$, where

$$\begin{aligned} y_1 &= a_{11}x_1 + a_{12}x_2 + \cdots + a_{1p}x_p \\ y_2 &= a_{21}x_1 + a_{22}x_2 + \cdots + a_{2p}x_p \\ &\vdots \\ y_p &= a_{p1}x_1 + a_{p2}x_2 + \cdots + a_{pp}x_p \end{aligned} \tag{16.1}$$

The coefficients defining each new variable are chosen so that the following conditions hold:

- The y variables (the principal components) are arranged in decreasing order of variance accounted for so that, for example, the first principal component accounts for as much as possible of the variation in the original data.
- The y variables are uncorrelated with one another.

The coefficients are found as the eigenvectors of the observed covariance matrix, **S**, although when the original variables are on very different scales, it is wiser to extract them from the observed correlation matrix, **R**, instead. The variances of the new variables are given by the eigenvectors of **S** or **R**.

The usual objective of this type of analysis is to assess whether the first few components account for a large proportion of the variation in the data, in which case they can be used to provide a convenient summary of the data for later analysis. Choosing the number of components adequate for summarizing a set of multivariate data is generally based on one or another of a number of relative ad hoc procedures:

- Retain just enough components to explain some specified large percentages of the total variation of the original variables. Values between 70% and 90% are usually suggested, although smaller values might be appropriate as the number of variables, p, or number of subjects, n, increases.
- Exclude those principal components whose eigenvalues are less than the average. When the components are extracted from the observed correlation matrix, this implies excluding components with eigenvalues less than 1.
- Plot the eigenvalues as a scree diagram and look for a clear "elbow" in the curve.

Principal component scores for an individual i with vector of variable values $\boldsymbol{x}_i'$ can be obtained from the equations

$$\begin{aligned} y_{i1} &= a_1'(\mathbf{x}_i - \bar{\mathbf{x}}) \\ &\vdots \\ y_{ip} &= a_p'(\mathbf{x}_i - \bar{\mathbf{x}}) \end{aligned} \tag{16.2}$$

where

$$\mathbf{a}_i' = \lfloor a_{i1}, a_{i2}, \ldots, a_{ip} \rfloor$$

$\bar{\boldsymbol{x}}$ is the mean vector of the observations

Full details of principal components analysis are given in Everitt and Dunn (2001).

16.2.2 Factor Analysis

Factor analysis is concerned with whether the covariances or correlations between a set of observed variables can be "explained" in terms of a smaller number of unobservable latent variables or common factors. "Explained" here means that the correlation between each pair of measured (manifest) variables arises because of their mutual association with the common factors. Consequently, the partial correlations between any pair of observed variables, given the values of the common factors, should be approximately zero.

The formal model linking manifest and latent variables is essentially that of multiple regression (see Chapter 7). In detail,

$$\begin{aligned} x_1 &= \lambda_{11} f_1 + \lambda_{12} f_2 + \cdots + \lambda_{1k} f_k + u_1 \\ x_2 &= \lambda_{21} f_1 + \lambda_{22} f_2 + \cdots + \lambda_{2k} f_k + u_2 \\ &\vdots \\ x_p &= \lambda_{p1} f_1 + \lambda_{p1} f_2 + \cdots + \lambda_{pk} f_k + u_p, \end{aligned} \tag{16.3}$$

where $f_1, f_2, \ldots, f_k$ are the latent variables (common factors), and $k < p$.

These equations may be written more concisely as

$$\mathbf{x} = \mathbf{\Lambda f} + \mathbf{u}, \tag{16.4}$$

where

$$\mathbf{\Lambda} = \begin{bmatrix} \lambda_{11} & \cdots & \lambda_k \\ \vdots & & \\ \lambda_{p1} & \cdots & \lambda_{pk} \end{bmatrix}, \quad \mathbf{f} = \begin{bmatrix} f_1 \\ \vdots \\ f_k \end{bmatrix}, \quad \mathbf{u} = \begin{bmatrix} u_1 \\ \vdots \\ u_p \end{bmatrix}$$

The residual terms (also known as specific variates), $u_1, \ldots, u_p$, are assumed uncorrelated with each other and with the common factors. The elements of $\mathbf{\Lambda}$ are usually referred to in this context as factor loadings.

Because the factors are unobserved, we can fix their location and scale arbitrarily, so we assume they are in standardized form with mean zero and standard deviation 1. (We will also assume they are uncorrelated although this is not an essential requirement.)

With these assumptions, the model in Equation 16.4 implies that the population covariance matrix of the observed variables, Σ, has the form

$$\mathbf{\Sigma} = \mathbf{\Lambda\Lambda}' + \mathbf{\Psi}, \tag{16.5}$$

where $\mathbf{\Psi}$ is a diagonal matrix containing the variances of the residual terms, $\psi_i = 1 \cdots p$.

The parameters in the factor analysis model may be estimated in a number of ways, including maximum likelihood, which also leads to a test for number of factors. The initial solution may be "rotated" as an aid to interpretation as described fully in Everitt and Dunn (2001). (Principal components may also be rotated but then the defining maximal proportion of variance property is lost.)

16.2.3 Factor Analysis and Principal Components Compared

Factor analysis, like principal components analysis, is an attempt to explain a set of data in terms of a smaller number of dimensions than one starts with, but the procedures used to achieve this goal are essentially quite different in the two methods. Factor analysis, unlike principal components analysis, begins with a hypothesis about the covariance (or correlational) structure of the variables. Formally, this hypothesis is that a covariance matrix $\mathbf{\Sigma}$, of order and rank p, can be partitioned into two matrices, $\mathbf{\Lambda\Lambda}'$ and $\mathbf{\Psi}$. The first is of order p but rank k (the number of common factors), whose off-diagonal elements are equal to those of $\mathbf{\Sigma}$. The second is a diagonal matrix of full rank p, whose elements when added to the diagonal elements of $\mathbf{\Lambda\Lambda}'$ give the diagonal elements of Σ. In other words, the hypothesis is that a set of k latent variables exists ($k < p$), and these are adequate to account for the inter-relationships of the variables although not for their full variances. Principal components analysis, however, is merely a transformation of the data, and no assumptions are made about the form of the covariance matrix from which the data arise. This type of analysis has no part corresponding to the specific variates of factor analysis. Consequently, if the factor model holds but the variances of the specific variables are small, we would expect both forms of analysis to give similar results. If, however, the specific variances are large, they will be absorbed into all the principal components, both retained and rejected, whereas factor analysis makes special provision for them. It should be remembered that both forms of analysis are similar in one important respect, namely that they are both pointless if the observed variables are almost uncorrelated, factor analysis because it has nothing to explain and principal components analysis because it would simply lead to components that are similar to the original variables.

16.3 Analysis Using SAS

16.3.1 Olympic Decathlon

The file olympic.dat contains the data shown in Table 16.1, with the names in the first 13 columns. They can be read with the following data step:

```
data decathlon;
  infile 'c:\handbook3\datasets\olympic.dat';
  input name $ 1-13 run100 Ljump shot Hjump run400 hurdle
    discus polevlt javelin run1500 score;
run;
```

Before undertaking a principal components analysis of the data, it is advisable to check them in some way for outliers. Here we shall examine the distribution of the total score assigned to each competitor with proc univariate and produce a box plot with proc sgplot.

```
proc univariate data=decathlon;
  var score;
  id name;
run;
proc sgplot data=decathlon;
  vbox score/datalabel=name;
run;
```

Details of the proc univariate were given in Chapter 2. The id statement is used here to label the extreme observations with the athlete's name. The proc sgplot step

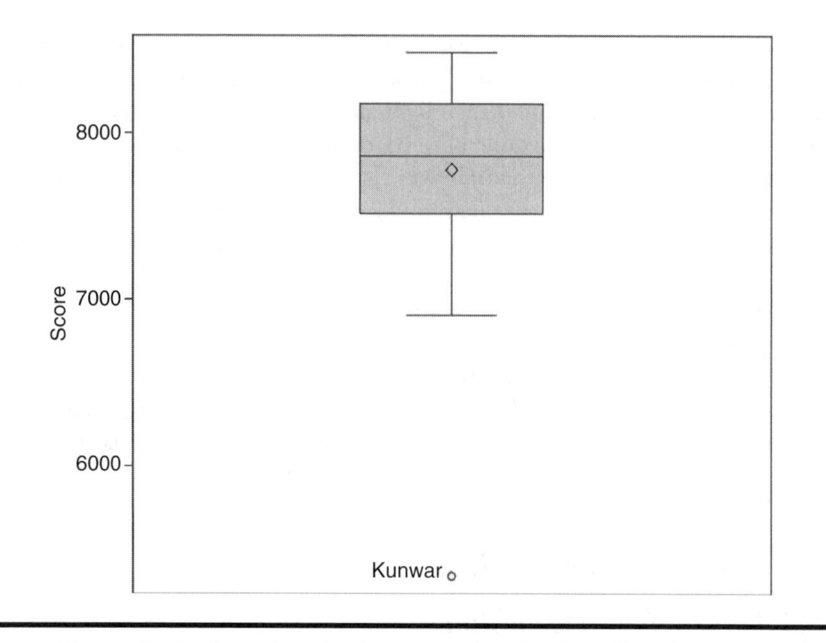

Figure 16.1 Distribution of total scores for Olympic decathlon data.

shows that it can be used for an overall box plot for the entire sample. The output of proc univariate is given in Table 16.3 and the box plot in Figure 16.1.

The athlete, Kunwar, with the lowest score, is very clearly an outlier and will now be removed from the data set before further analysis. And it will help in interpreting results if all events are "scored" in the same direction, so we take negative values for the four running events. In this way, all 10 events are such that small values represent a poor performance and large values the reverse.

```
data decathlon;
  set decathlon;
  if score > 6000;
  run100 = run100*-1;
  run400 = run400*-1;
  hurdle = hurdle*-1;
  run1500 = run1500*-1;
run;
```

A principal components can now be applied using proc princomp.

```
proc princomp data = decathlon out = pcout;
  var run100 -- run1500;
run;
```

The out= option in the proc statement names a data set that will contain the principal component scores plus all the original variables. The analysis is applied to the correlation matrix by default.

The output is shown as Table 16.4. Notice first that the components as given are scaled so that the sums of squares of their elements are equal to 1. To rescale them so that they represent correlations between variables and components, they would need to be multiplied by the square root of the corresponding eigenvalue. The coefficients defining the first component are all positive and it is clearly a measure of overall performance (see later). This component has variance 3.42 and

Table 16.4 Numerical and Graphical Output from Proc Princomp for Olympic Decathlon Data

Observations	33
Variables	10

Simple Statistics					
	run100	**Ljump**	**shot**	**Hjump**	**run400**
Mean	-11.19636364	7.133333333	13.97636364	1.982727273	-49.27666667
StD	0.24332101	0.304340133	1.33199056	0.093983799	1.06966019

Simple Statistics					
	hurdle	**discus**	**polevlt**	**javelin**	**run1500**
Mean	-15.04878788	42.35393939	4.739393939	59.43878788	-276.0384848
StD	0.50676522	3.71913123	0.334420575	5.49599841	13.6570975

Correlation Matrix										
	run100	**Ljump**	**shot**	**Hjump**	**run400**	**hurdle**	**discus**	**polevlt**	**javelin**	**run1500**
run100	1.0000	0.5396	0.2080	0.1459	0.6059	0.6384	0.0472	0.3891	0.0647	0.2610
Ljump	0.5396	1.0000	0.1419	0.2731	0.5153	0.4780	0.0419	0.3499	0.1817	0.3956
shot	0.2080	0.1419	1.0000	0.1221	-.0946	0.2957	0.8064	0.4800	0.5977	-.2688
Hjump	0.1459	0.2731	0.1221	1.0000	0.0875	0.3067	0.1474	0.2132	0.1159	0.1141
run400	0.6059	0.5153	-.0946	0.0875	1.0000	0.5460	-.1422	0.3187	-.1204	0.5873
hurdle	0.6384	0.4780	0.2957	0.3067	0.5460	1.0000	0.1105	0.5215	0.0628	0.1433
discus	0.0472	0.0419	0.8064	0.1474	-.1422	0.1105	1.0000	0.3440	0.4429	-.4023
polevlt	0.3891	0.3499	0.4800	0.2132	0.3187	0.5215	0.3440	1.0000	0.2742	0.0315
javelin	0.0647	0.1817	0.5977	0.1159	-.1204	0.0628	0.4429	0.2742	1.0000	-.0964
run1500	0.2610	0.3956	-.2688	0.1141	0.5873	0.1433	-.4023	0.0315	-.0964	1.0000

Eigenvalues of the Correlation Matrix				
	Eigenvalue	Difference	Proportion	Cumulative
1	3.41823814	0.81184501	0.3418	0.3418
2	2.60639314	1.66309673	0.2606	0.6025
3	0.94329641	0.06527516	0.0943	0.6968
4	0.87802124	0.32139459	0.0878	0.7846
5	0.55662665	0.06539914	0.0557	0.8403
6	0.49122752	0.06063230	0.0491	0.8894
7	0.43059522	0.12379709	0.0431	0.9324
8	0.30679812	0.03984871	0.0307	0.9631
9	0.26694941	0.16509526	0.0267	0.9898
10	0.10185415		0.0102	1.0000

Eigenvectors										
	Prin1	Prin2	Prin3	Prin4	Prin5	Prin6	Prin7	Prin8	Prin9	Prin10
run100	0.415882	-.148808	-.267472	-.088332	-.442314	-.030712	0.254398	0.663713	-.108395	0.109480
Ljump	0.394051	-.152082	0.168949	0.244250	-.368914	-.093782	-.750534	-.141264	-.046139	-.055804
shot	0.269106	0.483537	-.098533	0.107763	0.009755	0.230021	0.110664	-.072506	-.422476	-.650737
Hjump	0.212282	0.027898	0.854987	-.387944	0.001876	0.074544	0.135124	0.155436	0.102065	-.119412
run400	0.355847	-.352160	-.189496	0.080575	0.146965	0.326929	0.141339	-.146839	0.650762	-.336814
hurdle	0.433482	-.069568	-.126160	-.382290	-.088803	-.210491	0.272530	-.639004	-.207239	0.259718
discus	0.175792	0.503335	-.046100	-.025584	-.019359	0.614912	-.143973	-.009400	0.167241	0.534503
polevlt	0.384082	0.149582	-.136872	-.143965	0.716743	-.347760	-.273266	0.276873	0.017664	0.065896
javelin	0.179944	0.371957	0.192328	0.600466	-.095582	-.437444	0.341910	-.058519	0.306196	0.130932
run1500	0.170143	-.420965	0.222552	0.485642	0.339772	0.300324	0.186870	0.007310	-.456882	0.243118

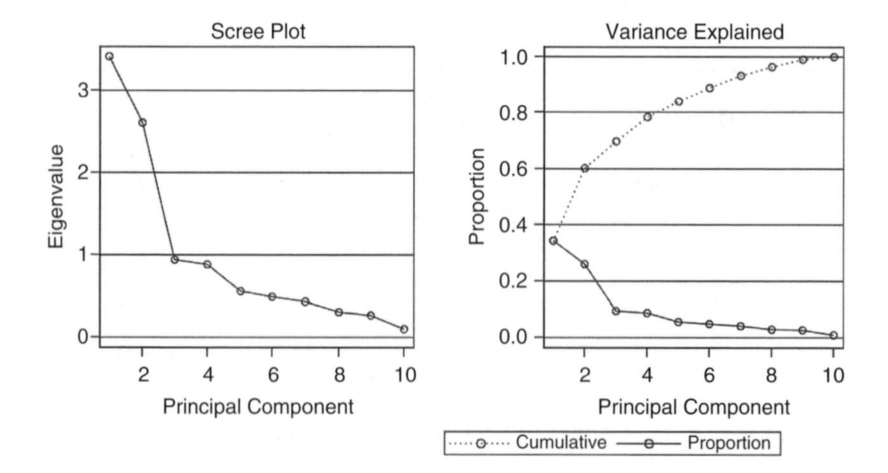

accounts for 34% of the total variation in the data. The second component contrasts performance on the "power" events such as shot and discuss with the only really "stamina" event, the 1500 m. The second component has variance 2.61, so between them the first two components account for 60% of the total variance.

Only the first two components have eigenvalues greater than 1, suggesting the first two principal component scores for each athlete provide an adequate and parsimonious description of the data.

We can use the first two principal component scores to produce a useful plot of the data, particularly if we label the points in an informative manner. As an example, we label the plot of the principal component scores with the athlete's overall position in the event.

```
proc rank data = pcout out = pcout descending;
  var score;
  ranks posn;
proc sgplot data = pcout;
  scatter y = prin1 x = prin2/markerchar = posn;
run;
```

Proc rank is used to calculate the finishing position in the event. The variable score is ranked in descending order and the ranks stored in the variable posn.

In the sgplot step, the markerchar option substitutes the value of posn for the plotting symbols. The resulting plot is shown in Figure 16.2. We will comment on this plot later.

Next we can plot the total score achieved by each athlete in the competition against each of the first two principal component scores and also find the corresponding correlations. Plots of the overall score against the first two principal components are shown in Figure 16.3 and the correlations in Table 16.5.

```
proc sgscatter data = pcout;
  compare y = score x = (prin1 prin2);
run;
proc corr data = pcout;
  var score prin1 prin2;
run;
```

Figure 16.3 shows the very strong relationship between total score and first principal component score – the correlation of the two variables is found from Table 16.5 to be 0.96 which is, of course, highly significant. But total score does not appear to be related to the second principal component score ($r = 0.16$).

And returning to Figure 16.2, the first principal component score is seen to largely rank the athletes in finishing position, confirming its interpretation as an overall measure of performance.

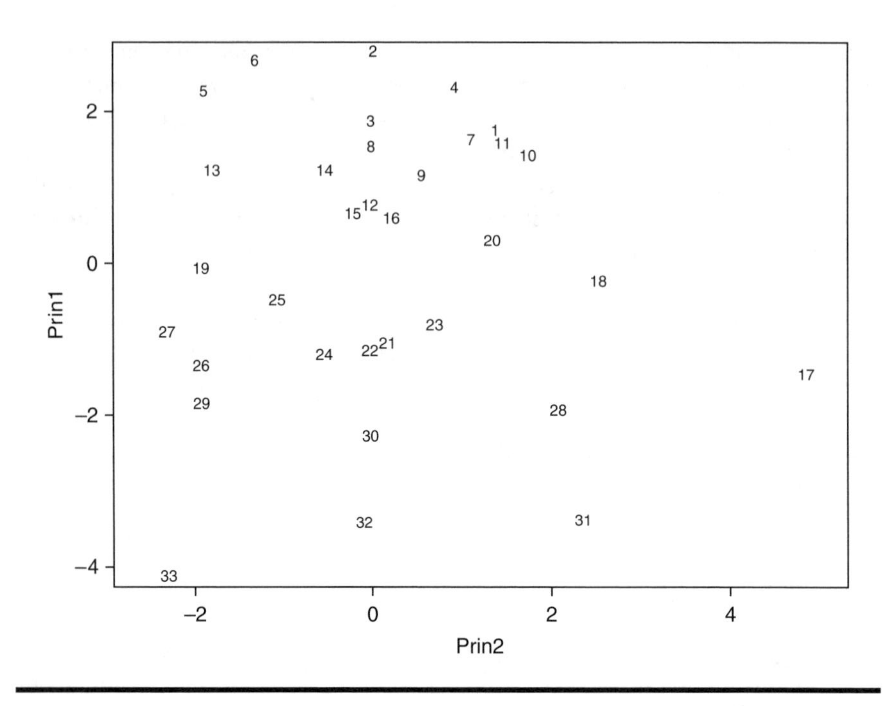

Figure 16.2 Plot of the first two principal component scores for the Olympic decathlon data.

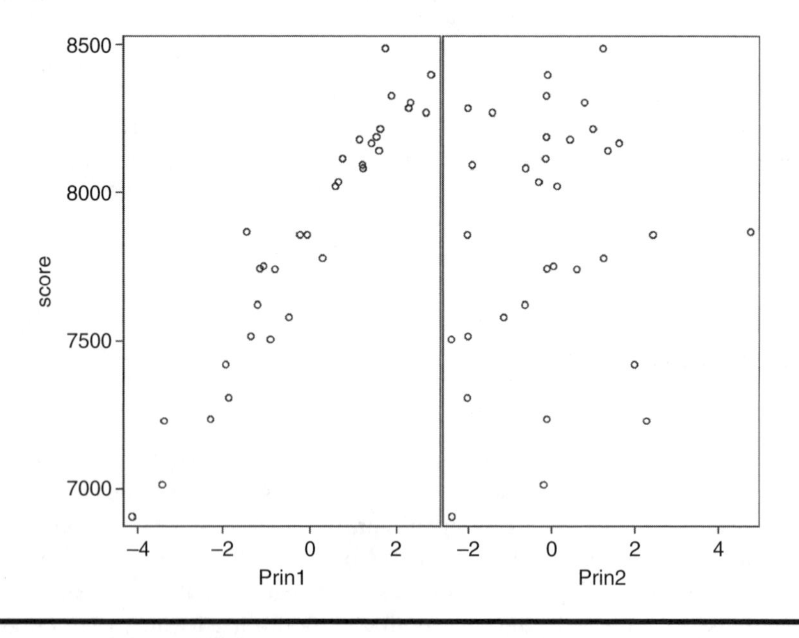

Figure 16.3 Plots of total score in Olympic decathlon against the first two principal component scores.

Table 16.5 Correlations between Overall Score in the Olympic Decathlon and the First Two Principal Component Scores

3 Variables:	score	Prin1	Prin2

Simple Statistics						
Variable	**N**	**Mean**	**Std Dev**	**Sum**	**Minimum**	**Maximum**
score	33	7857	415.06945	259278	6907	8488
Prin1	33	0	1.84885	0	-4.12390	2.78724
Prin2	33	0	1.61443	0	-2.40788	4.77144

Pearson Correlation Coefficients, N = 33
Prob > |r| under H0: Rho=0

	score	**Prin1**	**Prin2**
score	1.00000	0.96158 <.0001	0.16194 0.3679
Prin1	0.96158 <.0001	1.00000	0.00000 1.0000
Prin2	0.16194 0.3679	0.00000 1.0000	1.00000

16.3.2 Statements about Pain

The SAS procedure proc factor can accept data in the form of a correlation, or covariance matrix, as well as in the normal rectangular data matrix. To analyse a correlation or covariance matrix, the data need to be read into a special SAS data set with type = corr or type = cov. The correlation matrix shown in Table 16.3 was edited into the form shown in Table 16.6 and read in as follows:

```
data pain (type = corr);
infile 'c:\handbook3\datasets\pain.dat' expandtabs missover;
input _type_ $ _name_ $ p1-p9;
run;
```

The type = corr option on the data statement specifies the type of SAS data set being created. The value of the _type_ variable indicates what type of information the observation holds. When _type_ = "CORR", the values of the variables are correlation coefficients. When _type_ = "N", the values are the sample sizes. Only the correlations are necessary but the sample sizes have been entered because they will be used by the maximum likelihood method for the test of the number of factors.

Table 16.6 Correlations between Pain Statements

CORR	p1	1.0								
CORR	p2	−.0385	1.0							
CORR	p3	.6066	−.0693	1.0						
CORR	p4	.4507	−.1167	.5916	1.0					
CORR	p5	.0320	.4881	.0317	−.0802	1.0				
CORR	p6	−.2877	.4271	−.1336	−.2073	.4731	1.0			
CORR	p7	−.2974	.3045	−.2404	−.1850	.4138	.6346	1.0		
CORR	p8	.4526	−.3090	.5886	.6286	−.1397	−.1329	−.2599	1.0	
CORR	p9	.2952	−.1704	.3165	.3680	−.2367	−.1541	−.2893	.4047	1.0
N	N	123	123	123	123	123	123	123	123	123

The _name_ variable identifies the variable whose correlations are in that row of the matrix. The missover option on the infile statement obviates the need to enter the data for the upper triangle of the correlation matrix.

Both principal components analysis and maximum likelihood factor analysis might be applied to the pain statement data using proc factor. The following, however, specifies a maximum likelihood factor analysis extracting three factors and requesting a scree plot, often useful in selecting the appropriate number of components. A varimax rotation of the factors is also requested. The output is shown in Table 16.7.

```
proc factor data=pain method=ml n=3 rotate=varimax;
  var p1-p9;
run;
```

First the test for number factors indicates that a three factor solution provides an adequate description of the observed correlations. We can try to identify the three

Table 16.7 Output from Proc Factor Applied to Correlations between Pain Statements

Initial Factor Method: Maximum Likelihood

Prior Communality Estimates: SMC

p1	p2	p3	p4	p5	p6	p7	p8	p9
0.46369858	0.37626982	0.54528471	0.51155233	0.39616724	0.55718109	0.48259656	0.56935053	0.25371373

Preliminary Eigenvalues: Total = 8.2234784 Average = 0.91371982

	Eigenvalue	Difference	Proportion	Cumulative
1	5.85376325	3.10928282	0.7118	0.7118
2	2.74448043	1.96962348	0.3337	1.0456
3	0.77485695	0.65957907	0.0942	1.1398
4	0.11527788	0.13455152	0.0140	1.1538
5	-.01927364	0.13309824	-0.0023	1.1515
6	-.15237189	0.07592411	-0.0185	1.1329
7	-.22829600	0.10648720	-0.0278	1.1052
8	-.33478320	0.19539217	-0.0407	1.0645
9	-.53017537		-0.0645	1.0000

3 factors will be retained by the NFACTOR criterion.

Iteration	Criterion	Ridge	Change	Communalities								
1	0.1604994	0.0000	0.2170	0.58801	0.43948	0.66717	0.54503	0.55113	0.77414	0.52219	0.75509	0.24867
2	0.1568974	0.0000	0.0395	0.59600	0.47441	0.66148	0.54755	0.51168	0.81079	0.51814	0.75399	0.25112
3	0.1566307	0.0000	0.0106	0.59203	0.47446	0.66187	0.54472	0.50931	0.82135	0.51377	0.76242	0.24803
4	0.1566095	0.0000	0.0029	0.59192	0.47705	0.66102	0.54547	0.50638	0.82420	0.51280	0.76228	0.24757
5	0.1566078	0.0000	0.0008	0.59151	0.47710	0.66101	0.54531	0.50612	0.82500	0.51242	0.76293	0.24736

Convergence criterion satisfied.

Significance Tests Based on 123 Observations

Test	DF	Chi-Square	Pr > ChiSq
H0: No common factors	36	400.8045	<.0001
HA: At least one common factor			
H0: 3 Factors are sufficient	12	18.1926	0.1100
HA: More factors are needed			

Chi-Square without Bartlett's Correction	19.106147
Akaike's Information Criterion	-4.893853
Schwarz's Bayesian Criterion	-38.640066
Tucker and Lewis's Reliability Coefficient	0.949075

Squared Canonical Correlations

Factor1	Factor2	Factor3
0.90182207	0.83618918	0.60884385

Eigenvalues of the Weighted Reduced Correlation Matrix: Total = 15.8467138 Average = 1.76074598				
	Eigenvalue	Difference	Proportion	Cumulative
1	9.18558880	4.08098588	0.5797	0.5797
2	5.10460292	3.54807912	0.3221	0.9018
3	1.55652380	1.26852906	0.0982	1.0000
4	0.28799474	0.10938119	0.0182	1.0182
5	0.17861354	0.08976744	0.0113	1.0294
6	0.08884610	0.10414259	0.0056	1.0351
7	-.01529648	0.16841933	-0.0010	1.0341
8	-.18371581	0.17272798	-0.0116	1.0225
9	-.35644379		-0.0225	1.0000

Factor Pattern			
	Factor1	Factor2	Factor3
p1	0.60516	0.29433	0.37238
p2	-0.45459	0.29155	0.43073
p3	0.61386	0.49738	0.19172
p4	0.62154	0.39877	-0.00365
p5	-0.40635	0.45042	0.37154
p6	-0.67089	0.59389	-0.14907
p7	-0.62525	0.34279	-0.06302
p8	0.68098	0.47418	-0.27269
p9	0.44944	0.16166	-0.13855

Variance Explained by Each Factor		
Factor	Weighted	Unweighted
Factor1	9.18558880	3.00788644
Factor2	5.10460292	1.50211187
Factor3	1.55652380	0.61874873

Final Communality Estimates and Variable Weights		
Total Communality: Weighted = 15.846716 Unweighted = 5.128747		
Variable	Communality	Weight
p1	0.59151181	2.44807030
p2	0.47717797	1.91240023
p3	0.66097328	2.94991222
p4	0.54534606	2.19927836
p5	0.50603810	2.02479887
p6	0.82501333	5.71444465
p7	0.51242072	2.05095025
p8	0.76294154	4.21819901
p9	0.24732424	1.32865993

Rotation Method: Varimax

Orthogonal Transformation Matrix			
	1	2	3
1	0.72941	-0.56183	-0.39027
2	0.68374	0.61659	0.39028
3	0.02137	-0.55151	0.83389

Rotated Factor Pattern			
	Factor1	Factor2	Factor3
p1	0.65061	-0.36388	0.18922
p2	-0.12303	0.19762	0.65038
p3	0.79194	-0.14394	0.11442
p4	0.72594	-0.10131	-0.08998
p5	0.01951	0.30112	0.64419
p6	-0.08648	0.82532	0.36929
p7	-0.22303	0.59741	0.32525
p8	0.81511	0.06018	-0.30809
p9	0.43540	-0.07642	-0.22784

Variance Explained by Each Factor		
Factor	**Weighted**	**Unweighted**
Factor1	7.27423715	2.50415379
Factor2	5.31355675	1.34062697
Factor3	3.25892162	1.28396628

Final Communality Estimates and Variable Weights		
Total Communality: Weighted = 15.846716 Unweighted = 5.128747		
Variable	**Communality**	**Weight**
p1	0.59151181	2.44807030
p2	0.47717797	1.91240023
p3	0.66097328	2.94991222
p4	0.54534606	2.19927836
p5	0.50603810	2.02479887
p6	0.82501333	5.71444465
p7	0.51242072	2.05095025
p8	0.76294154	4.21819901
p9	0.24732424	1.32865993

common factors by examining the rotated loading in Table 16.7. The first factor loads highly on statements 1, 3, 4 and 8. These statements attribute pain relief to the control of doctors, and so we might label the factor, doctors control of pain. The second factor has its highest loadings on statements 6 and 7. These statements associated the cause of pain as one's own actions, and the factor might be labelled individual's responsibility for pain. The third factor has high loadings on statements 2 and 5. Again both involve an individual's own responsibility for his pain but now specifically because of things he has not done; the factor might be labelled lifestyle responsibility for pain.

Exercises

1 Olympic Decathlon

1. Repeat the principal components analysis of the Olympic decathlon data without removing the athlete who finished last in the competition.
2. How do the results compare with those reported in the chapter?

2 Pain Data

1. Run a principal components analysis on the pain data and compare the results with those from the maximum likelihood factor analysis.

3 Life Expectancies

Keyfitz and Flieger (1972) give life expectancies in years by country, age, and sex. (The figures given are for the 1960s.)

The variables in life.dat are

- **Country:** The country involved
- **M0:** Male life expectancy at birth
- **M25:** Male life expectancy at age 25
- **M50:** Male life expectancy at age 50
- **M75:** Male life expectancy at age 75
- **F0:** Female life expectancy at birth
- **F25:** Female life expectancy at age 25
- **F50:** Female life expectancy at age 50
- **F75:** Female life expectancy at age 75

1. Carry out a principal components analysis of the eight life expectancies using the covariance matrix.
2. Examine the scree plot and decide how many components are needed to adequately describe the data.
3. Use the chosen principal components from 2 to construct a graphical representation of the data.
4. Interpret your results.

4 Measurements on Criminals

The correlation matrix below has been calculated from the measurements of 7 physical characteristics in each of 3000 criminals.

1	1.000						
2	0.402	1.000					
3	0.396	0.618	1.000				
4	0.301	0.150	0.321	1.000			
5	0.305	0.135	0.289	0.846	1.000		
6	0.339	0.206	0.363	0.759	0.797	1.000	
7	0.340	0.183	0.345	0.661	0.800	0.736	1.000

The variables are as follows:

- Head length
- Head breadth
- Face breadth
- Left finger length
- Left forearm length
- Left foot length
- Height

1. Find the principal components of the correlation matrix.
2. Can you interpret the derived components?
3. Compare the PCA results with those from a maximum likelihood factor analysis of the correlations.

Chapter 17

Cluster Analysis: Air Pollution in the United States

17.1 Introduction

The data to be analysed in this chapter relate to air pollution in 41 U.S. cities. The data for five cities are given in Table 17.1 (they also appear in *SDS* as Table 26). Seven variables are recorded for each of the cities:

1. SO_2 content of air in micrograms per cubic metre
2. Average annual temperature in Fahrenheit
3. Number of manufacturing enterprises employing 20 or more workers
4. Population size (1970 census) in thousands
5. Average annual wind speed in miles per hour
6. Average annual precipitation in inches
7. Average number of days with precipitation per year

In this chapter we will use variables 2 to 7 in a cluster analysis of the data to investigate whether there is any evidence of distinct groups of cities. The resulting clusters will then be assessed in terms of their air pollution levels as measured by SO_2 content.

Table 17.1 Data on 41 Cities in the United States

Cities	*1*	*2*	*3*	*4*	*5*	*6*	*7*
Phoenix	10	70.3	213	582	6.0	7.05	36
Little Rock	13	61.0	91	132	8.2	48.52	100
San Francisco	12	56.7	453	716	8.7	20.66	67
Denver	17	51.9	454	515	9.0	12.95	86
Hartford	56	49.1	412	158	9.0	43.37	127

17.2 Cluster Analysis

Cluster analysis is a generic term for a large number of techniques that have the common aim of determining whether a (usually) multivariate data set contains distinct groups or clusters of observations and, if so, finding which of the observations belong in the same cluster. A detailed account of cluster analysis is given in Everitt et al. (2001).

The most commonly used class of clustering methods consists of those that lead to a series of nested or hierarchical classifications of the observations, beginning at the stage in which each observation is regarded as forming a single member "cluster" and ending at the stage in which all the observations are in a single group. The complete hierarchy of solutions can be displayed as a tree diagram known as a *dendrogram*. In practice, most users will be interested in choosing a particular partition of the data, that is, a particular number of groups which is optimal in some sense. This will entail "cutting" the dendrogram at some particular level.

Most hierarchical methods operate not on the raw data but on an inter-individual distance matrix calculated from the raw data. The most commonly used distance measure is Euclidean and is defined as

$$d_{ij} = \sqrt{\sum_{k=1}^{p} (x_{ik} - x_{jk})^2}, \qquad (17.1)$$

where x_{ik} and x_{jk} are the values of the kth variable for observations i and j, respectively, and p is the number of variables.

The different members of the class of hierarchical clustering techniques arise because of the variety of ways in which the distance between a cluster containing several observations and a single observation, or between two clusters, can be defined. The inter-cluster distances used by three commonly applied hierarchical clustering techniques are

1. *Single linkage clustering*: Distance between the closest observations.
2. *Complete linkage clustering*: Distance between the most remote observations.
3. *Average linkage clustering*: Average of distances between all pairs of observations, where members of a pair are in different groups.

Important issues that often need to be considered when using clustering in practice include how to scale the variables before calculating the distance matrix, which particular method of cluster analysis to use and how to decide on the appropriate number of groups in the data. All these and the many other practical problems of clustering are discussed in Everitt et al. (2001).

17.3 Analysis Using SAS

The data set for Table 26 in *SDS* does not contain the city names shown in Table 17.1. So we have edited the data set so that they occupy the first 16 columns. The resulting data set can be read in as follows:

```
data usair;
  infile 'c:\handbook3\datasets\usair2.dat' expandtabs;
  input city $16. so2 temperature factories population windspeed rain
     rainydays;
run;
```

The names of the cities are read into the variable city with a $16. format as several of them contain spaces and are longer than the default length of eight characters. The numeric data are read in with list input.

We begin by examining the distributions of the six variables to be used in the cluster analysis.

```
proc univariate data=usair;
  var temperature--rainydays;
  id city;
run;
```

The univariate procedure was described in Chapter 2. The id statement has the effect of labelling the extreme observations by city name rather than simply by observation number.

The output for factories and population is shown in Table 17.2. Chicago is clearly an outlier both in terms of manufacturing enterprises and population size. Although less extreme, Phoenix has the lowest value on all three climate variables (relevant output not given to save space). Both cities will therefore be excluded from the data set to be analysed.

```
data usair2;
  set usair;
  if city not in('Chicago','Phoenix');
run;
```

Table 17.2 Output from Proc Univariate for the Factories and Population Variables in the U.S. Cities Data

The UNIVARIATE Procedure

Variable: factories

Moments			
N	41	Sum Weights	41
Mean	463.097561	Sum Observations	18987
Std Deviation	563.473948	Variance	317502.89
Skewness	3.75488343	Kurtosis	17.403406
Uncorrected SS	21492949	Corrected SS	12700115.6
Coeff Variation	121.674998	Std Error Mean	87.9998462

Basic Statistical Measures			
Location		Variability	
Mean	463.0976	Std Deviation	563.47395
Median	347.0000	Variance	317503
Mode	.	Range	3309
		Inter-quartie Rangel	281.00000

Tests for Location: Mu0=0				
Test	Statistic		p Value	
Student's t	t	5.262481	Pr > ltl	<.0001
Sign	M	20.5	Pr >= lMl	<.0001
Signed Rank	S	430.5	Pr >= lSl	<.0001

Quantiles (Definition 5)	
Quantile	Estimate
100% Max	3344
99%	3344
95%	1064
90%	775
75% Q3	462
50% Median	347
25% Q1	181
10%	91
5%	46
1%	35
0% Min	35

Extreme Observations					
Lowest			Highest		
Value	city	Obs	Value	city	Obs
35	Charleston	40	775	St. Louis	21
44	Albany	24	1007	Cleveland	27
46	Albuquerque	23	1064	Detroit	18
80	Wilmington	6	1692	Philadelphia	29
91	Little Rock	2	3344	Chicago	11

The UNIVARIATE Procedure

Variable: population

Moments			
N	41	Sum Weights	41
Mean	608.609756	Sum Observations	24953
Std Deviation	579.113023	Variance	335371.894
Skewness	3.16939401	Kurtosis	12.9301083
Uncorrected SS	28601515	Corrected SS	13414875.8
Coeff Variation	95.1534243	Std Error Mean	90.4422594

Basic Statistical Measures			
Location		Variability	
Mean	608.6098	Std Deviation	579.11302
Median	515.0000	Variance	335372
Mode	.	Range	3298
		Inter-quartile Range	418.00000

Tests for Location: Mu0=0				
Test	Statistic		p Value	
Student's t	t	6.729263	Pr > |t|	<.0001
Sign	M	20.5	Pr >= |M|	<.0001
Signed Rank	S	430.5	Pr >= |S|	<.0001

Quantiles (Definition 5)	
Quantile	**Estimate**
100% Max	3369
99%	3369
95%	1513
90%	905
75% Q3	717
50% Median	515
25% Q1	299
10%	158
5%	116
1%	71
0% Min	71

Extreme Observations					
	Lowest			**Highest**	
Value	**city**	**Obs**	**Value**	**city**	**Obs**
71	Charleston	40	905	Baltimore	17
80	Wilmington	6	1233	Houston	35
116	Albany	24	1513	Detroit	18
132	Little Rock	2	1950	Philadelphia	29
158	Hartford	5	3369	Chicago	11

A single linkage cluster analysis and corresponding dendrogram can be obtained as follows:

```
proc cluster data = usair2 method = single simple ccc std outtree = single;
  var temperature--rainydays;
  id city;
  copy so2;
run;
goptions htext = .8;
axis1 label = (a = 90);
proc tree horizontal vaxis = axis1;
run;
```

The method = option on the proc statement is self-explanatory. The simple option provides information about the distribution of the variables used in the clustering. The ccc option includes the cubic clustering criterion in the output, which may be useful for indicating number of groups (Sarle, 1983). The std option standardizes the clustering variables to zero mean and unit variance, and the outtree = option names the data set that contains the information to be used in the dendrogram.

The var statement specifies which variables are to be used to cluster the observations and the id statement specifies the variable to be used to label the observations in the printed output and in the dendrogram. Variables mentioned in a copy statement are included in the outtree data set. Those mentioned on the var and id statements are included by default.

Proc tree produces the dendrogram using the outtree data set. It uses the traditional graphics system, and the goptions and axis statements are used to improve the appearance. The htext = .8 option reduces the text size on the dendrograms. The axis1 statement rotates the axis label 90° anticlockwise. This is then used for the vertical axis (vaxis = axis1). The horizontal (hor) option specifies the orientation which is vertical by default. The data set to be used by proc tree is left implicit and so will be the most recently created data set, that is, single.

The printed results are shown in Table 17.3 and the dendrogram in Figure 17.1. We see that Atlanta and Memphis are joined first to form a two number group.

Table 17.3 Results from Proc Cluster for Single Linkage Applied to the U.S. Cities Data

The CLUSTER Procedure

Single Linkage Cluster Analysis

Variable	Mean	Std Dev	Skewness	Kurtosis	Bimodality
temperature	55.5231	6.9762	0.9101	0.7883	0.4525
factories	395.6	330.9	1.9288	5.2670	0.5541
population	538.5	384.0	1.7536	4.3781	0.5341
windspeed	9.5077	1.3447	0.3096	0.2600	0.3120
rain	37.5908	11.0356	-0.6498	1.0217	0.3328
rainydays	115.7	23.9760	-0.1314	0.3393	0.2832

Eigenvalues of the Correlation Matrix				
	Eigenvalue	Difference	Proportion	Cumulative
1	2.09248727	0.45164599	0.3487	0.3487
2	1.64084127	0.36576347	0.2735	0.6222
3	1.27507780	0.48191759	0.2125	0.8347
4	0.79316021	0.67485359	0.1322	0.9669
5	0.11830662	0.03817979	0.0197	0.9866
6	0.08012683		0.0134	1.0000

The data have been standardized to mean 0 and variance 1

Root-Mean-Square Total-Sample Standard Deviation	1

Mean Distance Between Observations	3.21916

Cluster History

NCL	Clusters Joined		FREQ	SPRSQ	RSQ	ERSQ	CCC	Norm Min Dist	Tie
38	Atlanta	Memphis	2	0.0007	.999	.	.	0.1709	
37	Jacksonville	New Orleans	2	0.0008	.998	.	.	0.1919	
36	Des Moines	Omaha	2	0.0009	.998	.	.	0.2023	
35	Nashville	Richmond	2	0.0009	.997	.	.	0.2041	
34	Pittsburgh	Seattle	2	0.0013	.995	.	.	0.236	
33	Louisville	CL35	3	0.0023	.993	.	.	0.2459	
32	Washington	Baltimore	2	0.0015	.992	.	.	0.2577	
31	Columbus	CL34	3	0.0037	.988	.	.	0.2673	
30	CL32	Indianapolis	3	0.0024	.985	.	.	0.2823	
29	CL33	CL31	6	0.0240	.961	.	.	0.3005	
28	CL38	CL29	8	0.0189	.943	.	.	0.3191	
27	CL30	St. Louis	4	0.0044	.938	.	.	0.322	
26	CL27	Kansas City	5	0.0040	.934	.	.	0.3348	
25	CL26	CL28	13	0.0258	.908	.	.	0.3638	
24	Little Rock	CL25	14	0.0178	.891	.	.	0.3651	
23	Minneapolis	Milwaukee	2	0.0032	.887	.	.	0.3775	
22	Hartford	Providence	2	0.0033	.884	.	.	0.3791	
21	CL24	Cincinnati	15	0.0104	.874	.	.	0.3837	
20	CL21	CL36	17	0.0459	.828	.	.	0.3874	
19	CL37	Miami	3	0.0050	.823	.	.	0.4093	
18	CL20	CL22	19	0.0152	.808	.	.	0.4178	
17	Denver	Salt Lake City	2	0.0040	.804	.	.	0.4191	
16	CL18	CL19	22	0.0906	.713	.	.	0.421	
15	CL16	Wilmington	23	0.0077	.705	.	.	0.4257	
14	San Francisco	CL17	3	0.0083	.697	.	.	0.4297	
13	CL15	Albany	24	0.0184	.679	.	.	0.4438	
12	CL13	Norfolk	25	0.0084	.670	.	.	0.4786	
11	CL12	Wichita	26	0.0457	.625	.	.	0.523	
10	CL14	Albuquerque	4	0.0097	.615	.	.	0.5328	
9	CL23	Cleveland	3	0.0100	.605	.	.	0.5329	

Cluster History

NCL	Clusters Joined		FREQ	SPRSQ	RSQ	ERSQ	CCC	Norm Min Dist	Tie
8	CL11	Charleston	27	0.0314	.574	.	.	0.5662	
7	Dallas	Houston	2	0.0078	.566	.731	-6.1	0.5861	
6	CL8	CL9	30	0.1032	.463	.692	-7.6	0.6433	
5	CL6	Buffalo	31	0.0433	.419	.644	-7.3	0.6655	
4	CL5	CL10	35	0.1533	.266	.580	-8.2	0.6869	
3	CL4	CL7	37	0.0774	.189	.471	-6.6	0.6967	
2	CL3	Detroit	38	0.0584	.130	.296	-4.0	0.7372	
1	CL2	Philadelphia	39	0.1302	.000	.000	0.00	0.7914	

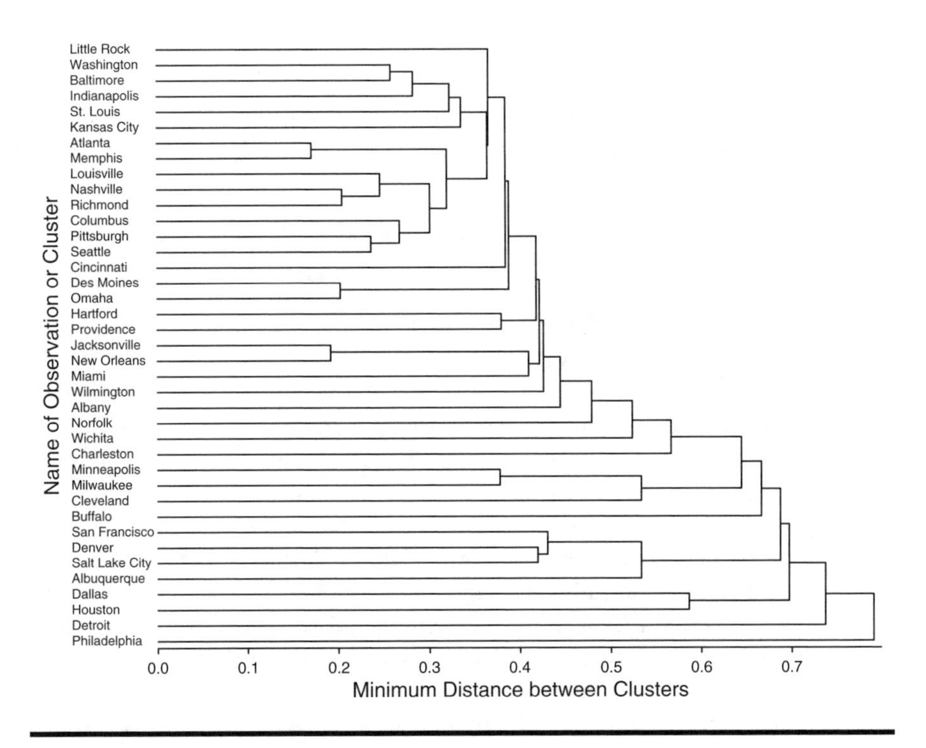

Figure 17.1 TREE procedure single linkage cluster analysis.

Then a number of other two member groups are produced. The first three member group involves Louisville, Nashville, and Richmond.

First, in Table 17.3, information is provided about the distribution of each variable in the data set. Of particular interest in the clustering context is the bimodality index which is the following function of skewness and kurtosis:

$$b = \frac{(m_3^2 + 1)}{m_4 + \frac{3(n-1)^2}{(n-2)(n-3)}}, \tag{17.2}$$

where m_3 is skewness, m_4 is kurtosis.

Values of b greater than 0.55 (the value for a uniform population) may indicate bimodal or multimodal marginal distributions. Here both factories and population have values very close to 0.55 suggesting possible clustering in the data.

The FREQ column of the cluster history simply gives the number of observations in each cluster at each stage of the process. The next two columns SPRSQ (semi-partial *R*-squared) and RSQ (*R*-squared) multiple correlation are defined as

$$\text{Semi-partial}R^2 = B_{kl}/T, \tag{17.3}$$

$$R^2 = 1 - P_g/T, \tag{17.4}$$

where $B_{kl} = W_m - W_k - W_l$, with m being the cluster formed from fusing clusters k and l, and W_k is the sum of the distances from each observation in the cluster to the cluster mean, that is,

$$W_k = \sum_{i \in C_k} \| x_i - \bar{x}_k \|^2 . \tag{17.5}$$

Finally, $P_g = \Sigma W_j$, where summation is over the number of clusters at the gth level of hierarchy.

The single linkage dendrogram in Figure 17.1 displays the "chaining" effect typical of this method of clustering. This phenomenon, though somewhat difficult to define formally, refers to the tendency of the technique to incorporate observations into existing clusters, rather than to initiate new ones.

Re-submitting the SAS code with method = complete, outree = complete and omitting the simple option yields the printed results in Table 17.4 and dendrogram in Figure 17.2. Then substituting average for complete and resubmitting give the results shown in Table 17.5 with the corresponding dendrogram in Figure 17.3.

We see in Table 17.4 that complete linkage clustering begins with the same initial "fusions" of pairs of cities as single linkage, but eventually begins to join different sets of observations. The corresponding dendrogram in Figure 17.2 shows a little more structure, although the number of groups is difficult to assess both from the dendrogram and using the CCC criterion.

The average linkage results in Table 17.5 are more similar to those of complete linkage rather than single linkage, and again the dendrogram (Figure 17.3) suggests more evidence of structure, without making the optimal number of groups obvious.

It is often useful to display the solutions given by a clustering technique by plotting the data in the space of the first two or three principal components and labelling the points by the cluster to which they have been assigned. The number of groups that we should use for these data is not clear from the previous analyses, but to illustrate we will show the four-group solution obtained from complete linkage. It will also be useful to look at the means of each variable in the four clusters. To begin we get the cluster labels for each city.

```
proc tree data=complete out=clusters n=4 noprint;
  copy city so2 temperature--rainydays;
run;
```

As well as producing a dendrogram, proc tree can also be used to create a data set containing a variable, cluster, that indicates which of a specified number of clusters each observation belongs to. The number of clusters is specified with the n= option. The copy statement transfers the named variables to this data set.

Table 17.4 Results from Proc Cluster for Complete Linkage Applied to the U.S. Cities Data

The CLUSTER Procedure

Complete Linkage Cluster Analysis

Eigenvalues of the Correlation Matrix				
	Eigenvalue	**Difference**	**Proportion**	**Cumulative**
1	2.09248727	0.45164599	0.3487	0.3487
2	1.64084127	0.36576347	0.2735	0.6222
3	1.27507780	0.48191759	0.2125	0.8347
4	0.79316021	0.67485359	0.1322	0.9669
5	0.11830662	0.03817979	0.0197	0.9866
6	0.08012683		0.0134	1.0000

The data have been standardized to mean 0 and variance 1

Root-Mean-Square Total-Sample Standard Deviation	1

Mean Distance Between Observations	3.21916

Cluster History

NCL	Clusters Joined		FREQ	SPRSQ	RSQ	ERSQ	CCC	Norm Max Dist	Tie
38	Atlanta	Memphis	2	0.0007	.999	.	.	0.1709	
37	Jacksonville	New Orleans	2	0.0008	.998	.	.	0.1919	
36	Des Moines	Omaha	2	0.0009	.998	.	.	0.2023	
35	Nashville	Richmond	2	0.0009	.997	.	.	0.2041	
34	Pittsburgh	Seattle	2	0.0013	.995	.	.	0.236	
33	Washington	Baltimore	2	0.0015	.994	.	.	0.2577	
32	Louisville	Columbus	2	0.0021	.992	.	.	0.3005	
31	CL33	Indianapolis	3	0.0024	.989	.	.	0.3391	
30	Minneapolis	Milwaukee	2	0.0032	.986	.	.	0.3775	
29	Hartford	Providence	2	0.0033	.983	.	.	0.3791	
28	Kansas City	St. Louis	2	0.0039	.979	.	.	0.412	
27	Little Rock	CL35	3	0.0043	.975	.	.	0.4132	
26	CL32	Cincinnati	3	0.0042	.970	.	.	0.4186	
25	Denver	Salt Lake City	2	0.0040	.967	.	.	0.4191	
24	CL37	Miami	3	0.0050	.962	.	.	0.4217	
23	Wilmington	Albany	2	0.0045	.957	.	.	0.4438	
22	CL31	CL28	5	0.0045	.953	.	.	0.4882	
21	CL38	Norfolk	3	0.0073	.945	.	.	0.5171	
20	CL36	Wichita	3	0.0086	.937	.	.	0.5593	
19	Dallas	Houston	2	0.0078	.929	.	.	0.5861	
18	CL29	CL23	4	0.0077	.921	.	.	0.5936	
17	CL25	Albuquerque	3	0.0090	.912	.	.	0.6291	
16	CL30	Cleveland	3	0.0100	.902	.	.	0.6667	
15	San Francisco	CL17	4	0.0089	.893	.	.	0.6696	
14	CL26	CL34	5	0.0130	.880	.	.	0.6935	
13	CL27	CL21	6	0.0132	.867	.	.	0.7053	
12	CL16	Buffalo	4	0.0142	.853	.	.	0.7463	
11	Detroit	Philadelphia	2	0.0142	.839	.	.	0.7914	
10	CL18	CL14	9	0.0200	.819	.	.	0.8754	

Cluster History

NCL	Clusters Joined		FREQ	SPRSQ	RSQ	ERSQ	CCC	Norm Max Dist	Tie
9	CL13	CL22	11	0.0354	.783	.	.	0.938	
8	CL10	Charleston	10	0.0198	.763	.	.	1.0649	
7	CL15	CL20	7	0.0562	.707	.731	-1.1	1.134	
6	CL9	CL8	21	0.0537	.653	.692	-1.6	1.2268	
5	CL24	CL19	5	0.0574	.596	.644	-1.9	1.2532	
4	CL6	CL12	25	0.1199	.476	.580	-3.2	1.5542	
3	CL4	CL5	30	0.1296	.347	.471	-3.2	1.8471	
2	CL3	CL7	37	0.1722	.174	.296	-3.0	1.9209	
1	CL2	CL11	39	0.1744	.000	.000	0.00	2.3861	

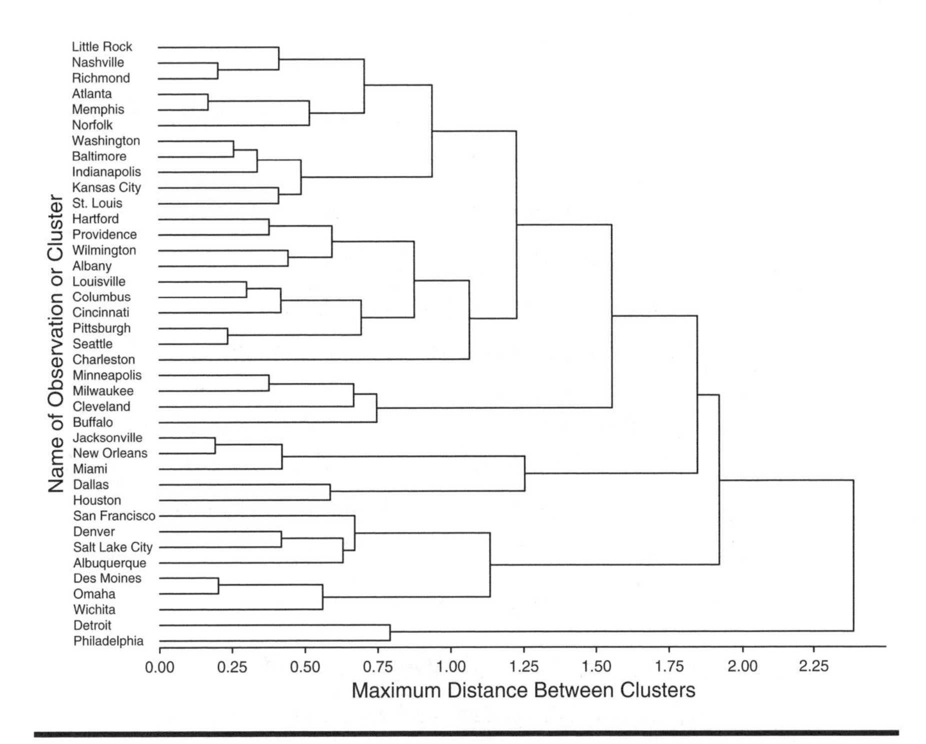

Figure 17.2 TREE procedure complete linkage cluster analysis.

The mean vectors of the four groups are obtained as follows and the output is shown in Table 17.6.

```
proc sort data=clusters;
  by cluster;
proc means data=clusters;
  var temperature--rainydays;
  by cluster;
run;
```

A plot of the first two principal components showing cluster membership can be produced as follows. The result is shown in Figure 17.4.

```
proc princomp data=clusters n=2 out=pcout noprint;
  var temperature--rainydays;
run;
proc sgplot data=pcout;
  scatter y=prinl x=prin2/markerchar=cluster;
run;
```

Table 17.5 Results from Proc Cluster for Average Linkage Applied to the U.S. Cities Data

The CLUSTER Procedure

Average Linkage Cluster Analysis

Eigenvalues of the Correlation Matrix				
	Eigenvalue	**Difference**	**Proportion**	**Cumulative**
1	2.09248727	0.45164599	0.3487	0.3487
2	1.64084127	0.36576347	0.2735	0.6222
3	1.27507780	0.48191759	0.2125	0.8347
4	0.79316021	0.67485359	0.1322	0.9669
5	0.11830662	0.03817979	0.0197	0.9866
6	0.08012683		0.0134	1.0000

The data have been standardized to mean 0 and variance 1

Root-Mean-Square Total-Sample Standard Deviation	1

Root-Mean-Square Distance Between Observations	3.464102

Cluster History

NCL	Clusters Joined		FREQ	SPRSQ	RSQ	ERSQ	CCC	Norm RMS Dist	Tie
38	Atlanta	Memphis	2	0.0007	.999	.	.	0.1588	
37	Jacksonville	New Orleans	2	0.0008	.998	.	.	0.1783	
36	Des Moines	Omaha	2	0.0009	.998	.	.	0.188	
35	Nashville	Richmond	2	0.0009	.997	.	.	0.1897	
34	Pittsburgh	Seattle	2	0.0013	.995	.	.	0.2193	
33	Washington	Baltimore	2	0.0015	.994	.	.	0.2395	
32	Louisville	CL35	3	0.0023	.992	.	.	0.2721	
31	CL33	Indianapolis	3	0.0024	.989	.	.	0.2899	
30	Columbus	CL34	3	0.0037	.985	.	.	0.342	
29	Minneapolis	Milwaukee	2	0.0032	.982	.	.	0.3508	
28	Hartford	Providence	2	0.0033	.979	.	.	0.3523	
27	CL31	Kansas City	4	0.0041	.975	.	.	0.3607	
26	CL27	St. Louis	5	0.0042	.971	.	.	0.3733	
25	CL32	Cincinnati	4	0.0050	.965	.	.	0.3849	
24	CL37	Miami	3	0.0050	.961	.	.	0.3862	
23	Denver	Salt Lake City	2	0.0040	.957	.	.	0.3894	
22	Wilmington	Albany	2	0.0045	.952	.	.	0.4124	
21	CL38	Norfolk	3	0.0073	.945	.	.	0.463	
20	CL28	CL22	4	0.0077	.937	.	.	0.4682	
19	CL36	Wichita	3	0.0086	.929	.	.	0.5032	
18	Little Rock	CL21	4	0.0082	.920	.	.	0.5075	
17	San Francisco	CL23	3	0.0083	.912	.	.	0.5228	
16	CL20	CL30	7	0.0166	.896	.	.	0.5368	
15	Dallas	Houston	2	0.0078	.888	.	.	0.5446	
14	CL18	CL25	8	0.0200	.868	.	.	0.5529	
13	CL29	Cleveland	3	0.0100	.858	.	.	0.5608	
12	CL17	Albuquerque	4	0.0097	.848	.	.	0.5675	
11	CL14	CL26	13	0.0347	.813	.	.	0.6055	
10	CL11	CL16	20	0.0476	.766	.	.	0.6578	

Cluster History

NCL	Clusters Joined		FREQ	SPRSQ	RSQ	ERSQ	CCC	Norm RMS Dist	Tie
9	CL13	Buffalo	4	0.0142	.752	.	.	0.6666	
8	Detroit	Philadelphia	2	0.0142	.737	.	.	0.7355	
7	CL10	Charleston	21	0.0277	.710	.731	-.97	0.8482	
6	CL12	CL19	7	0.0562	.653	.692	-1.6	0.8873	
5	CL7	CL24	24	0.0848	.569	.644	-2.9	0.9135	
4	CL5	CL6	31	0.1810	.388	.580	-5.5	1.0514	
3	CL4	CL9	35	0.1359	.252	.471	-5.3	1.0973	
2	CL3	CL15	37	0.0774	.174	.296	-3.0	1.1161	
1	CL2	CL8	39	0.1744	.000	.000	0.00	1.5159	

We see that this solution contains three clusters containing only a few observations each and is perhaps not ideal. Nevertheless, we shall continue to use it and look at differences between the derived clusters in terms of differences in air pollution, as measured by their SO_2 values. Box plots of SO_2 values for each of the four clusters can be found as follows:

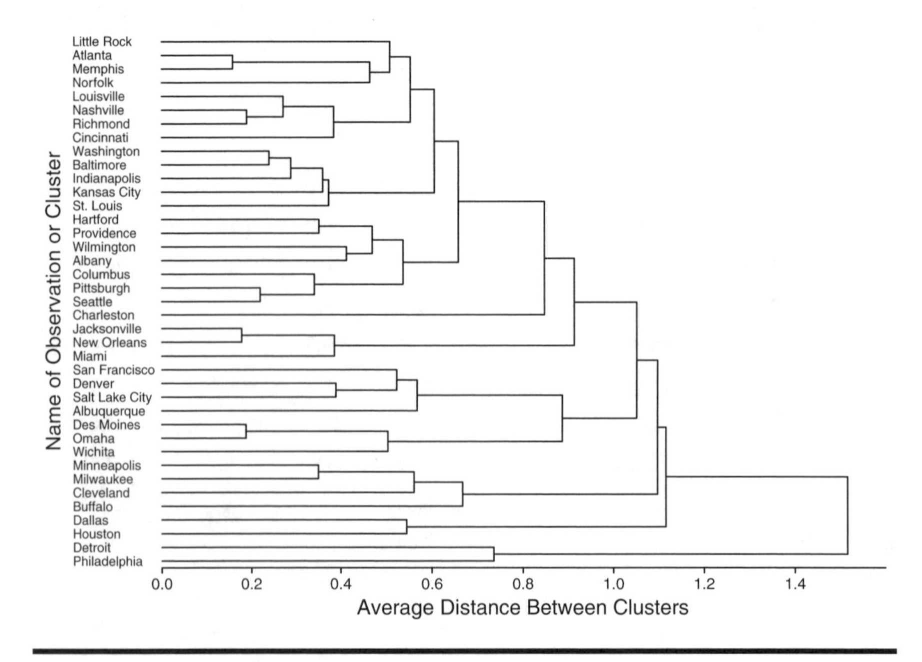

Figure 17.3 TREE procedure average linkage cluster analysis.

Table 17.6 Cluster Means for Four Group Solution Obtained by Applying Complete Linkage to the U.S. Cities Data

The MEANS Procedure

CLUSTER=1

Variable	N	Mean	Std Dev	Minimum	Maximum
temperature	25	53.6040000	5.0301160	43.5000000	61.6000000
factories	25	370.6000000	233.7716122	35.0000000	1007.00
population	25	470.4400000	243.8555037	71.0000000	905.0000000
windspeed	25	9.3240000	1.3690873	6.5000000	12.4000000
rain	25	39.6728000	5.6775101	25.9400000	49.1000000
rainydays	25	125.8800000	18.7401530	99.0000000	166.0000000

CLUSTER=2

Variable	N	Mean	Std Dev	Minimum	Maximum
temperature	5	69.4600000	3.5317135	66.2000000	75.5000000
factories	5	381.8000000	276.0556103	136.0000000	721.0000000
population	5	660.4000000	378.9364063	335.0000000	1233.00
windspeed	5	9.5800000	1.1798305	8.4000000	10.9000000
rain	5	51.0340000	9.4534084	35.9400000	59.8000000
rainydays	5	107.6000000	18.7962762	78.0000000	128.0000000

CLUSTER=3

Variable	N	Mean	Std Dev	Minimum	Maximum
temperature	7	53.3571429	3.2572264	49.0000000	56.8000000
factories	7	214.2857143	168.3168780	46.0000000	454.0000000
population	7	353.7142857	195.8466358	176.0000000	716.0000000
windspeed	7	10.0142857	1.5879007	8.7000000	12.7000000
rain	7	21.1657143	9.5465436	7.7700000	30.8500000
rainydays	7	83.2857143	16.0801564	58.0000000	103.0000000

CLUSTER=4

Variable	N	Mean	Std Dev	Minimum	Maximum
temperature	2	52.2500000	3.3234019	49.9000000	54.6000000
factories	2	1378.00	444.0630586	1064.00	1692.00
population	2	1731.50	309.0056634	1513.00	1950.00
windspeed	2	9.8500000	0.3535534	9.6000000	10.1000000
rain	2	35.4450000	6.3427478	30.9600000	39.9300000
rainydays	2	122.0000000	9.8994949	115.0000000	129.0000000

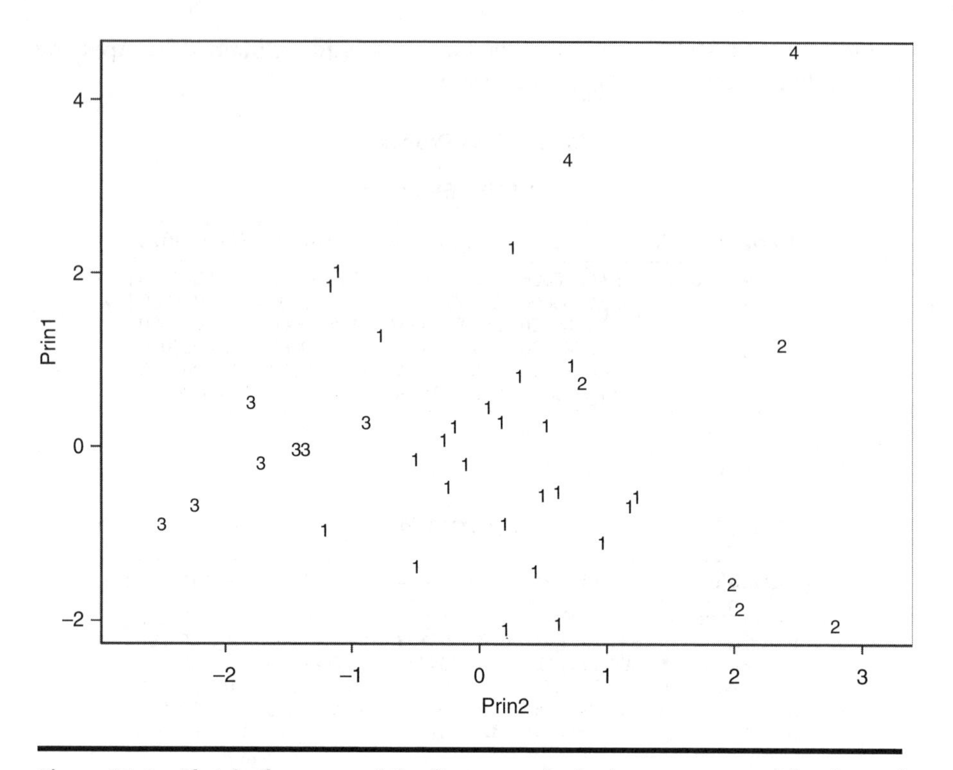

Figure 17.4 Plot, in the space of the first two principal components of the data, of the four-group solution from complete linkage applied to the U.S. cities data.

```
proc sgplot data=clusters;
  vbox so2/category=cluster;
run;
```

The plot is shown in Figure 17.5. The box plots suggest that the average sulphur dioxide concentration may differ in the four clusters, being higher in clusters 1 and 4 than in clusters 2 and 3. More formally, we might test for a cluster difference using a one way analysis of variance. Here we shall use proc glm for the analysis. The output is shown in Table 17.7. The F-test shows a highly significant difference that might be investigated in more detail using a suitable multiple comparison procedure.

```
proc glm data=clusters;
  class cluster;
  model so2=cluster;
run;
```

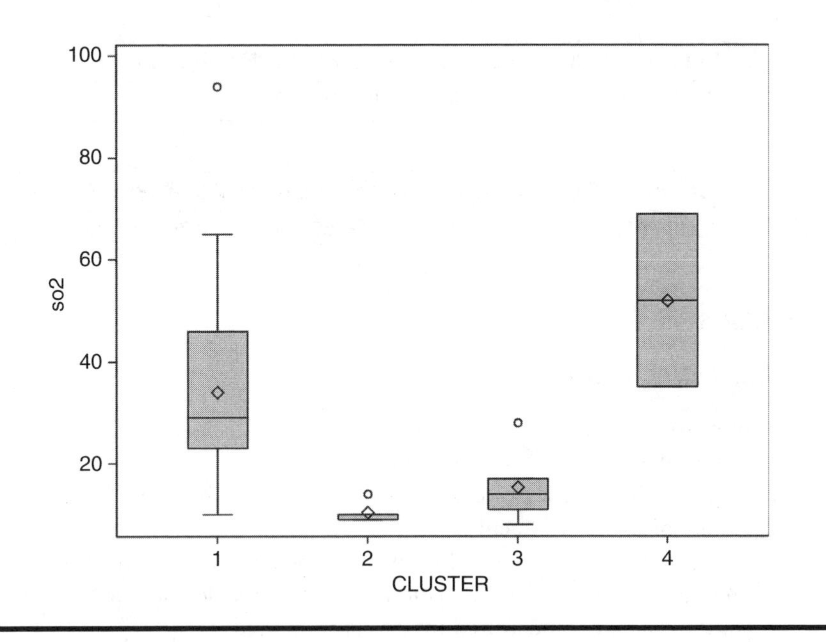

Figure 17.5 Box plots of SO_2 concentration in four clusters.

Table 17.7 Testing for Difference in SO_2 Levels between Groups in the Four Group Solution from Complete Linkage Clustering

Class Level Information		
Class	Levels	Values
CLUSTER	4	1 2 3 4

Number of Observations Read	39
Number of Observations Used	39

Dependent Variable: so2

Source	DF	Sum of Squares	Mean Square	F Value	Pr > F
Model	3	2.50666667	0.83555556	4.75	0.0070
Error	35	6.16000000	0.17600000		
Corrected Total	38	8.66666667			

R-Square	Coeff Var	Root MSE	so2 Mean
0.289231	125.8571	0.419524	0.333333

Source	DF	Type I SS	Mean Square	F Value	Pr > F
CLUSTER	3	2.50666667	0.83555556	4.75	0.0070

Source	DF	Type III SS	Mean Square	F Value	Pr > F
CLUSTER	3	2.50666667	0.83555556	4.75	0.0070

Exercises

1 U.S. Cities Data

1. Explore some of the other clustering procedures available in SAS, for example, proc modeclus, on the air pollution data.
2. Repeat the cluster analyses described in this chapter but now including the data on Chicago and Phoenix.
3. In the cluster analysis described in the text, the data were standardized before clustering to unit standard deviation for each variable. Repeat the analyses when the data are standardized by the range instead.

2 Romano–British Pottery

The data in pottery.dat show the chemical composition of 46 specimens of Romano–British pottery, determined by atomic absorption spectrophotometry, for nine oxides, and in addition the label of the kiln site at which the pottery was found (Tubb et al., 1980).

The variables in pottery.dat are

- Specimen ID
- Kiln site label 1 to 5
- Aluminium oxide
- Ferrous oxide
- Magnesium oxide
- Calcium oxide
- Sodium oxide
- Potassium oxide
- Titanium oxide
- Manganese oxide
- Barium oxide

1. Apply average linkage clustering to the data using the nine oxide values after suitable standardization.
2. Examine the resulting dendrogram to see if it is suggestive of a particular number of groups.
3. Investigate whether there is any association between the distinct compositional groups found by the cluster analysis and the kiln site.

Chapter 18

Discriminant Function Analysis: Classifying Tibetan Skulls

18.1 Description of Data

In the 1920s, Colonel Waddell collected 32 skulls in the southwestern and eastern districts of Tibet. The collection comprised skulls of two types:

1. *Type A*: Seventeen skulls from graves in Sikkim and neighbouring areas of Tibet.
2. *Type B*: Fifteen skulls picked up on a battlefield in the Lhausa district and believed to be those of native soldiers from the eastern province of Kharis.

It was postulated that Tibetans from Kharis might be survivors of a particular fundamental human type, unrelated to the Mongolian and Indian types that surrounded them.

A number of measurements were made on each skull, and Table 18.1 shows 5 of these for some of the 32 skulls collected by Colonel Waddell. (The data are given in Table 144 of *SDS*.) Of interest here is whether the two types of skulls can be accurately classified from the five measurements recorded, and which of these five measurements are the most informative in this classification exercise.

Table 18.1 Measurements on Tibetan Skulls of Two Types

Type A Skulls				
X1	*X2*	*X3*	*X4*	*X5*
190.5	152.5	145.0	73.5	136.5
172.5	132.0	125.5	63.0	121.0
167.0	130.0	125.5	69.5	119.5
169.5	150.5	133.5	64.5	128.0
175.0	138.5	126.0	77.5	135.5
⋮				
170.0	126.5	134.5	66.0	118.5
Type B Skulls				
X1	*X2*	*X3*	*X4*	*X5*
182.5	136.0	138.5	76.0	134.0
179.5	135.0	128.5	74.0	132.0
191.0	140.5	140.5	72.5	131.5
184.5	141.5	134.5	76.5	141.5
181.0	142.0	132.5	79.0	136.5
⋮				
182.5	131.0	135.0	68.5	136.0

Note: X1, greatest length of skull; X2, greatest horizontal breadth of skull; X3, height of skull; X4, upper faceheight; X5, facebreadth between outermost points of cheek bones.

18.2 Discriminant Function Analysis

Discriminant analysis is concerned with deriving helpful rules for allocating observations to one or other of a set of a priori defined classes in some optimal way, using the information provided by a series of measurements made of each sample member. The technique is used in situations where the investigator has one set of observations, the training sample, for which group membership is known with certainty a priori, and a second set, the test sample, consisting of the observations for which group membership is unknown and which we require to allocate to one of the known groups with as few misclassifications as possible.

An initial question that might be asked is since the members of the training sample can be classified with certainty, why not apply the procedure used in their classification to the test sample? Reasons are not difficult to find. In medicine, for example, it might be possible to diagnose a particular condition with certainty only as a result of a post-mortem examination. Clearly, for patients still alive and in need of treatment, a different diagnostic procedure would be useful.

Several methods for discriminant analysis are available, but here we shall concentrate on the one proposed by Fisher (1936) as a method for classifying an observation into one of the two possible groups using measurements, $x_1, x_2, \ldots, x_p$. Fisher's approach to the problem was to seek a linear function, z, of the variables

$$z = a_1x_1 + a_2x_2 + \cdots + a_px_p \tag{18.1}$$

such that the ratio of the between-groups variance of z to its within-group variance is maximized. This implies that the coefficients $\mathbf{a}' = [a_1, \ldots, a_p]$ have to be chosen so that V given by

$$V = \frac{\mathbf{a}'\mathbf{Ba}}{\mathbf{a}'\mathbf{Sa}} \tag{18.2}$$

is maximized. In Equation 18.2, $\mathbf{S}$ is the pooled within-group covariance matrix, that is,

$$\mathbf{S} = \frac{(n_1 - 1)\mathbf{S}_1 + (n_2 - 1)\mathbf{S}_2}{n_1 + n_2 - 2}, \tag{18.3}$$

where

$\mathbf{S}_1$ and $\mathbf{S}_2$ are the covariance matrices of the two groups
n_1 and n_2 the group sample sizes

The matrix $\mathbf{B}$ in Equation 18.2 is the covariance matrix of the group means.

The vector $\mathbf{a}$ that maximizes V is given by the solution of the equation

$$(\mathbf{B} - \lambda\mathbf{S})\mathbf{a} = 0. \tag{18.4}$$

In the two-group situation the single solution can be shown to be

$$\mathbf{a} = \mathbf{S}^{-1}(\bar{\mathbf{x}}_1 - \bar{\mathbf{x}}_2), \tag{18.5}$$

where $\bar{\mathbf{x}}_1$ and $\bar{\mathbf{x}}_2$ are the mean vectors of the measurements for the observations in each group.

The assumptions under which Fisher's method is optimal are

1. Data in both groups have a multivariate normal distribution.
2. Covariance matrices of each group are the same.

If the covariance matrices are not the same, but the data are multivariate normal, a quadratic discriminant function may be required. If the data are not multivariate normal, an alternative such as *logistic discrimination* (Everitt and Dunn, 2001) may be more useful, although Fisher's method is known to be relatively robust against departures from normality (Hand, 1981).

Assuming $\bar{z}_1 > \bar{z}_2$, where $\bar{z}_1$ and $\bar{z}_2$ are the discriminant function score means in each group, the classification rule for an observation with discriminant score z_i is

Assign to group 1 if $z_i - z_c < 0$,
Assign to group 2 if $z_i - z_c \geq 0$,

where

$$z_c = \frac{\bar{z}_1 + \bar{z}_2}{2}. \tag{18.6}$$

Subsets of variables most useful for discrimination can be identified by using procedures similar to the stepwise methods described in Chapter 7.

A question of some importance about a discriminant function is how well does it perform? One possible method of evaluating performance would be to apply the derived classification rule to the training set data and calculate the misclassification rate; this is known as the resubstitution estimate. But estimating misclassification rates in this way, although simple, is known in general to be optimistic (in some case wildly so). Better estimates of misclassification rates in discriminant analysis may be defined in a variety of ways (see Hand, 1997). A method that is commonly used is the so-called leaving-one-out method, in which the discriminant function is first derived from only $n - 1$ sample members, and then used to classify the observation not included. The procedure is repeated n times, each time omitting a different observation.

18.3 Analysis Using SAS

The data from Table 18.1 may be read in as follows:

```
data skulls;
  infile 'c:\handbook3\datasets\tibetan.dat' expandtabs;
  input length width height faceheight facewidth;
  if _n_<18 then type='A';
  else type='B';
run;
```

A parametric discriminant analysis can be specified as follows:

```
proc discrim data=skulls pool=test simple manova wcov
  crossvalidate;
  class type;
  var length--facewidth;
run;
```

The option pool=test provides a test of the equality of the within-group covariance matrices. If the test is significant beyond a level specified by slpool, then a quadratic rather than a linear discriminant function is derived. The default value of slpool is 0.1.

The manova option provides a test of the equality of the mean vectors of the two groups. Clearly if there is no difference, a discriminant analysis is largely a waste of time. The simple option provides useful summary statistics both overall and within groups; wcov gives the within-group covariance matrices. The crossvalidate option will be discussed later in the chapter. The class statement names the variable that defines the groups, and the var statement names the variables to be used to form the discriminant function.

The output is shown in Table 18.2. The results for the test of the equality of the within-group covariance matrices are also shown in Table 18.2. The Chi-squared test of

Table 18.2 Results from Proc Discrim Applied to the Tibetan Skull Data

The DISCRIM Procedure

Total Sample Size	32	**DF Total**	31
Variables	5	**DF Within Classes**	30
Classes	2	**DF Between Classes**	1

Number of Observations Read	32
Number of Observations Used	32

Class Level Information

type	**Variable Name**	**Frequency**	**Weight**	**Proportion**	**Prior Probability**
A	A	17	17.0000	0.531250	0.500000
B	B	15	15.0000	0.468750	0.500000

The DISCRIM Procedure

Within-Class Covariance Matrices

type = A, DF = 16

Variable	**length**	**width**	**height**	**faceheight**	**facewidth**
length	45.52941176	25.22242647	12.39062500	22.15441176	27.97242647
width	25.22242647	57.80514706	11.87500000	7.51930147	48.05514706
height	12.39062500	11.87500000	36.09375000	-0.31250000	1.40625000
faceheight	22.15441176	7.51930147	-0.31250000	20.93566176	16.76930147
facewidth	27.97242647	48.05514706	1.40625000	16.76930147	66.21139706

type = B, DF = 14					
Variable	length	width	height	faceheight	facewidth
length	74.42380952	-9.52261905	22.73690476	17.79404762	11.12500000
width	-9.52261905	37.35238095	-11.26309524	0.70476190	9.46428571
height	22.73690476	-11.26309524	36.31666667	10.72380952	7.19642857
faceheight	17.79404762	0.70476190	10.72380952	15.30238095	8.66071429
facewidth	11.12500000	9.46428571	7.19642857	8.66071429	17.96428571

Simple Statistics

Total-Sample					
Variable	N	Sum	Mean	Variance	Standard Deviation
length	32	5758	179.93750	87.70565	9.3651
width	32	4450	139.06250	46.80242	6.8412
height	32	4266	133.29688	36.99773	6.0826
faceheight	32	2334	72.93750	29.06048	5.3908
facewidth	32	4279	133.70313	55.41709	7.4443

type = A					
Variable	N	Sum	Mean	Variance	Standard Deviation
length	17	2972	174.82353	45.52941	6.7475
width	17	2369	139.35294	57.80515	7.6030
height	17	2244	132.00000	36.09375	6.0078
faceheight	17	1187	69.82353	20.93566	4.5756
facewidth	17	2216	130.35294	66.21140	8.1370

type = B					
Variable	N	Sum	Mean	Variance	Standard Deviation
length	15	2786	185.73333	74.42381	8.6269
width	15	2081	138.73333	37.35238	6.1117
height	15	2022	134.76667	36.31667	6.0263
faceheight	15	1147	76.46667	15.30238	3.9118
facewidth	15	2063	137.50000	17.96429	4.2384

Within Covariance Matrix Information		
type	Covariance Matrix Rank	Natural Log of the Determinant of the Covariance Matrix
A	5	16.16370
B	5	15.77333
Pooled	5	16.72724

Test of Homogeneity of Within Covariance Matrices

Chi-Square	DF	Pr > ChiSq
18.370512	15	0.2437

Since the Chi-Square value is not significant at the 0.1 level, a pooled covariance matrix will be used in the discriminant function.

Reference: Morrison, D.F. in Multivariate Statistical Methods, 1976, p252.

Generalized Squared Distance to type		
From type	A	B
A	0	3.50144
B	3.50144	0

Multivariate Statistics and Exact F Statistics S=1 M=1.5 N=12					
Statistic	Value	F Value	Num DF	Den DF	Pr > F
Wilks' Lambda	0.51811582	4.84	5	26	0.0029
Pillai's Trace	0.48188418	4.84	5	26	0.0029
Hotelling-Lawley Trace	0.93007040	4.84	5	26	0.0029
Roy's Greatest Root	0.93007040	4.84	5	26	0.0029

Linear Discriminant Function for type		
Variable	A	B
Constant	-514.26257	-544.72605
length	1.46831	1.55762
width	2.36106	2.20528
height	2.75219	2.74696
faceheight	0.77530	0.95250
facewidth	0.19475	0.37216

Classification Summary for Calibration Data: WORK.SKULLS

Resubstitution Summary using Linear Discriminant Function

Number of Observations and Percent Classified into type			
From type	**A**	**B**	**Total**
A	14 82.35	3 17.65	17 100.00
B	3 20.00	12 80.00	15 100.00
Total	17 53.13	15 46.88	32 100.00
Priors	0.5	0.5	

Error Count Estimates for type			
	A	**B**	**Total**
Rate	0.1765	0.2000	0.1882
Priors	0.5000	0.5000	

Classification Summary for Calibration Data: WORK.SKULLS

Cross-validation Summary using Linear Discriminant Function

Number of Observations and Percent Classified into type			
From type	**A**	**B**	**Total**
A	12 70.59	5 29.41	17 100.00
B	6 40.00	9 60.00	15 100.00
Total	18 56.25	14 43.75	32 100.00
Priors	0.5	0.5	

Error Count Estimates for type			
	A	**B**	**Total**
Rate	0.2941	0.4000	0.3471
Priors	0.5000	0.5000	

the equality of the two covariance matrices is not significant at the 0.1 level and so a linear discriminant function will be derived. The results of the multivariate analysis of variance are also shown in Table 18.2. Because there are only two groups here, all four test criteria lead to the same *F* value, which is significant well beyond the 5% level.

The results defining the discriminant function are also given in Table 18.2. The two sets of coefficients given need to be subtracted to give the discriminant function in the form described in the previous section. This leads to

$$\mathbf{a}' = [-0.0893,\ 0.1158,\ 0.0052,\ -0.1772,\ -0.1774]. \quad (18.7)$$

The group means on the discriminant function are $\bar{z}_1 = -28.713$, $\bar{z}_2 = -32.214$, leading to a value of $\bar{z}_c = -30.463$.

So, for example, a skull having a vector of measurements, $\mathbf{x}' = [185,\ 142,\ 130,\ 72,\ 133]$ has a discriminant score of -30.07, and $z_1 - z_c$ in this case is therefore 0.39 and the skull should be assigned to group 1.

The resubstitution approach to estimating the misclassification rate of the derived allocation rule is seen from Table 18.2 to be 18.82%. But the leaving-out-one (cross-validation) approach increases this to a more realistic 34.71%.

To identify the most important variables for discrimination, proc stepdisc can be used as follows:

```
proc stepdisc data = skulls sle = .05 sls = .05;
  class type;
  var length--facewidth;
run;
```

The output is shown in Table 18.3. The significance levels required for variables to enter and be retained are set with the sle (slentry) and sls (slstay) options,

Table 18.3 Results from Proc Stepdisc Applied to the Tibetan Skull Data

The STEPDISC Procedure

The Method for Selecting Variables is STEPWISE			
Total Sample Size	32	**Variable(s) in the Analysis**	5
Class Levels	2	**Variable(s) will be Included**	0
		Significance Level to Enter	0.05
		Significance Level to Stay	0.05

Number of Observations Read	32
Number of Observations Used	32

Class Level Information				
type	Variable Name	Frequency	Weight	Proportion
A	A	17	17.0000	0.531250
B	B	15	15.0000	0.468750

The STEPDISC Procedure

Stepwise Selection: Step 1

Statistics for Entry, DF = 1, 30				
Variable	R-Square	F Value	Pr > F	Tolerance
length	0.3488	16.07	0.0004	1.0000
width	0.0021	0.06	0.8029	1.0000
height	0.0532	1.69	0.2041	1.0000
faceheight	0.3904	19.21	0.0001	1.0000
facewidth	0.2369	9.32	0.0047	1.0000

Variable faceheight will be entered.

Variable(s) that have been Entered
faceheight

Multivariate Statistics					
Statistic	Value	F Value	Num DF	Den DF	Pr > F
Wilks' Lambda	0.609634	19.21	1	30	0.0001
Pillai's Trace	0.390366	19.21	1	30	0.0001
Average Squared Canonical Correlation	0.390366				

The STEPDISC Procedure

Stepwise Selection: Step 2

Statistics for Removal, DF = 1, 30			
Variable	R-Square	F Value	Pr > F
faceheight	0.3904	19.21	0.0001

No variables can be removed.

Statistics for Entry, DF = 1, 29				
Variable	**Partial R-Square**	**F Value**	**Pr > F**	**Tolerance**
length	0.0541	1.66	0.2081	0.4304
width	0.0162	0.48	0.4945	0.9927
height	0.0047	0.14	0.7135	0.9177
facewidth	0.0271	0.81	0.3763	0.6190

No variables can be entered.

No further steps are possible.

The STEPDISC Procedure

Stepwise Selection Summary								
Step	**Number In**	**Entered**	**Removed**	**Partial R-Square**	**F Value**	**Pr > F**	**Wilks' Lambda**	**Pr < Lambda**
1	1	faceheight		0.3904	19.21	0.0001	0.60963388	0.0001

Step	**Number In**	**Entered**	**Removed**	**Average Squared Canonical Correlation**	**Pr > ASCC**
1	1	faceheight		0.39036612	0.0001

respectively. The default value for both is $p = .15$. By default, a "stepwise" procedure is used (other options can be specified using a method = statement). Variables are chosen to enter or leave the discriminant function according to one of the two criteria:

- The significance level of an F test from an analysis of covariance, where the variables already chosen act as covariates and the variable under consideration is the dependent variable
- The squared multiple correlation for predicting the variable under consideration from the class variable controlling for the effects of the variables already chosen

The significance level and the squared partial correlation criteria select variables in the same order although they may select different numbers of variables.

Increasing the sample size tends to increase the number of variables selected when using significance levels, but has little effect on the number selected when using squared partial correlations.

At step 1 in Table 18.3, the variable faceheight has the highest R^2 value and is the first variable selected. At step 2, none of the partial R^2 values of the other variables meets the criterion for inclusion and the process therefore ends. The tolerance shown for each variable is 1 – the squared multiple correlation of the variable, with the other variables already selected. A variable can only be entered if its tolerance is above a value specified in the singular statement. The value set by default is 1.0E–8.

Details of the "discriminant function" using only faceheight are found as follows:

```
proc discrim data=skulls crossvalidate;
  class type;
  var faceheight;
run;
```

The output is shown in Table 18.4. Here the coefficients of faceheight in each class are simply the mean of the class on faceheight divided by the pooled within-group variance of the variable. Here the resubstitution and leaving-one-out methods of estimating the misclassification rate give the same value of 24.71%.

Table 18.4 Results from Proc Discrim Applied to the Tibetan Skull Data Using Only faceheight

The DISCRIM Procedure

Total Sample Size	32	**DF Total**	31
Variables	1	**DF Within Classes**	30
Classes	2	**DF Between Classes**	1

Number of Observations Read	32
Number of Observations Used	32

Class Level Information

type	Variable Name	Frequency	Weight	Proportion	Prior Probability
A	A	17	17.0000	0.531250	0.500000
B	B	15	15.0000	0.468750	0.500000

Pooled Covariance Matrix Information	
Covariance Matrix Rank	Natural Log of the Determinant of the Covariance Matrix
1	2.90727

The DISCRIM Procedure

Generalized Squared Distance to type		
From type	A	B
A	0	2.41065
B	2.41065	0

Linear Discriminant Function for type		
Variable	A	B
Constant	-133.15615	-159.69891
faceheight	3.81408	4.17695

Classification Summary for Calibration Data: WORK.SKULLS

Re-substitution Summary using Linear Discriminant Function

Number of Observations and Percent Classified into type			
From type	A	B	Total
A	12 70.59	5 29.41	17 100.00
B	3 20.00	12 80.00	15 100.00
Total	15 46.88	17 53.13	32 100.00
Priors	0.5	0.5	

Error Count Estimates for type			
	A	B	Total
Rate	0.2941	0.2000	0.2471
Priors	0.5000	0.5000	

Exercises

1 Tibetan Skulls

1. Use the posterr options in proc discrim to estimate error rates for the discriminant functions derived for the skull data. Compare these with those given in Tables 18.2 and 18.4.
2. Investigate the use of the non-parametric discriminant methods available in proc discrim for the skull data. Compare the results with those for the simple linear discriminant function given in the text.

2 Sudden Infant Deaths

The data in sids.dat were collected by Spicer et al. (1987) in an investigation of sudden infant death syndrome (SIDS). The two groups here consist of 16 SIDS victims and 49 controls. All the infants have a gestational age of 37 weeks or more and were regarded as full term.

The variables in sids.dat are as follows:

- **Group:** 1 = control, 2 = SIDS
- **HR:** Heart rate, bpm
- **BW:** Birthweight, grams
- **Factor 68:** A value arising from spectral analysis of 24 h recordings of electrocardiograms and respiratory movements made on each child
- **Gestational age:** Weeks

1. Construct Fisher's linear discriminant function using only the FACTOR68 and BIRTHWEIGHT variables. Show the derived discriminant function on a scatterplot of the data.
2. Construct the discriminant function based on all four variables and find an appropriate estimate of the misclassification rate.
3. How would you incorporate prior probabilities into your discriminant function?

Chapter 19

Correspondence Analysis: Smoking and Motherhood, Sex and the Single Girl, and European Stereotypes

19.1 Description of Data

Three sets of data will be considered in this chapter, all of which arise in the form of two-dimensional contingency tables, as met previously in Chapter 3. The three data sets are given in Tables 19.1 through 19.3, and their details are as follows:

- Data in Table 19.1 involve the association between a girl's age and her relationship with her boyfriend.
- Data in Table 19.2 show the distribution of birth outcomes by age of mother, length of gestation and whether the mother smoked during the prenatal period. We shall consider the data as a two-dimensional contingency table with four row categories and four column categories.
- Data in Table 19.3 were obtained by asking a large number of people in the United Kingdom which of 13 characteristics they would associate with the nationals of the United Kingdom's partner countries in the European Community. Entries in the table give the percentages of respondents agreeing that the nationals of a particular country possess the particular characteristic.

Table 19.1 Girl's Age and Relationship with Boyfriend

	Age Group				
	Under 16	*16–17*	*17–18*	*18–19*	*19–20*
No boyfriend	21	21	14	13	8
Boyfriend/no sexual intercourse	8	9	6	8	2
Boyfriend/sexual intercourse	2	3	4	10	10

Table 19.2 Age and Smoking Behaviour of Mother and Gestational Age of Child

	Premature		*Full Term*	
	Died in First Year	*Alive at Year 1*	*Died in First Year*	*Alive at Year 1*
Young mothers				
Non-smokers	50	315	24	4012
Smokers	9	40	6	459
Old mothers				
Non-smokers	41	147	14	1594
Smokers	4	11	1	124

Table 19.3 Perceived Characteristics by U.K. Nationals of Nationals from Other Countries in The European Union

	Characteristic												
Country	*1*	*2*	*3*	*4*	*5*	*6*	*7*	*8*	*9*	*10*	*11*	*12*	*13*
France	37	29	21	19	10	10	8	8	6	6	5	2	1
Spain	7	14	8	9	27	7	3	7	3	23	12	1	3
Italy	30	12	19	10	20	7	12	6	5	13	10	1	2
U.K.	9	14	4	6	27	12	2	13	26	16	29	6	25
Ireland	1	7	1	16	30	3	10	9	5	11	22	2	27
Holland	5	4	2	2	15	2	0	13	24	1	28	4	6
Germany	4	48	1	12	3	9	2	11	41	1	38	8	8

Note: Characteristics: 1, stylish; 2, arrogant; 3, sexy; 4, devious; 5, easy-going; 6, greedy; 7, cowardly; 8, boring; 9, efficient; 10, lazy; 11, hard working; 12, clever; 13, courageous.

19.2 Displaying Contingency Table Data Graphically Using Correspondence Analysis

Correspondence analysis is a technique for displaying the associations among a set of categorical variables in a type of scatterplot or map, thus allowing a visual examination of the structure or pattern of these associations. A correspondence analysis should ideally be seen as an extremely useful supplement to, rather than a replacement for, the more formal inferential procedures generally used with categorical data (see Chapters 3 and 8). The aim of using correspondence analysis is summarized well in the following quotation from Greenacre (1992):

> An important aspect of correspondence analysis which distinguishes it from more conventional statistical methods is that it is not a confirmatory technique, trying to prove a hypothesis, but rather an exploratory technique, trying to reveal the data content. One can say that it serves as a window onto the data, allowing researchers easier access to their numerical results and facilitating discussion of the data and possibly generating hypotheses which can be formally tested at a later stage.

Mathematically, correspondence analysis can be regarded as either

1. Method for decomposing the Chi-squared statistic for a contingency table into components corresponding to different dimensions of the heterogeneity between its rows and columns or
2. Method for simultaneously assigning a scale to rows and a separate scale to columns so as to maximize the correlation between the resulting pair of variables

Quintessentially, however, correspondence analysis is a technique for displaying multivariate categorical data graphically, by deriving coordinate values to represent the categories of the variables involved, which may then be plotted to provide a "picture" of the data.

In the case of two categorical variables forming a two-dimensional contingency table, the required coordinates are obtained from the singular value decomposition (Everitt and Dunn, 2001) of a matrix **E** with elements e_{ij} given by

$$e_{ij} = \frac{p_{ij} - p_{i+}p_{j+}}{\sqrt{p_{i+}p_{+j}}}, \tag{19.1}$$

where $p_{ij} = n_{ij}/n$ with n_{ij} being the number of observations in the *ij*th cell of the contingency table and n the total number of observations; the total number of observations in row i is represented by n_{i+}, and the corresponding value for column j, n_{+j}. Finally, $p_{i+} = n_{i+}/n$ and p_{+j}/n. The elements of $\boldsymbol{E}$ can be written in terms of the familiar "observed" (O) and "expected" (E) nomenclature used for contingency tables as

$$e_{ij} = \frac{1}{\sqrt{n}} \frac{\mathrm{O} - \mathrm{E}}{\sqrt{E}}. \tag{19.2}$$

Written this way, it is clear that the terms are a form of residual from fitting the independence model to the data.

The singular value decomposition of $\boldsymbol{E}$ consists of finding matrices, **U**, **V** and **Δ** (diagonal), such that

$$\mathbf{E} = \mathbf{U}\boldsymbol{\Delta}\mathbf{V}', \tag{19.3}$$

where

U contains the eigenvectors of $\mathbf{EE}'$
V the eigenvectors of $\mathbf{E}'\mathbf{E}$

The diagonal matrix $\boldsymbol{\Delta}$ contains the ranked singular values, δ_k, so that δ_k^2 are the eigenvalues (in decreasing) order of either $\mathbf{EE}'$ or $\mathbf{E}'\mathbf{E}$. The coordinate of the row *i*th category on the *k*th coordinate axis is given by $\delta_k u_{ik}/\sqrt{p_{i+}}$, and the coordinate of the *j*th column category on the same axis is given in $\delta_k v_{jk}/\sqrt{p_{+j}}$, where u_{ik} $i = 1 \cdots r$ and v_{jk}, $j = 1 \cdots c$ are, respectively, the elements of the *k*th column of **U** and the *k*th column of **V**.

To represent the table fully requires at most $R = \min(r, c) - 1$ dimensions, where *r* and *c* are the number of rows and columns of the table. *R* is the rank of the matrix $\boldsymbol{E}$. The eigenvalues, δ_k^2, are such that

$$\mathrm{Trace}(\mathbf{EE}') = \sum_{k=1}^{R} \delta_k^2 = \sum_{i=1}^{r} \sum_{j=1}^{c} e_{ij}^2 = \frac{X^2}{n}, \tag{19.4}$$

where X^2 is the usual Chi-squared test statistic for independence. In the context of correspondence analysis, X^2/n is known as inertia. Correspondence analysis produces a graphical display of the contingency table from the columns of **U** and **V**, in most cases from the first two columns, $\mathbf{u}_1$, $\mathbf{u}_2$, $\mathbf{v}_1$, $\mathbf{v}_2$, of each, since these give the "best" two-dimensional representation. It can be shown that the first two coordinates give the following approximation to the e_{ij}:

$$e_{ij} \approx u_{i1}v_{j1} + u_{i2}v_{j2}, \tag{19.5}$$

so that a large positive residual corresponds to u_{ik} and v_{jk} for $k = 1$ or 2 being large and of the same sign. A large negative residual corresponds to u_{ik} and v_{jk}, being large and of the opposite sign for each value of *k*. When u_{ik} and v_{jk} are small and their signs are not consistent for each k, the corresponding residual term will be small. The adequacy of the representation produced by the first two coordinates can be assessed informally by calculating the percentages of the inertia they account for, that is,

$$\text{Percentage inertia} = \frac{\delta_1^2 + \delta_2^2}{\sum_{k=1}^{R} \delta_k^2}. \tag{19.6}$$

Values of 60% and over usually mean that the two-dimensional solution gives a reasonable account of the structure in the table.

19.3 Analysis Using SAS

19.3.1 Boyfriends

The 15 cell counts shown in Table 19.1 are in an ASCII file, tab separated. So a suitable data set can be created as follows:

```
data boyfriends;
  infile 'c:\handbook3\datasets\boyfriends.dat' expandtabs;
  input c1-c5;
  if _n_=1 then rowid='NoBoy';
  if _n_=2 then rowid='NoSex';
  if _n_=3 then rowid='Both';
  label c1='under 16' c2='16-17' c3='17-18' c4='18-19' c5='19-20';
run;
```

The data are already in the form of a contingency table and can be simply read into a set of variables representing the columns of the table. The label statement is used to assign informative labels to these variables. More informative variable names could also have been used, but labels are more flexible in that they may begin with numbers and include spaces. It is also useful to label the rows, and here the SAS automatic variable, _n_, is used to set the values of a character variable, rowid.

A correspondence analysis of this table can be performed as follows:

```
proc corresp data=boyfriends out=coor;
  var c1-c5;
  id rowid;
run;
```

The out= option names the data set that will contain the coordinates of the solution. By default, two dimensions are used and the dimens= option is used to specify an alternative.

The var statement specifies the variables that represent the columns of the contingency table and the id statement specifies a variable to be used to label the rows. The latter is optional, but without it the rows will simply be labelled row 1, row 2, etc. The output appears in Table 19.4.

A plot of the results is produced automatically using ODS graphics and is shown in Figure 19.1. (The plotit macro used in earlier versions of SAS is still available if preferred.) Displaying the categories of a contingency table in a scatterplot in this way involves the concept of distance between the percentage profiles of row and column categories. The distance measure used in a correspondence analysis is known as the

Table 19.4 Output from Proc Corresp Applied to the Age and Relationship Data

The CORRESP Procedure

Inertia and Chi-Square Decomposition

Singular Value	Principal Inertia	Chi-Square	Percent	Cumulative Percent	19 38 57 76 95 ----+----+----+----+----+---
0.37596	0.14135	19.6473	95.36	95.36	************************
0.08297	0.00688	0.9569	4.64	100.00	*
Total	0.14823	20.6042	100.00		

Degrees of Freedom = 8

Row Coordinates

	Dim1	Dim2
NoBoy	-0.1933	-0.0610
NoSex	-0.1924	0.1425
Both	0.7322	-0.0002

Summary Statistics for the Row Points

	Quality	Mass	Inertia
NoBoy	1.0000	0.5540	0.1536
NoSex	1.0000	0.2374	0.0918
Both	1.0000	0.2086	0.7546

Partial Contributions to Inertia for the Row Points		
	Dim1	Dim2
NoBoy	0.1465	0.2996
NoSex	0.0622	0.7004
Both	0.7914	0.0000

Indices of the Coordinates that Contribute Most to Inertia for the Row Points			
	Dim1	Dim2	Best
NoBoy	2	2	2
NoSex	0	2	2
Both	1	0	1

Squared Cosines for the Row Points		
	Dim1	Dim2
NoBoy	0.9094	0.0906
NoSex	0.6456	0.3544
Both	1.0000	0.0000

Column Coordinates		
	Dim1	Dim2
under 16	-0.3547	-0.0550
16-17	-0.2897	0.0003
17-18	-0.1033	0.0001
18-19	0.2806	0.1342
19-20	0.7169	-0.1234

Summary Statistics for the Column Points			
	Quality	Mass	Inertia
under 16	1.0000	0.2230	0.1939
16-17	1.0000	0.2374	0.1344
17-18	1.0000	0.1727	0.0124
18-19	1.0000	0.2230	0.1455
19-20	1.0000	0.1439	0.5137

Partial Contributions to Inertia for the Column Points		
	Dim1	**Dim2**
under 16	0.1985	0.0981
16-17	0.1410	0.0000
17-18	0.0130	0.0000
18-19	0.1242	0.5837
19-20	0.5232	0.3183

Indices of the Coordinates that Contribute Most to Inertia for the Column Points			
	Dim1	**Dim2**	**Best**
under 16	1	0	1
16-17	1	0	1
17-18	0	0	1
18-19	0	2	2
19-20	1	1	1

Squared Cosines for the Column Points		
	Dim1	**Dim2**
under 16	0.9765	0.0235
16-17	1.0000	0.0000
17-18	1.0000	0.0000
18-19	0.8137	0.1863
19-20	0.9712	0.0288

Chi-squared distance. The calculation of this distance can be illustrated using the proportions of girls in age groups 1 and 2 for each relationship type in Table 19.1.

$$\text{Chi-squared distance} = \sqrt{\frac{(0.68-0.64)^2}{0.55} + \frac{(0.26-0.27)^2}{0.24} + \frac{(0.06-0.09)^2}{0.21}}$$

$$= 0.09.$$

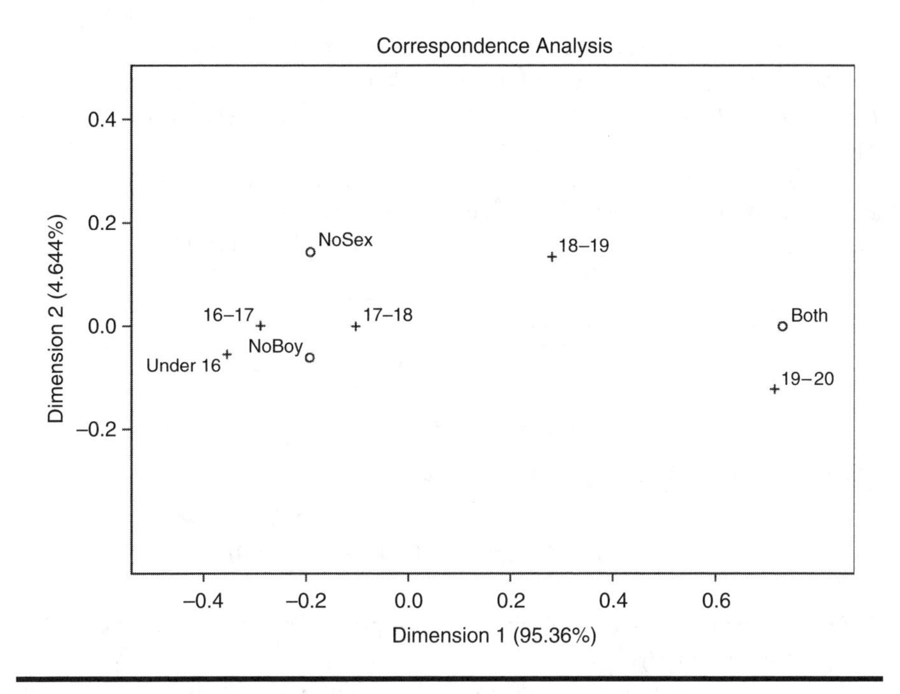

Figure 19.1 Plot of correspondence analysis solution for the age and type of relationship data.

This is similar to ordinary "straight line" or Pythagorean distance, but differs by dividing each term by the corresponding average proportion; in this way the procedure effectively compensates for the different levels of occurrence of the categories. (More formally, the chance of the Chi-squared distance for measuring relationships between profiles can be justified as a way of standardizing variables under a multinomial or Poisson distributional assumption—see Greenacre, 1992.)

The complete set of Chi-squared distances for all pairs of the five age groups, calculated as shown above, can be arranged in a matrix as follows:

$$
\begin{array}{c}
\\
1\\
2\\
3\\
4\\
5
\end{array}
\begin{pmatrix}
1 & 2 & 3 & 4 & 5\\
0.00 & & & & \\
0.09 & 0.00 & & & \\
0.26 & 0.19 & 0.00 & & \\
0.66 & 0.59 & 0.41 & 0.00 & \\
1.07 & 1.01 & 0.83 & 0.51 & 0.00
\end{pmatrix}
$$

The points representing the age groups in Figure 19.1 give the two-dimensional representation of these distances, the Euclidean distance between two points representing the Chi-squared distance between the corresponding age groups (similar for the point representing type of relationship). For a contingency table with r rows and c columns, it can be shown that the Chi-squared distances can be represented exactly in $\min\{r-1, c-1\}$ dimensions; here, since $r=3$ and $c=5$, this means that the coordinates in Figure 19.1 will lead to Euclidean distances that are identical to the Chi-squared distances given above. For example, the correspondence analysis coordinates for age groups 1 and 2 taken from Equation 16.4 are

Age Group	x	y
1	−0.355	−0.055
2	−0.290	0.000

The corresponding Euclidean distance is calculated as

$$\sqrt{(-0.355 + 0.290)^2 + (-0.055 - 0.000)^2},$$

that is, a value of 0.09—agreeing with the Chi-squared distance between the two age groups given previously.

Of most interest in correspondence analysis solutions, such as that graphed in Figure 19.1, is the joint interpretation of the points representing the row and column categories. It can be shown that row and column coordinates that are large and of the same sign correspond to a large positive residual term in the contingency table. Row and column coordinates that are large but of opposite signs imply a cell in the table with a large negative residual. Finally, small coordinate values close to the origin correspond to small residuals. In, for example, age group 5 and boyfriend/sexual intercourse, both have large negative coordinate values on the first dimension; consequently, the corresponding cell in the table will have a large positive residual. Again, age group 5 and boyfriend/no sexual intercourse have coordinate values with opposite signs on both dimensions, implying a negative residual for the corresponding cell in the table.

19.3.2 Smoking and Motherhood

The cell counts of Table 19.2 are in an ASCII file, births.dat, tab separated, and may be read in as follows:

```
data births;
  infile 'c:\handbook3\datasets\births.dat' expandtabs;
  input c1-c4;
  length rowid $12.;
  select(_n_);
```

```
    when(1) rowid='Young NS';
    when(2) rowid='Young Smoker';
    when(3) rowid='Old NS';
    when(4) rowid='Old Smoker';
  end;
  label c1='Prem Died' c2='Prem Alive' c3='FT Died' c4='FT
    Alive';
run;
```

As with the previous example, the data are read into a set of variables corresponding to the columns of the contingency table, and labels are assigned to them. A character variable, rowid, is assigned appropriate values, using the automatic SAS variable, _n_, to label the rows. This is explicitly declared as a 12-character variable with the length statement. Where a character variable is assigned values as part of the data step, rather than reading them from a data file, the default length is determined from its first occurrence. In this example, that would have been from rowid='Young NS'; and its length would have been 8, with longer values truncated. This example also shows the use of the select group as an alternative to multiple if-then statements. The expression in parentheses on the select statement is compared to those on the when statements and the rowid variable set accordingly. The end statement terminates the select group.

The correspondence analysis and plot are produced in the same way as for the first example. The inertia table is shown in Table 19.5 and the plot in Figure 19.2.

```
proc corresp data=births out=coor;
  var c1-c4;
  id rowid;
run;
```

Table 19.5 Inertia Table from Proc Corresp Applied to the Age/Smoking Behaviour of Mother and Gestational Age of Child Data

The CORRESP Procedure

Inertia and Chi-Square Decomposition

Singular Value	Principal Inertia	Chi-Square	Percent	Cumulative Percent	18 36 54 72 90 ----+----+----+----+----+---
0.05032	0.00253	17.3467	90.78	90.78	************************
0.01562	0.00024	1.6708	8.74	99.52	**
0.00365	0.00001	0.0914	0.48	100.00	
Total	0.00279	19.1090	100.00		

Degrees of Freedom = 9

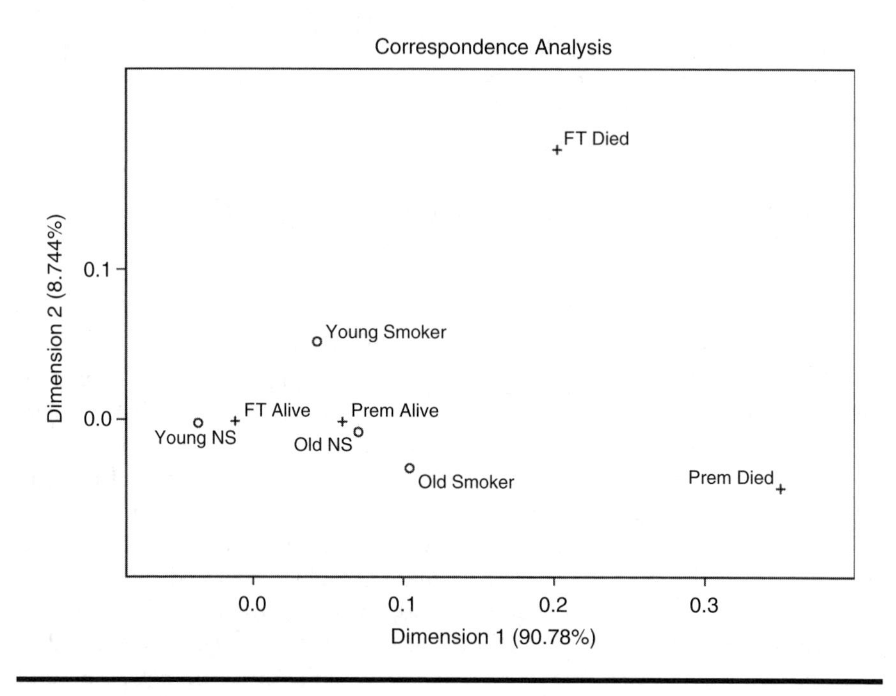

Figure 19.2 Correspondence analysis solution for the age/smoking behaviour of mother and gestational age of child data.

The Chi-squared statistic for these data is 19.1090, which with 9 degrees of freedom has an associated p value of 0.024. So it appears that "type" of mother is related to what happens to the newborn baby. The correspondence analysis of the data shows that the first two eigenvalues account for 99.5% of the inertia. Clearly, a two-dimensional solution provides an extremely good representation of the relationship between the two variables. The two-dimensional solution plotted in Figure 19.2 suggests that young mothers who smoke tend to produce more full-term babies who then die in the first year, and older mothers who smoke have rather more than expected premature babies who die in the first year. It does appear that smoking is a risk factor for death in the first year of the baby's life and that age is associated with length of gestation, with older mothers delivering more premature babies.

19.3.3 Are the Germans Really Arrogant?

The data on perceived characteristics of European nationals can be read in as follows:

```
data europeans;
  infile 'c:\handbook3\datasets\europeans.dat' expandtabs;
  input country $ c1-c13;
  label c1 = 'stylish'
        c2 = 'arrogant'
        c3 = 'sexy'
        c4 = 'devious'
        c5 = 'easy-going'
        c6 = 'greedy'
        c7 = 'cowardly'
        c8 = 'boring'
        c9 = 'efficient'
        c10 = 'lazy'
        c11 = 'hard working'
        c12 = 'clever'
        c13 = 'courageous';
run;
```

In this case, the name of the country is included in the data file, so that it can be read in with the cell counts and used to label the rows of the table. The correspondence analysis and plot are produced in the same way and the results are shown in Table 19.6 and Figure 19.3.

Table 19.6 Inertia Table from Proc Corresp Applied to the Perceived European Characteristics Data

The CORRESP Procedure

Inertia and Chi-Square Decomposition

Singular Value	Principal Inertia	Chi-Square	Percent	Cumulative Percent	10 20 30 40 50 ----+----+----+----+----+---
0.49161	0.24168	255.697	49.73	49.73	*************************
0.38474	0.14803	156.612	30.46	80.19	***************
0.20156	0.04063	42.985	8.36	88.55	****
0.19476	0.03793	40.133	7.81	96.36	****
0.11217	0.01258	13.313	2.59	98.95	*
0.07154	0.00512	5.415	1.05	100.00	*
Total	0.48597	514.153	100.00		

Degrees of Freedom = 72

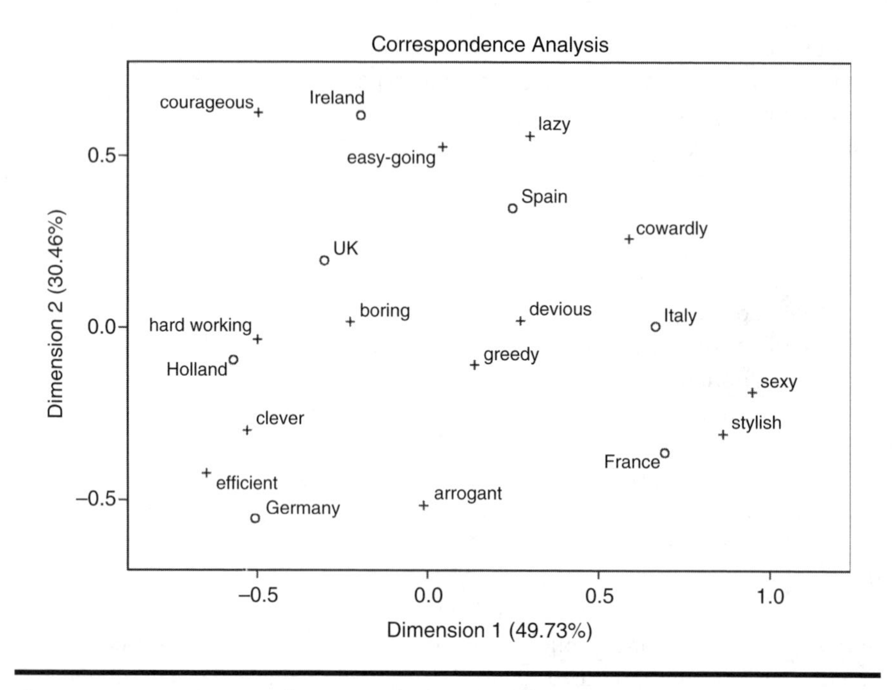

Figure 19.3 Correspondence analysis solution for the perceived European characteristics data.

```
proc corresp data=europeans out=coor;
  var c1-c13;
  id country;
run;
```

Here a two-dimensional representation accounts for about 80% of the inertia. The two-dimensional solution plotted in Figure 19.3 is left for detailed interpretation to the reader, noting here only that it largely fits the authors' own prejudices about perceived national stereotypes.

A comprehensive account of correspondence analysis is given in Greenacre (2007).

Exercises

1 *European Stereotypes*

Construct a scatterplot matrix of the first four correspondence analysis coordinates of the European stereotypes data.

2 *Smoking and Motherhood Data*

Calculate the Chi-squared distances for both the row and column profiles of the smoking and motherhood data, and compare them with the corresponding Euclidean distances in Figure 19.2.

3 *Hair Colour and Eye Colour*

The data below give the hair colour and eye colour of a large number of people.

	Fair	*Red*	*Medium*	*Dark*	*Black*
Light	688	116	584	188	4
Blue	326	38	241	110	3
Medium	343	84	909	412	26
Dark	98	48	403	681	81

1. Find the two-dimensional correspondence analysis solution for the data and plot the results.

4 *Murders*

The data below show the methods by which victims of persons convicted of murder between 1970 and 1977 had been killed.

1. How many dimensions would be needed for an exact correspondence analysis of these data?
2. Use the first three correspondence analysis coordinates to plot a 3×3 scatterplot matrix.
3. Interpret the results of the correspondence analysis.

	1970	*1970*	*1972*	*1973*	*1974*	*1975*	*1976*	*1977*
Shooting	15	15	31	17	42	49	38	27
Stabbing	95	113	94	125	124	126	148	127
Blunt instrument	23	16	34	34	35	33	41	41
Poison	9	4	8	3	5	3	1	4
Manual violence	47	60	54	70	69	66	70	60
Stangulation	43	45	43	53	51	63	47	51
Smothering/ drowning	26	16	20	24	15	15	15	15

Appendix

Answers to Selected Exercises

Chapter 2

Exercise 1(1)

```
proc sort data=water;
  by location;
run;
proc boxplot data=water;
  plot (mortal hardness)*location;
run;
```

Exercise 1(4)

```
proc sort data=water;
  by location;
run;
proc kde data=water;
  bivar mortal hardness /plots=surface;
  by location;
run;
```

Exercise 1(5)

```
proc sgplot data=water;
  reg y=mortal x=hardness /group=location;
run;
```

Chapter 3

Exercise 2(1)

```
proc freq data=pistons order=data noprint;
  tables machine*site / out=tabout outexpect outpct;
  weight n;
run;
data resids;
  set tabout;
  r=(count-expected)/sqrt(expected);
  radj=r/sqrt((1-percent/pct_row)*(1-percent/pct_col));
run;
proc tabulate data=resids;
  class machine site;
  var r radj;
  table machine,
    site*r;
  table machine,
    site*radj;
run;
```

Chapter 4

Exercise 3(1)

```
proc sgplot data=cortisol;
  vbox cortisol /category=group;
run;
data cortisol;
  set cortisol;
  logcortisol=log(cortisol);
run;
proc glm data=cortisol;
  class group;
  model logcortisol=group / solution;
run;
```

Chapter 5

Exercise 2(1)

```
proc tabulate data=genotypes f=5.1;
  class gl gm;
  var weight;
```

```
  table gl,
    gm*weight*(mean std n);
run;
```

Exercise 2(3)

```
proc glm data=genotypes;
  class gl gm;
  model weight=gl|gm / ss1;
run;
proc glm data=genotypes;
  class gl gm;
  model weight=gm|gl / ss1;
run;
```

Chapter 6

Exercise 2(2)

```
data universe;
  set universe;
  distsqd=distance**2;
run;
proc reg data=universe;
  model velocity= distance distsqd / noint;
  output out=universe predicted=qpred;
run;
proc sort data=universe;
  by distance;
run;
proc sgplot data=universe noautolegend;
  scatter y=velocity x=distance;
  series y=lpred x=distance;
  series y=qpred x=distance / lineattrs=(pattern=2);
  xaxis min=0;
  yaxis min=0;
run;
```

Exercise 4(1)

```
proc sgplot data=expend;
  scatter y=percent x=expend /datalabel=region;
run;
* or...;
```

```
proc gplot data=expend;
  plot percent*expend;
  symbol1 v=dot pointlabel=('#region' j=r position=middle);
run;
```

Chapter 7

Exercise 2(1)

```
proc sgplot data=peanuts;
  reg y=percent x=level;
run;
```

Chapter 8

Exercise 1(1)

```
proc logistic data=ghq;
  class sex;
  model cases/total=ghq sex ghq*sex;
  output out=ghq2 predicted=pr;
run;
proc sgplot data=ghq2;
  series y=pr x=ghq / group=sex;
run;
```

Exercise 4(1)

```
data roles;
  infile 'c:\handbook3\datasets\role.dat';
  input Years Agreem Disagreem Agreef Disagreef;
  mprob=agreem/(agreem+disagreem);
  fprob=agreef/(agreef+disagreef);
run;
proc sgplot data=roles;
  scatter y=mprob x=years/markerattrs=(symbol=circle);
  scatter y=fprob x=years;
run;
```

Chapter 9

Exercise 3(1)

```
data bladder;
  infile 'c:\handbook3\datasets\bladder.dat';
  input id time tum n;
```

```
  logtime = log(time);
run;
proc genmod data = bladder;
  model n = tum / dist = p offset = logtime;
run;
```

Chapter 12

Exercise 1(1)

```
data pndep2;
  set pndep2;
  zdep = dep;
run;
proc sort data = pndep2;
  by time;
run;
proc stdize data = pndep2 out = pndep2;
  var zdep;
  by time;
run;
proc sgpanel data = pndep2 noautolegend;
  panelby group / rows = 2;
  series y = zdep x = time /group = idno;
run;
* or. . . . ;
symbol1 i = join v = none l = 1 r = 27;
symbol2 i = join v = none l = 2 r = 34;
proc gplot data = pndep2;
  plot zdep*time = idno / skipmiss;
run;
```

Exercise 2(1)

```
data pip;
  infile 'c:\handbook3\datasets\phos.dat';
  input id group p0 p05 p1 p15 p2-p5;
run;
data pipl;
  set pip;
  array ps {*} p0--p5;
  array t{8} t1-t8 (0.5 1 1.5 2 3 4 5);
  do i = 1 to 8;
    time = t{i};
```

```
    pip = ps{i};
    output;
  end;
run;
proc sgpanel data = pipl noautolegend;
  panelby group / rows = 2;
  series y = pip x = time /group = id;
  colaxis type = linear values = (0 to 2 by 0.5 3 to 5 by 1);
run;
* or...;
proc sort data = pipl;
  by group id;
run;
proc gplot data = pipl uniform;
  plot pip*time = id /nolegend;
  by group;
  symbol1 i = join v = none r = 50;
run;
```

Chapter 13

Exercise 3(1)

```
proc mixed data = pndep2 covtest noclprint;
  class group idno;
  model dep = group|time /s ddfm = bw;
  random int time / sub = idno type = un;
run;
```

Chapter 14

Exercise 2(1)

```
data respw;
  infile 'c:\handbook3\datasets\resp.dat';
  input id centre treat sex age bl v1-v4;
run;
data respl;
  set respw;
  array vs {4} v1-v4;
  do time = 1 to 4;
  status = vs{time};
```

```
  output;
  end;
run;
proc genmod data=respl desc;
  class centre treat sex id;
  model status=centre treat sex age time bl / d=b;
  repeated subject=id / type=ind;
run;
```

Chapter 15

Exercise 2(1)

```
infile 'c:\handbook3\datasets\glioma.dat';
  input id age sex $ group $ event $ time;
  if event='FALSE' then censor=1;
    else censor=0;
run;
proc lifetest data=glioma plots=(s);
    time time*censor(1);
    strata group;
run;
```

Chapter 17

Exercise 2(1)

```
data pottery;
  infile 'c:\handbook3\datasets\pottery.dat' expandtabs;
  input id Kiln Al2O3 Fe2O3 MgO CaO Na2O K2O TiO2 MnO BaO;
run;
proc cluster data=pottery method=average ccc std outtree=pottree;
  var Al2O3--BaO;
  id id;
  copy kiln;
run;
proc tree horizontal vaxis=axis1;
axis1 label=none;
run;
```

Chapter 18

Exercise 2(1)

```
data sids;
  infile 'c:\handbook3\datasets\sids.dat' expandtabs;
  input Group HR BW Factor68 Gestage;
run;
proc discrim data=sids out=discout canonical;
  class group;
  var bw Factor68;
  ods output LinearDiscFunc=func;
run;
data func;
  set func;
  _3=_1-_2;
run;
proc print;
var variable _3;
format _3 12.8;
run;
data discout;
  set discout;
  pbw=((factor68*16.07705412)+0.50619161)/ 0.00194756;
run;
proc sort data=discout; by factor68; run;
proc sgplot data=discout;
  scatter y=bw x=factor68 / group=_into_;
  series y=pbw x= factor68;
run;
```

Chapter 19

Exercise 3(1)

```
proc corresp data=colour;
  var FairHair--BlackHair;
  id rowid;
run;
```

References

Agresti, A. (2007) *Introduction to Categorical Data Analysis*, 2nd Edition, Wiley, New York.

Aitkin, M. (1978) The analysis of unbalanced cross-classifications. *Journal of the Royal Statistical Society* A, 141, 195–223.

Beck, A., Steer, R. and Brown, G. (1996) *BDI-II Manual*, 2nd Edition, The Psychological Corporation, San Antonio, TX.

Berk, K.N. (1977) Tolerance and condition in regression computations. *Journal of the American Statistical Association*, 72, 863–866.

Caplehorn, J. and Bell, J. (1991) Methadone dosage and the retention of patients in maintenance treatment. *The Medical Journal of Australia*, 154, 195–199.

Carpenter, J., Pocock, S. and Lamm, C.J. (2002) Coping with missing data in clinical trials: A model-based approach to asthma trials. *Statistics in Medicine*, 21, 1043–1066.

Chambers, J.M. and Hastie, T.J. (1993) *Statistical Models in S*, CRC/Chapman and Hall, New York.

Chatterjee, S. and Price, B. (1999) *Regression Analysis by Example*, 3rd Edition, Wiley, New York.

Clayton, D. and Hills, M. (1993) *Statistical Models in Epidemiology*, Oxford University Press, Oxford.

Cleveland, W.S. (1979) Robust locally weighted regression and smoothing scatterplots. *Journal of the American Statistical Association*, 74, 829–836.

Collett, D. (2003a) *Modelling Binary Data*, 2nd Edition, Chapman & Hall/CRC, London.

Collett, D. (2003b) *Modelling Survival Data in Medical Research*, 2nd Edition, Chapman and Hall/CRC, London.

Cook, R.D. and Weisberg, S. (1982) *Residuals and Influence in Regression*, Chapman & Hall, London.

Cox, D.R. (1972) Regression models and life tables. *Journal of the Royal Statistical Society B*, 34, 187–220.

Davidson, M.L. (1972) Univariate versus multivariate tests in repeated measurements experiments. *Psychological Bulletin*, 77, 446–452.

Davis, C.S. (1991) Semiparametric and nonparametric methods for the analysis of repeated measurements with applications to clinical trials. *Statistics in Medicine*, 16, 1959–1980.

Davis, C.S. (2002) *Statistical Methods for the Analysis of Repeated Measures*, Springer, New York.

De Backer, M., De Vroey, C., Lesaffre, E., Scheys, I. and De Keyser, P. (1998) Twelve weeks of continuous oral therapy for toenail onchyomycosis caused by dermatophytes: A double-blind comparative trial of terbinafine 250 mg/day versus itraconazole 200 mg/day. *Journal of the American Academy of Dermatology*, 38, 57–63.

Diggle, P.J. (1998) Dealing with missing values in longitudinal studies. In B.S. Everitt and G. Dunn (Eds.), *Statistical Analysis of Medical Data*, Arnold, London.

Diggle, P.J. and Kenward, M.G. (1994) Informative drop-out in longitudinal analysis (with discussion), *Applied Studies*, 43, 49–93.

Diggle, P.J., Liang, K. and Zeger, S.L. (2002) *Analysis of Longitudinal Data*, 2nd Edition, Oxford University Press, Oxford.

Dizney, H. and Gromen, L. (1967) Predictive validity and differential achievement in three MLA comparative foreign language tests. *Educational and Psychological Measurement*, 27, 1127–1130.

Everitt, B.S. (1992) *The Analysis of Contingency Tables*, 2nd Edition, CRC/Chapman and Hall, London.

Everitt, B.S. (2001) *Statistics in Psychology: An Intermediate Course*, Laurence Erlbaum, Mahwah, NJ.

Everitt, B.S. (2006) *The Cambridge Dictionary of Statistics*, 3rd Edition, Cambridge University Press, Cambridge.

Everitt, B.S. and Dunn, G. (2001) *Applied Multivariate Data Analysis*, 2nd Edition, Arnold, London.

Everitt, B.S. and Pickles, A. (2004) *Statistical Aspects of the Design and Analysis of Clinical Trials*, 2nd Edition, Imperial College Press, London.

Everitt, B.S., Landau, S. and Leese, M. (2001) *Cluster Analysis*, 4th Edition, Arnold, London.

Fisher, L.D. and Van Belle, G. (1996) *Biostatistics: A Methodology for the Health Sciences*, 2nd Edition, Wiley, New York.

Fisher, R.A. (1936) The use of multiple measurement in taxonomic problems. *Annals of Eugenics*, 7, 179–184.

Fitzmaurice, G.M., Laird, N.M. and Ware, J.H. (2004) *Applied Longitudinal Analysis*, Wiley, New York.

Friedman, J.H. (1991) Multiple adaptive regression splines. *Annals of Statistics*, 19, 1–67.

Goldberg, D. (1972) *The Detection of Psychiatric Illness by Questionnaire*, Oxford University Press, Oxford.

Grambsch, P.M. and Therneau, T.M. (1994) Proportional hazard tests and diagnostics based on weighted residuals. *Biometrika*, 81, 515–526.

Grana, C., Chinol, M., Robertson, C., Mazzetta, C., Bartolemei, M., Cicco, C.D., Fiorenza, M., Gatti, M., Caliceti, P. and Paganell, G. (2002) Pretargeted adjuvant radioimmunotherapy with yttrium-90-biotin in malignant glioma patients: A pilot study. *British Journal of Cancer*, 86, 207–212.

Greenacre, M. (1992) Correspondence analysis in medical research. *Statistical Methods in Medical Research*, 1, 97–117.

Greenacre, M. (2007) *Correspondence Analysis in Practice*, 2nd Edition, Chapman and Hall/CRC, London.

Greenhouse, S.W. and Geisser, S. (1959) On methods in the analysis of profile data. *Psychometrika*, 24, 95–112.

Gregoire, A.J.P., Kumar, R., Everitt, B.S., Henderson, A.F. and Studd, J.W.W. (1996) Transdermal oestrogen for the treatment of severe post-natal depression. *The Lancet*, 347, 930–934.

Hancock, B.W., Aitkin, M., Martin, J., Dunsmore, I.R., Ross, C.M., Carr, I. and Emmanuel, I.G. (1979) Hodgkin's disease in Sheffield (1971–1976) with computer analysis of variables, *Clinical Oncology*, 5, 283–297.

Hand, D.J. (1981) *Discrimination and Classification*, Wiley, Chichester.

Hand, D.J. (1997) *Construction and Assessment of Classification Rules*, Wiley, Chichester.

Hand, D.J., Daly, F., Lunn, A.D., McConway, K.J. and Ostrowski, E. (1994) *A Handbook of Small Data Sets*, Chapman & Hall, London.

Hastie, T.J. and Tibshirani, R.J. (1990) *Generalized Additive Models*, Chapman and Hall/CRC, London.

Heitjan, D.F. (1997) Approaches to inputation, *American Journal of Public Health*, 87, 545–550.

Hosmer, D.W. and Lemeshow, S. (1999) *Applied Survival Analysis*, Wiley, New York.

Hotelling, H. (1933) Analysis of a complex of statistical variables into principal components. *Journal of Educational Psychology*, 24, 417–441.

Howell, D.C. (2002) *Statistical Methods for Psychology*, 5th Edition, Duxbury Press, Belmont, CA.

Huynh, H. and Feldt, L.S. (1976) Estimates of the Box correction for degrees of freedom for sample data in randomised block and split plot designs. *Journal of Educational Statistics*, 1, 69–82.

Kalbfleisch, J.D. and Prentice, J.L. (2002) *The Statistical Analysis of Failure Time Data*, 2nd Edition, Wiley, New York.

Kelsey, J.L. and Hardy, R.J. (1975) Driving of motor vehicles as a risk factor for acute herniated lumbar intervetebral disc. *American Journal of Epidemiology*, 102, 63–73.

Keyfitz, N. and Flieger, W. (1972) *Population: Facts and Methods of Demography*, W.H. Freeman, San Francisco, CA.

Lawless, J.F. (2002) *Statistical Models and Methods for Lifetime Data*, 2nd Edition, Wiley, New York.

Levene, H. (1960) Robust tests for the equality of variance. In O. Olkin (Ed.), *Contributions to Probability and Statistics*, Stanford University Press, Stanford, CA.

Liang, K.Y. and Zeger, S.L. (1986) Longitudinal data analysis using generalized linear models, *Biometrika*, 73, 13–22.

Longford, N.T. (1993) *Random Coefficient Models*, Oxford University Press, Oxford.

McCullagh, P. and Nelder, J.A. (1989) *Generalized Linear Models*, CRC/Chapman and Hall, London.

McKay, R.J. and Campbell, N.A. (1982a) Variable selection techniques in discriminant analysis. I. Description. *British Journal of Mathematical and Statistical Psychology*, 35, 1–29.

McKay, R.J. and Campbell, N.A. (1982b) Variable selection techniques in discriminant analysis. II. Allocation. *British Journal of Mathematical and Statistical Psychology*, 35, 30–41.

Mallows, C.L. (1973) Some comments on Cp. *Technometrics*, 15, 661–675.

Matthews, J.N.S. (1993) A refinement to the analysis of serial data using summary measures. *Statistics in Medicine*, 12, 27–37.

Maxwell, S.E. and Delaney, H.D. (2003) *Designing Experiments and Analysing Data*, 2nd Edition, Wadsworth, Belmont, CA.

Morrison, D.F. (1976) *Multivariate Statistical Methods*, McGraw-Hill, New York.

Murray, G.D. and Findlay, J.G. (1988) Correcting for bias caused by drop-outs in hypertension trials, *Statistics in Medicine*, 7, 941–946.

Nelder, J.A. (1977) A reformulation of linear models. *Journal of the Royal Statistical Society A*, 140, 48–63.

Osborne, J.F. (1979) *Statistical Exercises in Medical Research*, Blackwell Scientific Publications, Oxford.

Pearson, K. (1901) On lines and planes of closest fit to systems of points in space. *Philosophical Magazine*, 2, 559–572.

Piantadosi, S. (1997) *Clinical Trials: A Methodological Perspective*, Wiley, New York.

Pinheiro, J.C. and Bates, D.M. (2000) *Mixed-Effects Models in S and S-PLUS*, Springer, New York.

Proudfoot, J., Mann, A., Everitt, B.S., Marks, I., Goldberg, D. and Gray, J.A. (2002) Computerized interactive, multimedia cognitive-behavioural program for anxiety and depression in general practice, *Psychological Medicine*, 33, 217–227.

Quine, S. (1975) *Achievement Orientation of Aboriginal and White Adolescents*. Doctoral dissertation, Australian National University, Canberra.

Rothschild, A.J., Schatzberg, A.F., Rosenbaum, H.A., Stahl, J.B. and Cole, J.O. (1982) The dexamethasone suppression test as a discriminator among subtypes of psychotic patients. *British Journal of Psychiatry*, 141, 471–474.

Rubin, D. (1976) Inference and missing data. *Biometrika*, 63, 581–592.

Sarle, W.S. (1983) SAS Technical Report A-108 Cubic Clustering Criterion, SAS Institute Inc., Cary, NC.

Sartwell, P.E., Mazi, A.T., Aertles, F.G., Greene, G.R. and Smith, M.E. (1969) Thromboembolism and oral contraceptives: An epidemiological case-control study. *American Journal of Epidemiology*, 90, 365–375.

Satterthwaite, F.W. (1946) An approximate distribution of estimates of variance components. *Biometrics Bulletin*, 2, 110–114.

Sauerbrei, W. and Royston, P. (1999) Building multivariable prognostic and diagnostic models: Transformation of the predictors by using fractional polynomials. *Journal of the Royal Statistical Society, Series A*, 162, 71–94.

Schumacher, M., Basert, G., Bojar, H., Hübner, K., Olschewski, M., Sauerbrei, W., Schmoor, C., Beyerle, C., Neumann, R.L.A. and Rauschecker, H.F. (1994) Randomised 2×2 trial evaluating hormonal treatment and the duration of chemotherapy in node positive breast cancer patients. *Journal of Clinical Oncology*, 12, 2086–2093.

Seeber, G.U.H. (1998) Poisson regression. In P. Armitage and T. Colton (Eds.), *Encyclopedia of Biostatistics*, Wiley, Chichester.

Senn, S. (1997) *Statistical Issues in Drug Development*, Wiley, Chichester.

Shapiro, S.S. and Wilk, M.B. (1965) An analysis of variance test for normality. *Biometrika*, 52, 591–611.

Silverman, B.W. (1986) *Density Estimation in Statistics and Data Analysis*, CRC/Chapman and Hall, London.

Skevington, S.M. (1990) A standardised scale to measure beliefs about controlling pain (B.P.C.Q.); a preliminary study. *Psychological Health*, 4, 221–232.

Somes, G.W. and O'Brien, K.F. (1985) Mantel–Haenszel statistic. In S. Kotz, N.L. Johnson and C.B. Read (Eds.), *Encyclopedia of Statistical Sources*, Volume 5, Wiley, New York.

Spicer, C.C., Lawrence, G.J. and Southall, D.P. (1987) Statistical analysis of heart rates and subsequent victims of sudden infant death syndrome. *Statistics in Medicine*, 6, 159–166.

Thail, P.F. and Vail, S.C. (1990) Some covariance models for longitudinal count data with overdispersion. *Biometrics*, 46, 657–671.

Tubb, A., Parker, N.J. and Nickless, G. (1980) The analysis of Romano-British pottery by atomic absorption spectrophotometry. *Archaeometry*, 22, 153–171.

Vandaele, W. (1978) Participation in illegitimate activities: Erlich revisited. In A. Blumstein, J. Cohen and D. Nagin (Eds.), *Deterrence and Incapacitation*, Natural Academy of Sciences, Washington, DC.

Vanisma, F. and DeGreve, J.P. (1972) Close binary systems before and after mass transfer. *Astrophysics and Space Science*, 87, 377–401.

Wood, S.N. (2006) *Generalized Additive Models*, Chapman and Hall/CRC, London.

Woodley, W.L., Simpson, J., Biondrini, R.C. and Berkeley, J. (1977) Rainfall results 1970–1975: Florida area cumulus experiment. *Science*, 195, 735–742.

Zeger, S.L. and Liang, K.Y. (1986) Longitudinal data analysis for discrete and continuous outcomes. *Biometrics*, 42, 121–130.

Index

A

B

C

D

M

N

O

P

Q

R

S